Mar Gómez Gallego · Miguel A. Sierra

Organic Reaction Mechanisms

Springer-Verlag Berlin Heidelberg GmbH

Mar Gómez Gallego · Miguel A. Sierra

Organic Reaction Mechanisms

40 Solved Cases

With 432 figures, 116 in colour

Springer

Prof. Mar Gómez Gallego

margg@quim.ucm.es

Prof. Miguel A. Sierra

sierraor@quim.ucm.es

Departamento de Química Orgánica
Facultad de Química
Universidad Complutense
E-28040 Madrid, Spain

ISBN 978-3-642-62292-2   ISBN 978-3-642-18788-9 (eBook)
DOI 10.1007/978-3-642-18788-9

Library of Congress Cataloging-in-Publication-Data

A catalog record for this book is available from the Library of Congress.

Bibliographic information published by Die Deutsche Bibliothek.
Die Deutsche Bibliothek lists this publication in the Deutsche Nationalbibliographie;
detailed bibliographic data is available in the Internet at http://dnb.ddb.de

springeronline.com

© Springer-Verlag Berlin Heidelberg 2004
Originaly published by Springer-Verlag Berlin Heidelberg New York 2004
Softcover reprint of the hardcover 2nd edition 2004

Typesetting: D.A.S Büro, Angelika Schulz, Zülpich
Cover design: Künkel & Lopka, Heidelberg
Printed on acid free paper   02/3020/M - 5 4 3 2 1 0

Foreword

Teaching organic chemistry is a difficult task. One reason is the large gap between the observable properties of a given chemical species and its formal representation on a blackboard or on a computer screen. This concern is even more important in the case of the elucidation of reaction mechanisms, since it involves entities such as reaction intermediates or transition states which are difficult, if not impossible, to observe. The second (obvious) reason is that organic chemistry is an experimental science. We do not have in our discipline any "theorem of stereocontrol" for example, although we have experimental data and mechanistic models that provide us with conceptual tools to explain or even to predict the stereocontrol of a new experiment in organic chemistry. Therefore, according to Popper, we cannot demonstrate that a given mechanism is "true" although we can conclude that it is "false" if it is not compatible with the experimental observations. In addition, the concept of reaction mechanism can be misleading because in many textbooks it is implicitly assumed that there is a common mechanism for, let us say the Diels Alder-reaction although we know that any cycloaddition can be stepwise or concerned.

This book constitutes an innovative contribution within this context. The authors have collected very illustrative problems that cover the main topics of organic chemistry. However, they have not followed the usual admonitory way in which only the correct answers are commented. Instead, the authors show the possible alternative mechanisms and how the experimental data provided by carefully designed experiments can help to discard very plausible hypotheses.

In summary, the reader, after studying the problems presented and discussed in this book, will appreciate not only the beauty and diversity of organic chemistry, but also the necessity of being creative and (reasonably) unbiased as well as sceptical and cautious when confronted with novel mechanistic problems.

University of the Basque Country, June 2003   Fernando P. Cossío

Preface

Thes are the times of highly sophisticated computer programs that lead to pretty pictures about how an organic reaction works. Nevertheless, the knowledge of the mechanism of an organic reaction is still derived from very hard experimental work and the ability to process and to get information from the data obtained. The interpretation of the experimental data is a key point in the formation of organic chemists, independent of their future work. It is sometimes confusing to learn how to do this in a jumble of equations, theoretical studies, and physical-chemistry postulates. The approach we have chosen for this work is a practical one. Instead of formulating study mechanisms from the diverse theories and experimental methods in the classical way, the aim of this book is to discuss a series of selected examples of organic reaction mechanisms to understand how they have been proposed.

All the cases presented have been taken from recent literature and deal with the formulation and experimental determination of a mechanistic proposal. They are based on real cases that demonstrate how nowadays organic chemists are still deeply concerned about insights into a reaction. Through the book we use any type of information that the authors of the original work considered necessary in the elucidation of the reaction mechanism. This includes spectroscopic data, kinetic and thermodynamic data, isotopic labelling, theoretical studies ... and of course, the knowledge of the reactivity of organic compounds.

The examples have been ordered into three levels. Level 1 is dedicated to revising fundamental concepts regarding the elucidation of an organic reaction mechanism. The subjects discussed in this level will be the tools to be used later in levels 2 and 3. Level 1 starts with the most immediate source of information when studying an organic reaction mechanism, the structure and stereochemistry of the reaction products. Basic concepts regarding crossover experiments, neighboring group participation and classical/nonclassical carbocations are followed by some examples that illustrate the use of isotopic labelling in mechanistic studies. The importance of data obtained from kinetic measurements is discussed later and subjects like catalysis in solution, Hammett constants, activation parameters and kinetic isotope effects are revised. To complete this first part of the book, there are some examples dedicated to revising the different types of pericyclic reactions such as cycloadditions, electrocyclic ring closures and sigmatropic rearrangements.

Once the fundametal concepts have been established, they are all combined in the discussion of the collection of examples proposed in levels 2 and 3. Now the cases have been placed in order of increasing difficulty from easy to medium in level 2 and medium to difficult in level 3, mixing the different techniques and data to understand the different aspects of the example under discussion. All examples are treated in detail following the same methodology: introduction, experimental data and discussion. Key points, additional references and solved questions related to the main subject studied are also included.

This book is aimed to advanced undergraduates in chemistry but it also may be useful for the instructors who should be able to find in the text new examples to illustrate topics of advanced organic chemistry or physical organic chemistry. A good knowledge of organic chemistry, how to write an organic reaction mechanism, and a basic understanding of the basics of physical organic chemistry is presumed for the readers in order to get the most out of each case. The work done by the authors of the papers used to build up this book has made possible the discussion of mechanistic concepts from an organic chemistry basis. Moreover, their work demonstrates that experimental mechanistic studies are still alive and are necessary to understand the insights of an organic process.

June 2003   Mar Gómez Gallego
                                                             Miguel A. Sierra

Table of Contents

Level 1

Level 2

Level 3

Level 1 – Case 1
A Surprise in the Synthesis of Guanacastepene A

Key point: *Unexpected result in a well-known reaction*

Guanacastepene A

A Surprise in the Synthesis of Guanacastepene A

Sometimes a reaction with a well-established mechanism does not work as expected. In these cases the mechanism has to be reinvestigated and the careful analysis of the reaction products usually gives the key of the mechanistic pathway. This situation can be found in multistep synthetic sequences that have been designed to achieve a target molecule and that are based on the *predictable* outcome of each step. We have an example in the synthesis of guanacastepene A **1**, the parent member of a family of diterpene natural products obtained from a fungus. This compound is a synthetic target of current interest due to its novel structure and the possibility of exploring its activity against antibiotic-resistant bacteria.

1

One of the diverse approaches to the synthesis of **1** is based on the formation of the tricyclic core of guanacastepene **3** by means of a direct Knoevenagel cyclization of β-keto ester precursor **2** (Scheme 1.1).

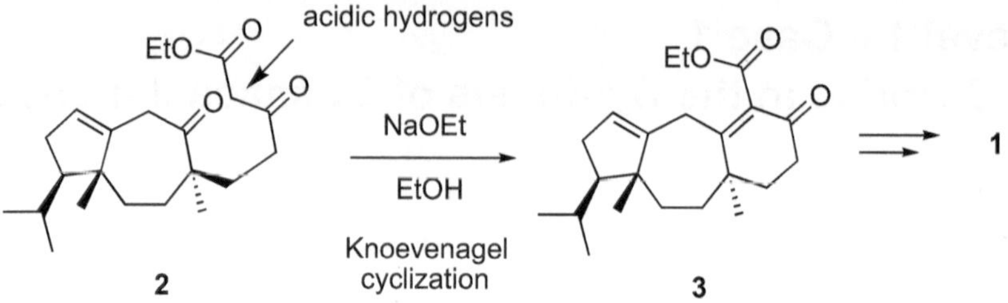

Scheme 1.1

Unfortunately, exposure of **2** to sodium ethoxide in ethanol at 60°C did not yield the expected cyclization product **3**, but a mixture of compounds. The major product was the ester **4** accompanied by two minor compounds that were identified as tricyclic alcohol **5** and tricyclic diketone **6** (this latter probably resulting from the oxidation of alcohol **5**) (Scheme 1.2).[1]

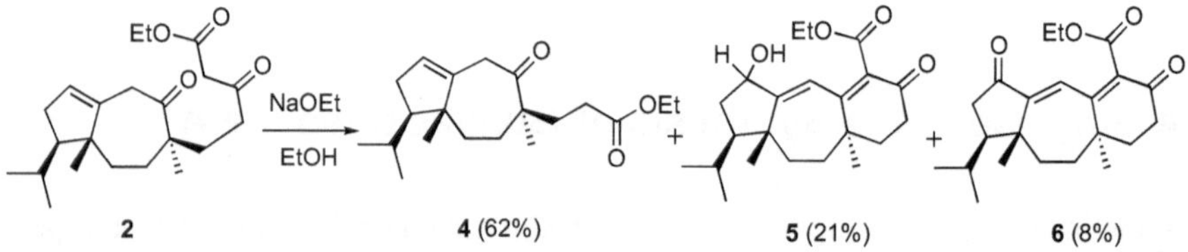

Scheme 1.2

*Suggest a reasonable explanation for the failure of the Knoevenagel cyclization of compound **2**. Propose a mechanism to justify how compounds **4** and **5** are obtained. To answer these questions use the information attained in the following experiments.*

Experimental Data

Exposure of the dioxolane-protected β-keto ester **7** to the Knoevenagel cyclization conditions did not result in the formation of the ester **8** (Scheme 1.3).

---

[1] The configuration of the alcohol in **5** was tentatively assigned as shown in Figure 1.1 upon NOESY experiments.

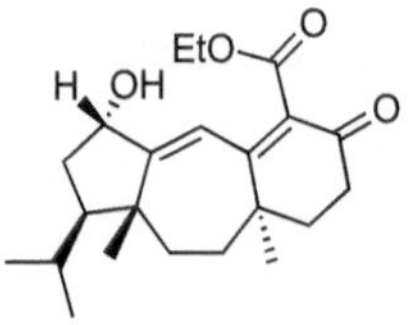

Figure 1.1

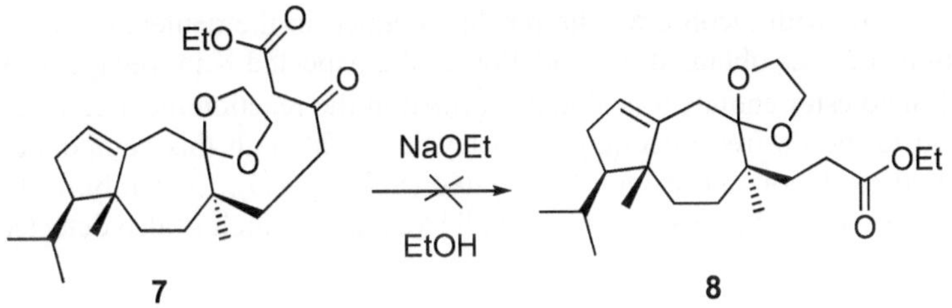

Scheme 1.3

The cyclization of **2** was carried out in deuterated solvent and base. When the reaction was performed with rigorous exclusion of air, the expected Knoevenagel product, deuterated in nine positions (**3**-d_9) could be detected by mass spectrometric analysis of the crude reaction mixture. Furthermore, octadeuterated tricyclic alcohol (**5**-d_8) was obtained as the only reaction product after the reaction mixture was exposed to air (Scheme 1.4).[2]

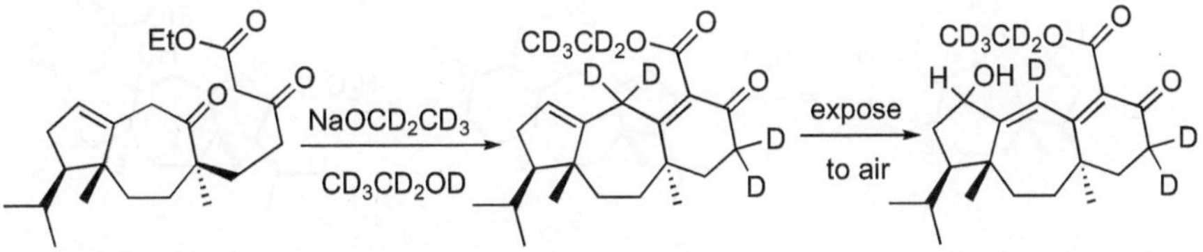

Scheme 1.4

Discussion

The starting point in the investigation of a reaction mechanism is always the analysis of the number of products obtained in the reaction and the determination of their structures. Particularly, from the study of the structure of the reaction products we can obtain valuable information about the bonds that have been broken and those that have been formed during the process. In this case, the analysis of the structure of the reaction products **4** and **5** is the key to understanding why an ideal substrate for a Knoevenagel condensation as β-keto ester **2** reacts in a different way.

Considering the structures of the reaction products **4** and **5** we notice that tricyclic alcohol **5** resembles the expected Knoevenagel product (although it has been oxidized), but in the case of **4** the structure is completely different. It is very unlikely that two products, so different from each other, would come from a single reaction pathway. Hence, a much more reasonable option is to consider *two competing reaction pathways* when **2** was treated under Knoevenagel conditions.

---

[2] Compound **5**-d_8 was identified by ^{1}H NMR. The methine (C*H*OH) proton (δ = 4.38 ppm) appeared only after the reaction mixture was exposed to air.

Let us start with alcohol **5**. The results obtained in the deuteration experiment confirm that **5** was obtained by oxidation of the expected Knoevenagel product **3**. Thus, β-keto ester enolate **9** is already formed in the reaction medium and does in fact lead to the desired unconjugated tricycle **3**, although this compound undergoes rapid oxidation and cannot be isolated (Scheme 1.5). Evidently, further oxidation of the alcohol **5** leads to the tricyclic ketone **6**, which is also detected in the reaction mixture.

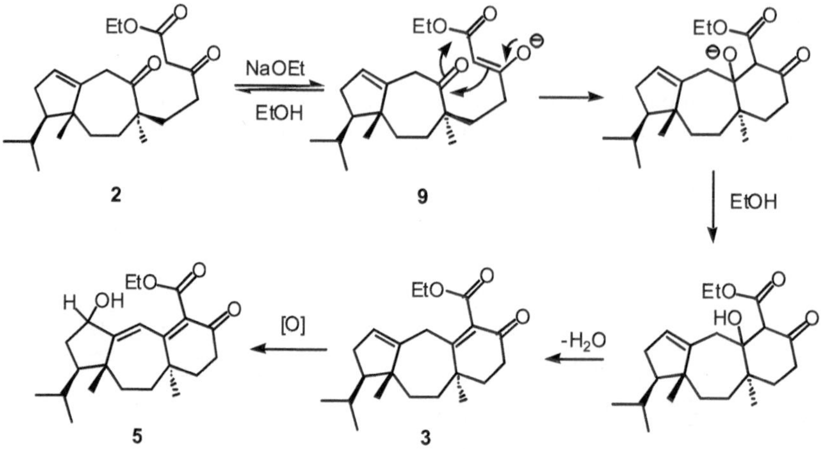

Scheme 1.5

Does this mechanism agree with the position of the deuterium labels in **5** when the reaction was carried out in deuterated base and solvent? The starting compound **2** has several acidic positions (α to the ester and α to the keto groups) that can exchange protons with the solvent. In addition, in the presence of deuterated ethoxide, transesterification of the carboxylate group can also occur. The starting material in the deuterated medium will be **10** rather than **2**. Following the Knoevenagel mechanism previously discussed, but now starting from **10**, compounds **3**-d_9 and **5**-d_8 are obtained (Scheme 1.6).

Competing with this route is the pathway leading to the major reaction product **4**. To understand how this compound is formed we should have in mind the failure of the reaction when the endocyclic keto group is protected as dioxolane (see Scheme 1.3). This data indicates that a *free* keto group is essential for the outcome of the reaction.

Compound **2** is distinguished from the typical Knoevenagel cyclization substrates by the enhanced acidity of the unconjugated enone α-protons. Thus, a reasonable alternative to the Knoevenagel mechanism could be to consider that under the reaction conditions, the formation of hydroazulenone enolate **11** would take place. Addition of the enolate oxygen to the keto group should form lactone **12**, which would yield bicyclic ester **4** by addition of the ethoxide to the lactone carbonyl group and subsequent lactone breakage (Scheme 1.7). As the endocyclic keto group in **2** is directly involved in the reaction, dioxolane **7** (which lacks the feature of a participating cyclic enolate) does not undergo this reaction.

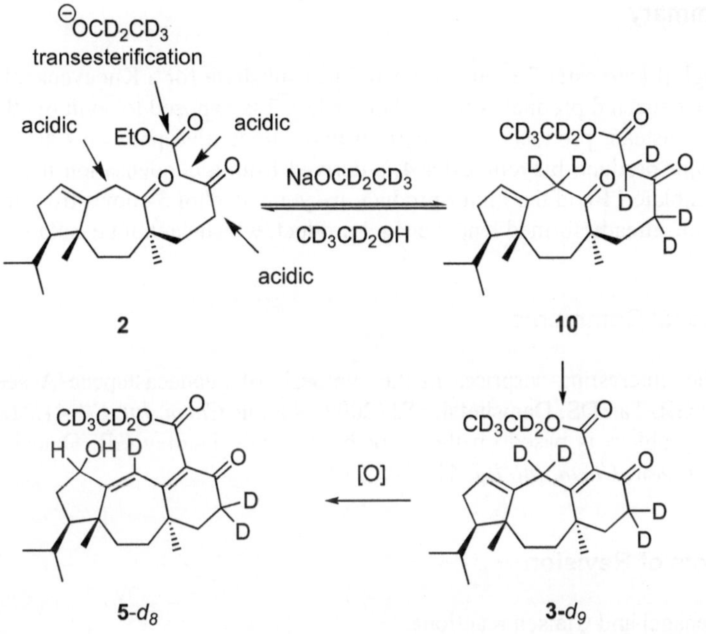

Scheme 1.6

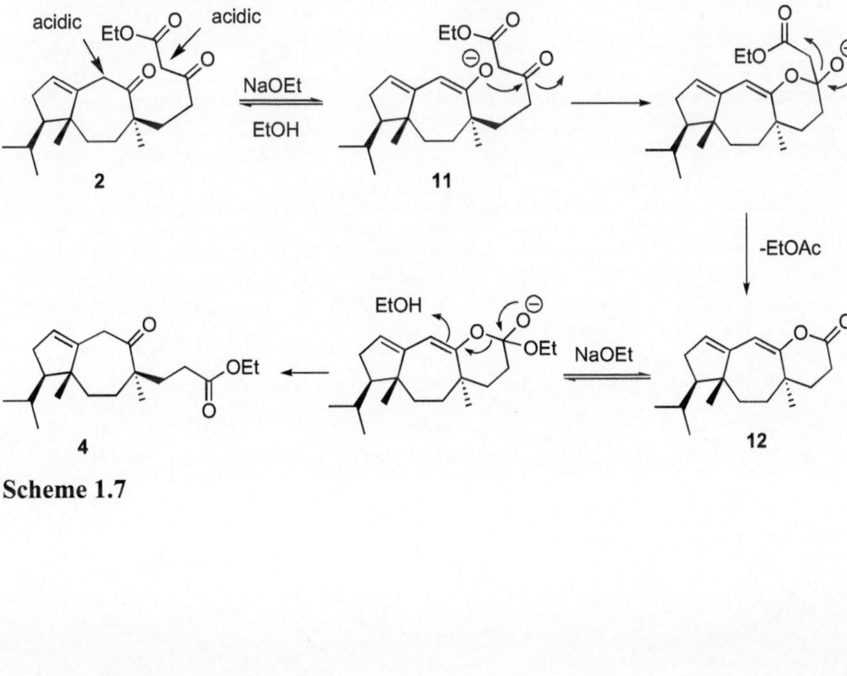

Scheme 1.7

In Summary

Although β-keto ester **2** seems to be an ideal substrate for a Knoevenagel cycliza-tion, the expected product is not isolated when **2** is exposed to sodium ethoxide in ethanol. Instead, products derived from two competitive pathways are observed. The major product, bicyclic ester **4** is derived from a condensation involving the cyclic enolate **11** and the minor product, tricyclic alcohol **5** comes from the oxida-tion of the already formed Knoevenagel product, which cannot be isolated.

Additional Comments

For other interesting surprises in the synthesis of guanacastepene A see Lin S, Dudley GB, Tan DS, Danishefsky SJ (2002) *Angew. Chem. Int. Ed.* 41:2188-2191.

This problem is based on the work by Tan DS, Dudley GB, Danishefsky SJ (2002) *Angew. Chem. Int. Ed.* 41:2185-2188.

Subjects of Revision

Knoevenagel and Claisen reactions.

Level 1– Case 2
Sulfenylation of Indole

Key point: *Structure of products and mechanism*

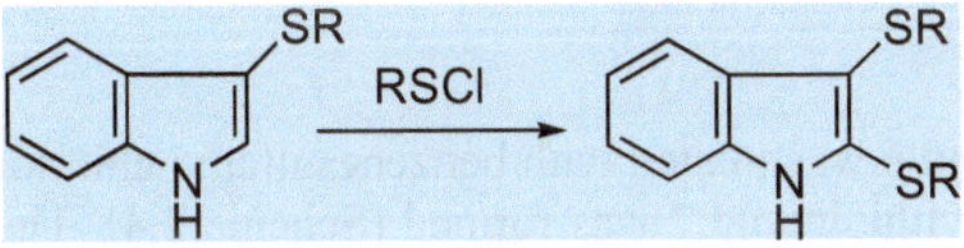

Sulfenylation of Indole

3-Indolyl sulfides **1** are easily prepared by sulfenylation of indoles with sulfenyl chlorides. The mechanism of the sulfenylation is well known and follows the usual S_EAr-type substitution pathway depicted in Scheme 2.1.

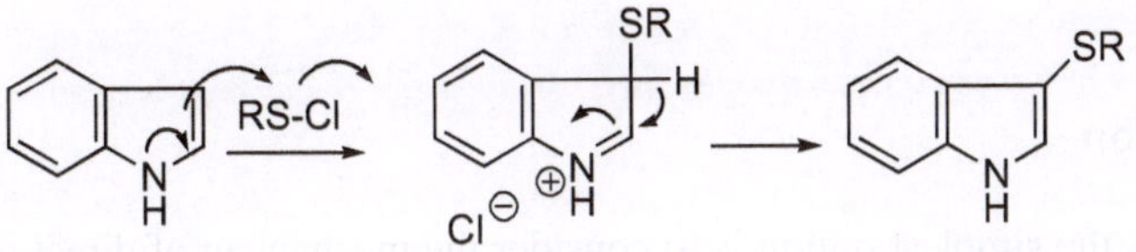

Scheme 2.1

One drawback to the use of sulfenyl chlorides for the sulfenylation of indoles is their exceptional reactivity. If the 2-position of the indole is unoccupied, the slightest excess of reagent leads to a second sulfenylation, and a full second equivalent leads to excellent yields of 2,3-indolyl bis-sulfides **2** (Scheme 2.2).

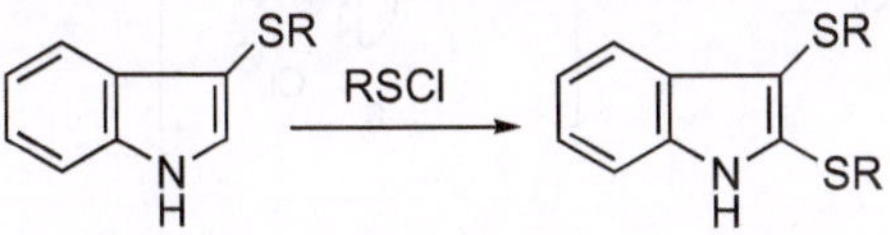

Scheme 2.2

The question to be resolved is the following: *is the second sulfide group introduced directly at the 2-position of the indole ring or is it necessary to consider an alternative mechanism for the formation of bis-sulfides **2**?*

Experimental Data

The second sulfenylation of 3-(phenylthio)indole **3** with methanesulfenyl chloride yields a mixture of the bis-sulfides **4** and **5** (Scheme 2.3).

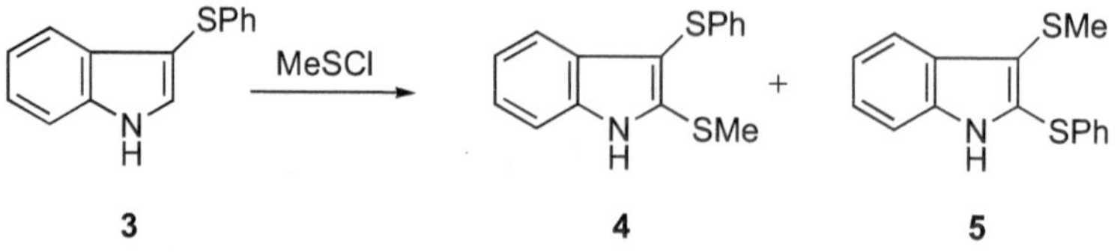

Scheme 2.3

When compound **6** was treated with benzenesulfenyl chloride at room temperature, 2,3-bis-phenylthioindole **7** was formed (Scheme 2.4). The reaction needs **at least** a full equivalent of sulfenyl chloride to go to completion.

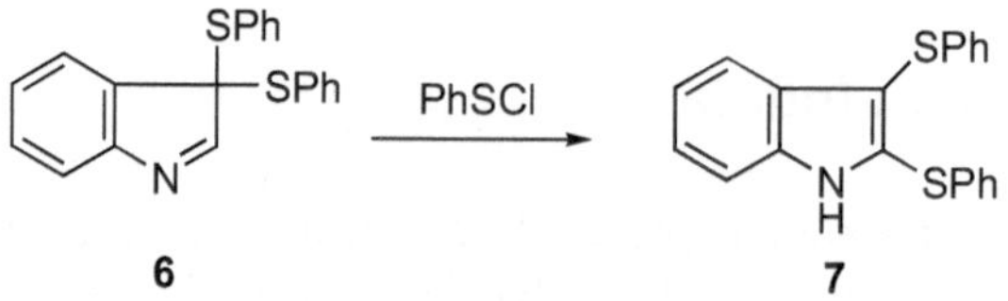

Scheme 2.4

Discussion

For a start, the simplest option is to consider the mechanism of direct introduction of the sulfenyl group into the 2-position (Scheme 2.5). The nucleophilic attack of the indole nucleus to the sulfenyl chloride requires a temporary disruption of the aromaticity of the ring system leading to intermediate **8**, which after aromatization leads to the final product **2**.

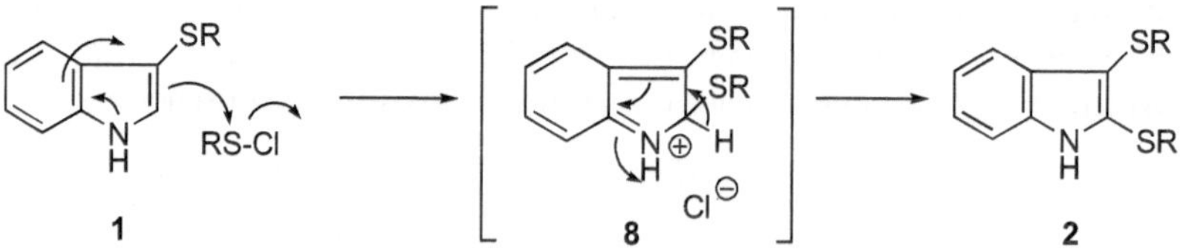

Scheme 2.5

Although reasonable, this mechanism cannot explain the mixture of isomers obtained in the sulfenylation of 3-(phenylthio)indole **3** with methanesulfenyl chloride. As represented in Scheme 2.6, under these conditions, **3** should exclusively yield bis-sulfide **4**. The direct sulfenylation at the 2-position of the indole ring should be thereafter discarded.

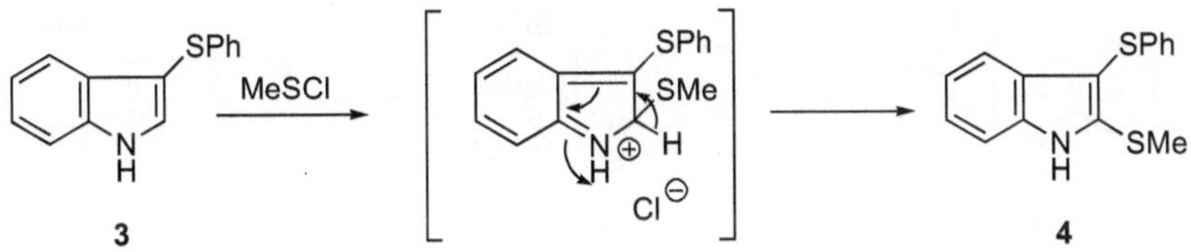

Scheme 2.6

As an alternative, we should consider the introduction of the second sulfide at the 3-position of the indole, and check if it is possible to obtain the reaction products this way. The appealing feature of this mechanism is that the aromaticity of the benzene ring remains undisturbed along the process.

The 3-position is the most nucleophilic site of the indole nucleus and considering the absence of steric effects, the presence of a sulfide group would not be expected to deactivate the 3-position toward further substitution. An intermediate like **9** should be formed in the first step of the reaction (Scheme 2.7). Migration of the SR group would probably occur through an episulfonium ion intermediate **10**. The cleavage of the three-memberered ring in **10** and subsequent aromatization should yield bis-sulfide **2**. The formation of an episulfonium intermediate is reasonable, considering that the sulfur atom is a good nucleophile and that the iminium group of the five-membered ring is highly electrophilic.

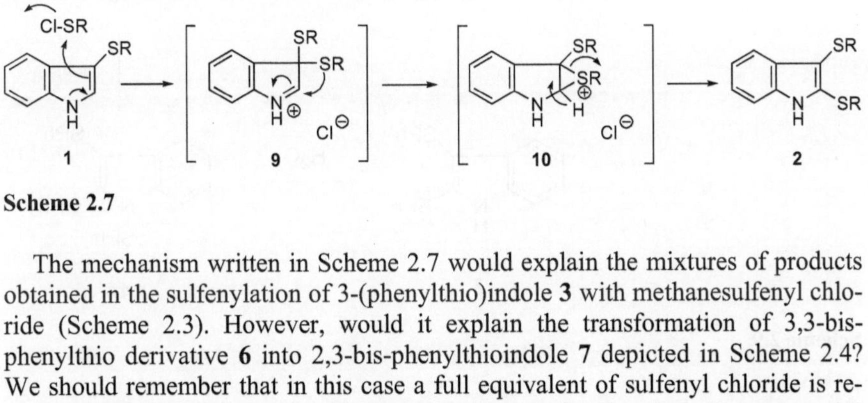

Scheme 2.7

The mechanism written in Scheme 2.7 would explain the mixtures of products obtained in the sulfenylation of 3-(phenylthio)indole **3** with methanesulfenyl chloride (Scheme 2.3). However, would it explain the transformation of 3,3-bis-phenylthio derivative **6** into 2,3-bis-phenylthioindole **7** depicted in Scheme 2.4? We should remember that in this case a full equivalent of sulfenyl chloride is required for the completion of the reaction. Considering this fact, if compound **7** were formed just by rearrangement of **6** no extra sulfenyl chloride would be needed. *What for compound* **6** *requires the extra equivalent of reagent?*

The sulfenylation-rearrangement mechanism for compound **6** would require the formation of intermediate **11** (Scheme 2.8). However, the C=N bond in **6** is less electrophilic than the iminium bond of the intermediate **9** in Scheme 2.7 and hence, the nucleophilic attack by the adjacent SPh group should be less favored. In addition, even if **11** is formed, the rearrangement to **7** by ring opening and protonation of the nitrogen, does not require (in principle) any extra amount of the reagent.

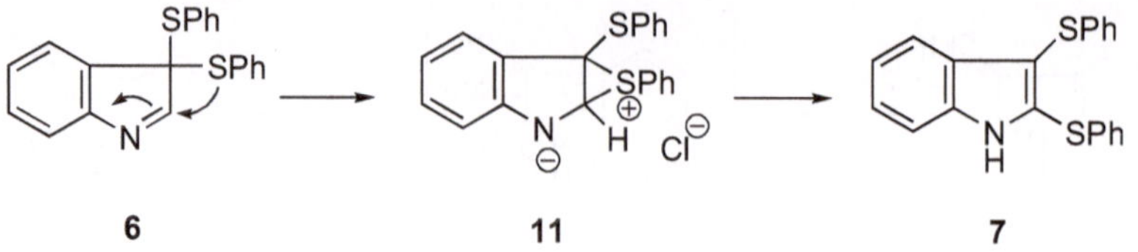

Scheme 2.8

It is more likely to assume that if sulfonyl chloride is present in the medium, the quaternization of the nitrogen in **6** is possible, leading to a *N*-sulfenyl intermediate **12** in the first step of the reaction. Now, the electrophilicity of the C=N bond is increased and a nucleophilic attack by the SPh group would form the episulfonium species **13** that by ring opening would lead to *N*-phenylsulfenyl 2,3-bis-sulfide **14**. Removal of the SPh group in **14** during the aqueous workup would yield the final bis-sulfide **7** (Scheme 2.9).

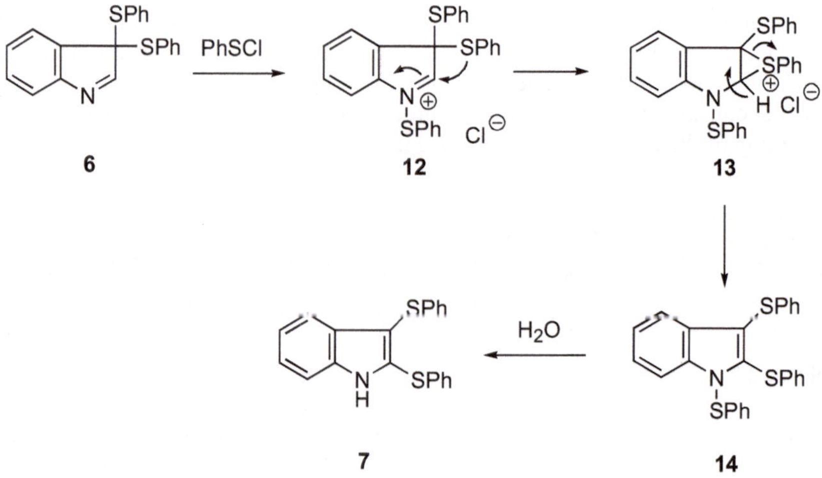

Scheme 2.9

In Summary

Sulfenylation of indole with sulfenyl chlorides occurs initially at the 3-position. In excess of the reagent, the second sulfide group is introduced at the 2-position by initial formation of an indolenium 3,3-bis-sulfide intermediate like **9**, followed by migration of one of the sulfide groups to the 2-position through an episulfonium species **10**.

Question

How could we get evidence in support of the proposed quaternization of the nitrogen atom in 3,3-bis(phenylthio)-3H-indole **6**?

Answer

If quaternization of **6** did occur, the rearrangement would also proceed in the presence of alkylating agents like methyl iodide, benzyl bromide or methyl trifluoromethanesulfonate. Alkyl halides would lead to N-alkylated derivatives **15**, which after rearrangement in the medium and aqueous workup would yield the 2,3-indolyl bis-sulfide **7** (Scheme 2.10).

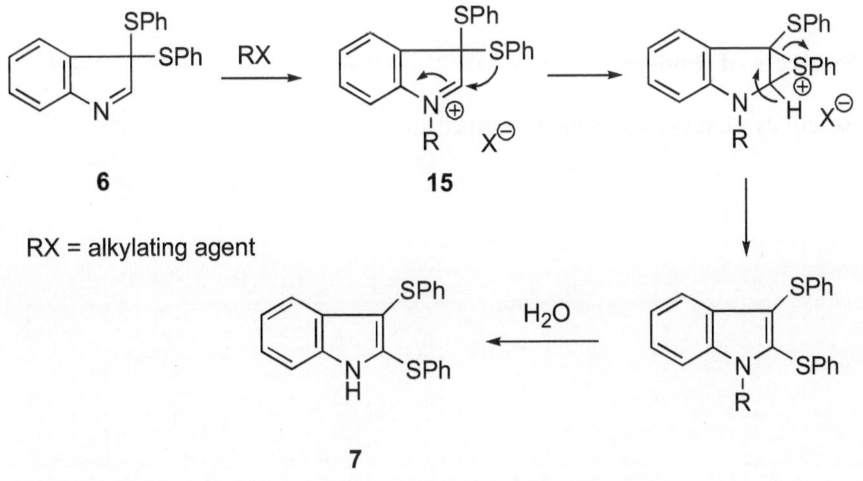

RX = alkylating agent

Scheme 2.10

The authors have tried the alkylation experiments with no success. An explanation for the failure of the rearrangement in these cases can be found in the different reactivity of indole towards sulfenyl halides and alkylating agents. Whereas the sulfenylation of indole is very facile and does not need the addition of base, it is well known that alkylation of indole requires the initial generation of the anion **16** to obtain 1- or 3-alkylated products (Scheme 2.11). Therefore, the failure in promoting the transformation of **6** into **7** by alkylating agents should be attributed to the failure of these reagents to alkylate the indole nitrogen atom under the reaction conditions.

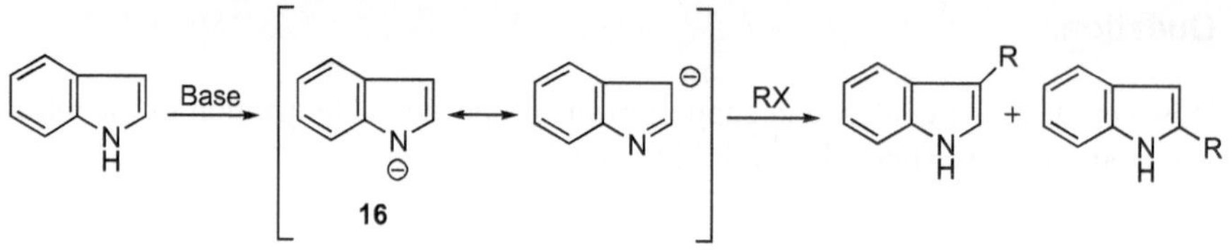

Scheme 2.11

Additional Comments

This problem is based on the work by Hamel P (2002) *J. Org. Chem.* 67:2854-2858 and by the work of Hamel P, Préville P (1996) *J. Org. Chem.* 61:1573-1577.

Subjects of Revision

Reactivity of indoles. Reaction intermediates.

Level 1– Case 3
Substrate Selective Reactions in the Presence of Lewis Acids

Key point: *Selectivity*

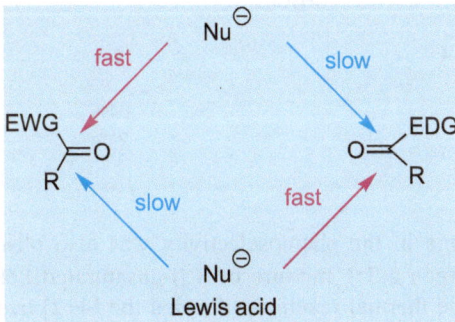

Substrate Selective Reactions in the Presence of Lewis Acids

For an organic chemist it is evident that more electrophilic aldehydes and ketones **1** react with nucleophiles much faster than less electrophilic analogues **2**. However, not as evident is the fact that the reverse situation occurs in the Lewis acid-mediated reactions. That is, more electrophilic aldehydes and ketones **1** react much slower than the less electrophilic analogues **2** in the presence of a Lewis acid (Scheme 3.1).

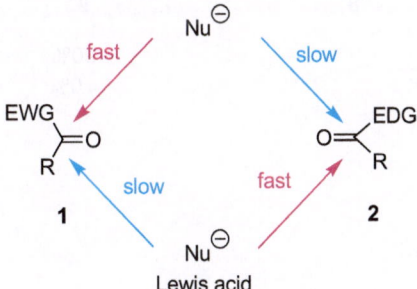

EWG = electron-withdrawing group
EDG = electron-donating group

Scheme 3.1

These statements are supported by the results obtained in the following examples. The reaction of a 1:1 mixture of *p*-trifluoromethylbenzaldehyde **3** and *p*-tolualdehyde **4** with one equivalent of allyltributylstannane gives the homoallylic alcohol **5** derived from **3**, and only trace amounts of the corresponding alcohol **6** derived from **4**. This is an expected result. However, when the reaction is carried out in the presence of $BF_3 \cdot OEt_2$, alcohol **6** is isolated as the major reaction product (Scheme 3.2).

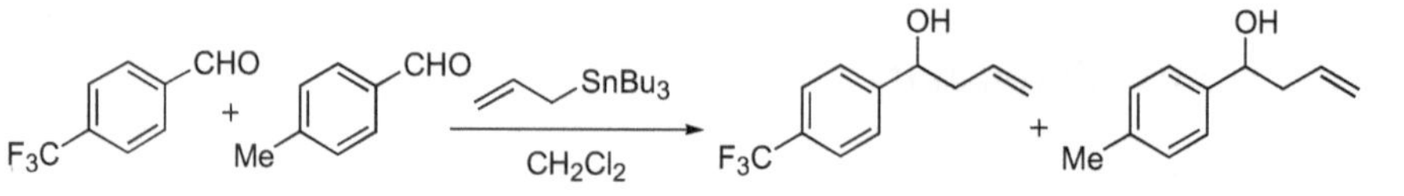

3	**4**		**5**	**6**
120 °C			64%	trace
$BF_3.OEt_2$, -100 °C			21%	63%

Scheme 3.2

A dramatic change in the chemoselectivity was also observed in the Diels-Alder reaction between a 1:1 mixture of α,β-unsaturated ketones **7** and **8** with cyclopentadiene. The thermal reactions afforded the [4+2] cycloadduct **9** derived from trifluoromethyl ketone **7** (90% yield) along with a small amount of adduct **10** derived from methyl ketone **8**. When the reaction was run in the presence of a Lewis acid ($BF_3 \cdot OEt_2$) cycloadduct **10** was the sole product, isolated in 49% yield (Scheme 3.3).

Rationalize all the affirmations made previously and justify the experimental results!

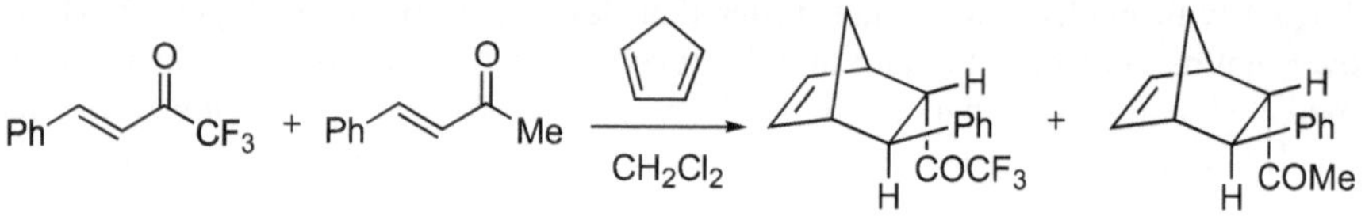

7	**8**		**9**	**10**
40 °C			90%	8%
$BF_3.OEt_2$, -78 °C			0%	49%

Scheme 3.3

Experimental Data

1. The *ab initio* study about the stability of the BF_3 complexes of acetaldehyde and trifluoroacetaldehyde (GAUSSIAN 94 program), has indicated that the $CF_3CHO \cdot BF_3$ complex **11** is 6.46 kcal mol^{-1} less stable than the $CH_3CHO \cdot BF_3$

complex **12** (Figure 3.1). In both cases the most stable conformers have been compared.

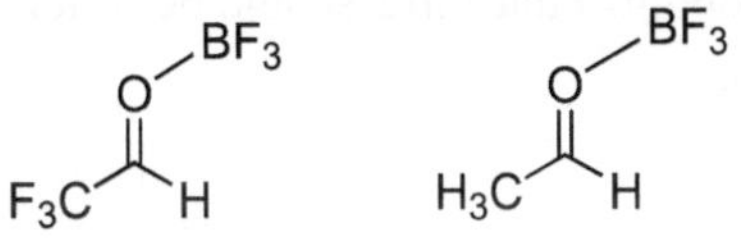

11

less stable

12

Figure 3.1

2. The ^{13}C-NMR signals of the carbonyl carbons ($-78°C$ in CD_2Cl_2) of a mixture of acetophenone **13** and trifluoroacetophenone **14** appear at 197.94 and 179.59 ppm, respectively. When the mixture was treated with $BF_3 \cdot OEt_2$ (up to two equivalents) signals at 214.65 and 179.59 ppm were observed (Figure 3.2).

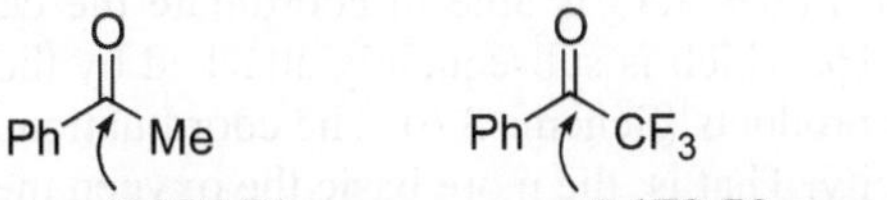

δ 197.94 ppm   δ 179.59 ppm

13   **14**

Figure 3.2

Discussion

The carbonyl carbon is electron deficient due to the electronegativity of the oxygen atom. In consequence, the nucleophilic attack to a C=O group strongly depends on the electrophilicity of the carbon atom in the carbonyl function. Electron-withdrawing substituents (R^1, R^2 = EWG) will make this position more electrophilic and hence, will favor the nucleophilic addition as indicated in Scheme 3.4.

Nucleophilic addition to a carbonyl group

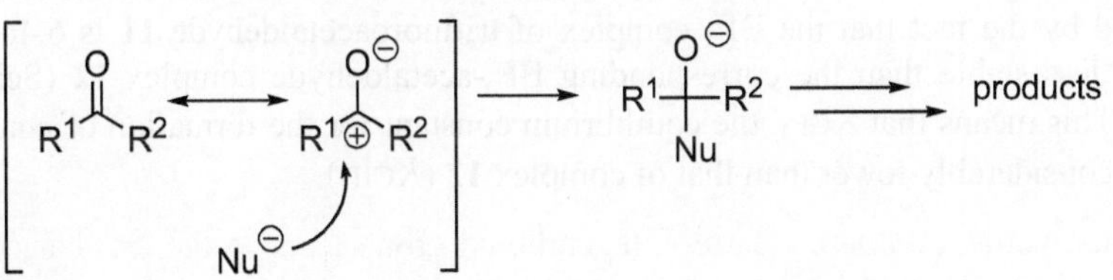

Scheme 3.4

These affirmations could explain the results obtained in the reaction between benzaldehydes **3** and **4** and allyltributylstannane. Thus, *p*-trifluoromethyl benzaldehyde **3** is more electrophilic than *p*-tolualdehyde **4**, as a result of the presence of the electron-withdrawing CF_3 group at the *para* position of the aromatic ring. In consequence, the nucleophilic attack in **3** should be faster than in **4** (Scheme 3.5).

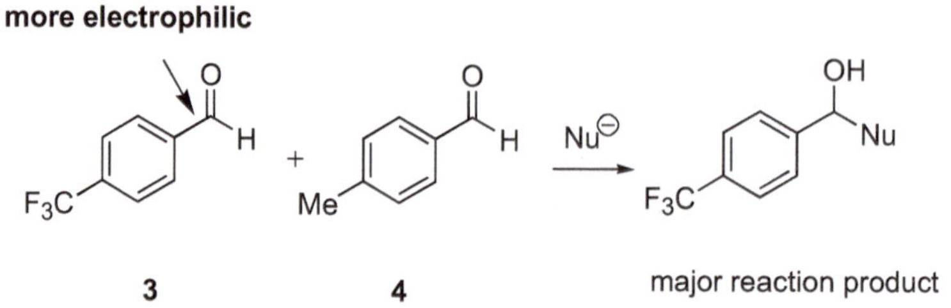

more electrophilic

3   **4**   major reaction product

Scheme 3.5

The situation is different when a Lewis acid is involved in the reaction. It is well known that the Lewis acid is able to coordinate the carbonyl oxygen atom forming a complex **15**, which is subsequently attacked by the nucleophile leading to the final reaction products (Scheme 3.6). The coordination ability of the oxygen depends on its basicity. That is, the more basic the oxygen the better the coordination complex **15** is formed. The presence of electron-withdrawing groups (R^1, R^2 = EWG) should decrease the basicity of the oxygen atom, making the formation of the complex with the Lewis acid much more difficult.

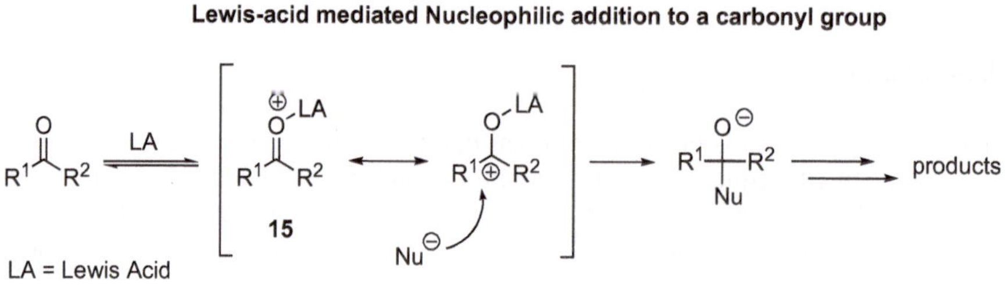

Lewis-acid mediated Nucleophilic addition to a carbonyl group

LA = Lewis Acid

Scheme 3.6

These arguments, based only on elementary electronic considerations, are confirmed by the fact that the BF_3 complex of trifluoroacetaldehyde **11** is 6.46 kcal mol^{-1} less stable than the corresponding BF_3-acetaldehyde complex **12** (Scheme 3.7). This means that K_{CF_3}, the equilibrium constant for the formation of complex **11** is considerably lower than that of complex **12** (K_{CH_3}).

In other words, the more electrophilic carbonyl group leads to the less basic oxygen and hence to the less stable complex with the Lewis acid.

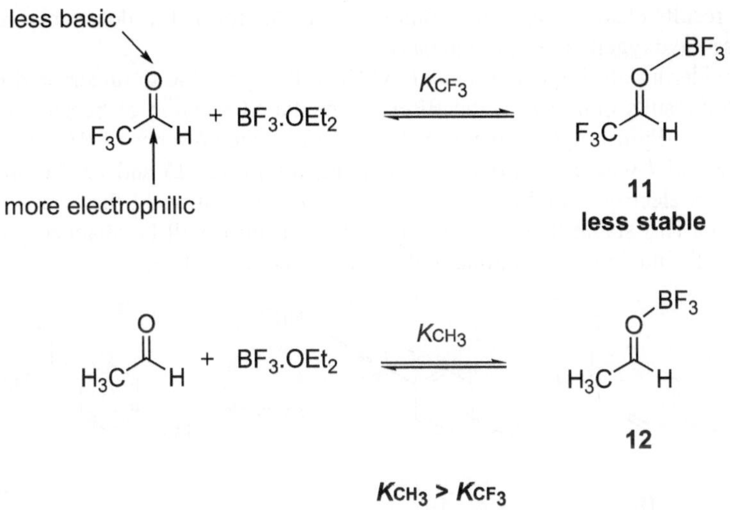

less basic

more electrophilic

K_{CF_3}

11
less stable

F_3C ... H + BF$_3$·OEt$_2$

K_{CH_3}

12

H_3C ... H + BF$_3$·OEt$_2$

$$K_{CH_3} > K_{CF_3}$$

Scheme 3.7

The preferred formation of complex **12** rather than **11** was also supported by the low-temperature ^{13}C-NMR data of acetophenone **13** and trifluoroacetophenone **14** shown in Figure 3.3. The ^{13}C-NMR chemical shift of the carbon bearing a positive charge (the carbonyl carbon atom in this case) correlates with the electron density on this position. If the carbon atom becomes more positive, the ^{13}C signal will be deshielded. As the coordination with a Lewis acid makes the carbonyl carbon more positive, the study of the ^{13}C-NMR spectra provides a useful source of information about the stability of the complexes formed.

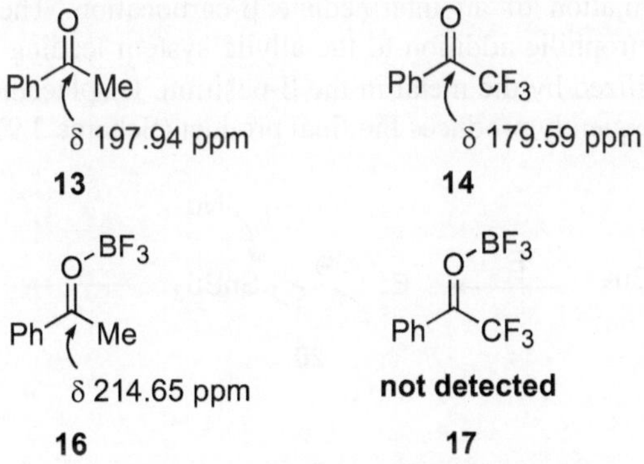

Ph ... Me
δ 197.94 ppm
13

Ph ... CF$_3$
δ 179.59 ppm
14

Ph ... Me
δ 214.65 ppm
16

Ph ... CF$_3$
not detected
17

Figure 3.3

When the ^{13}C-NMR study of the mixture of ketones **13** and **14** was run in the presence of BF$_3$·OEt$_2$, the original signal at 197.94 ppm of acetophenone **13** was replaced by other considerably deshielded at 214.65 ppm, assignable to the corresponding BF$_3$ complex **16**. In contrast, the signal at 179.59 ppm of trifluoroacetophenone **14** did not change, even when an excess of BF$_3$·OEt$_2$ was employed.

These results clearly imply that complex **17** is not formed in detectable amounts because the oxygen is not enough basic.

Considering all this experimental evidence it is possible to understand the *unexpected* results obtained in the $BF_3 \cdot OEt_2$-mediated reaction of benzaldehydes **3** and **4** with allyltributylstannnane (Scheme 3.8). In the presence of $BF_3 \cdot OEt_2$, aldehydes **3** and **4** would form the corresponding complexes **18** and **19**. Complex **18**, having an electron-withdrawing group at the *para* position of the aromatic ring, should be less stable than complex **19**, which in turn, will be attacked preferentially by the nucleophile, leading to the main reaction product.

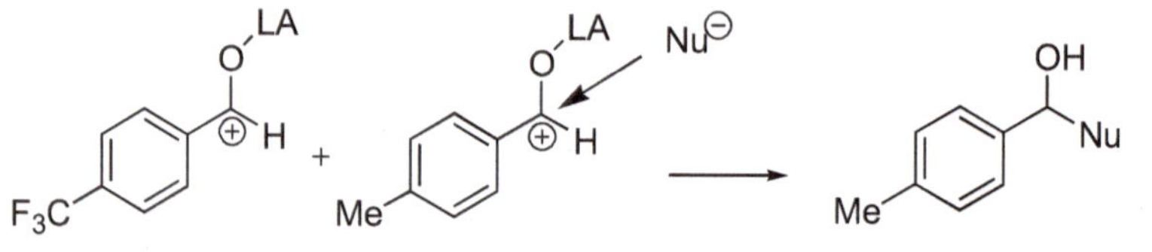

18	19	
LA = Lewis Acid	**more stable**	major reaction product

Scheme 3.8

At this point the student is encouraged to draw a reasonable mechanism to explain the addition of allyltributylstannane to aldehydes 3 and 4.

As help, we will remember how an allyltributylstannane reacts with an electrophile. Allylstannanes (like allylsilanes) have been extensively used as allyl anion equivalents. Considering the ability of the metal to stabilize a positive charge in the β-position by hyperconjugation, it seems reasonable to propose a mechanism involving the formation of an intermediate β-carbocation. Then, the first step would be the electrophilic addition to the allylic system leading to the formation of cation **20**, stabilized by the metal in the β-position. Displacement of the metallic fragment subsequently produces the final product (Scheme 3.9).

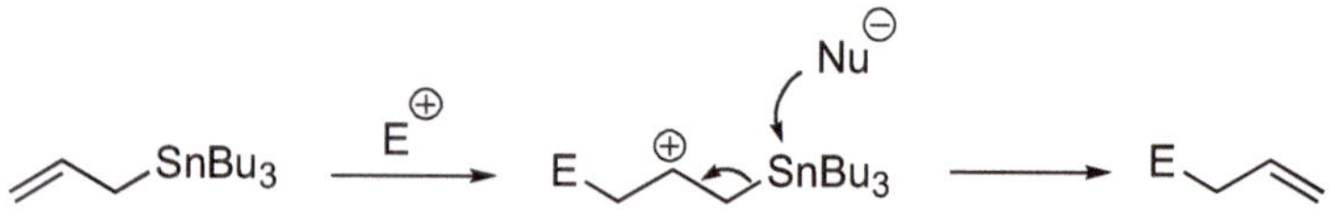

20

Scheme 3.9

Once we have discussed the results obtained in the reactions with aromatic aldehydes and allylstannanes, we will focus our attention to the results obtained in the Diels-Alder reaction.

Clearly, the outcome of the Diels-Alder reaction between compounds **7** and **8** and cyclopentadiene is not surprising since it is well known that **7**, having an electron-withdrawing CF_3 group, must be a better dienophile than **8** (Scheme 3.10). We should remember that *normal* Diels-Alder reactions are HOMO diene-LUMO dienophile controlled and that electron-withdrawing substituents lower the LUMO energies of dienophiles, accelerating the reaction rate.

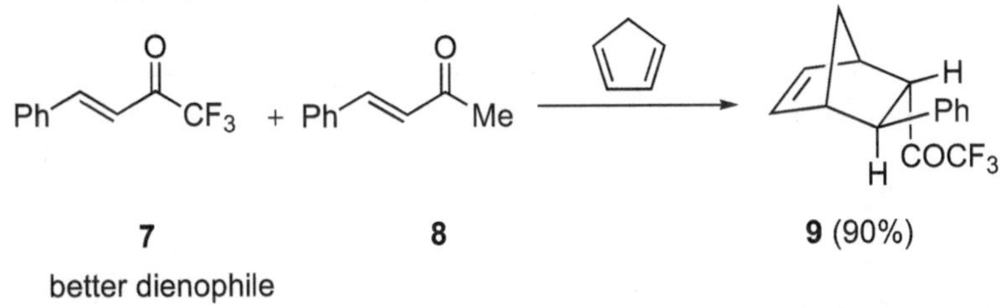

7   **8**   **9** (90%)

better dienophile

Scheme 3.10

The reversal of the selectivity during the reaction in the presence of a Lewis acid could be interpreted (as we have discussed previously), in terms of **the different stabilities of the Lewis acid-carbonyl complexes formed in each case.**

The Lewis acid-catalyzed Diels-Alder reactions have been studied in depth, particularly when α,β-unsaturated aldehydes and ketones are involved as dienophiles. It has been established that the effect of the catalyst is to lower the LUMO energy of the dienophile, by coordination with the carbonyl group. It is evident that the more stable complex will cause a larger lowering in the LUMO energy. In this case, in the presence of the Lewis acid, complexes **21** and **22** should be formed. However, as we have discussed above, the coordination ability of more electrophilic carbonyl groups to a Lewis acid is weaker than that of their less electrophilic analogues. In fact, we have commented previously that trifluoroacetophenone-BF$_3$ complex **17** has not been detected by NMR. Consequently, it is reasonable to consider that in the presence of a Lewis acid only **22** is formed in the reaction medium (Scheme 3.11).

Therefore, in the presence of BF$_3$·OEt$_2$, compound **8** would form a coordination complex with the Lewis acid **22**, which is a better dienophile (lower LUMO) than **7**. The cycloaddition between **22** and cyclopentadiene affords the observed reaction product (Scheme 3.11).

7

21   **22**   **10**

not formed   reaction product

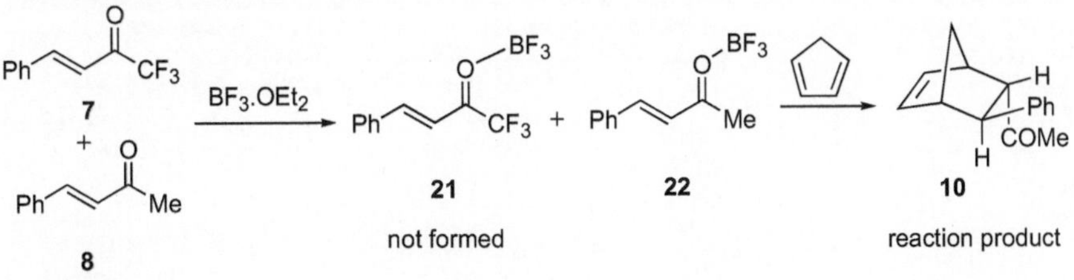

Scheme 3.11

In Summary

The chemoselectivity of the reaction of carbonyl groups and nucleophiles can be reversed in the presence of a Lewis acid in the reaction medium. Whereas the nucleophilic attack to a carbonyl group depends mainly on the electrophilicity of the carbon atom, in the Lewis acid promoted reactions the chemoselectivity of the reaction seems to be determined by the stability of the complex formed between the Lewis acid and the carbonyl group.

Additional Comments

This problem is based on the work by Asao N, Asano T, Yamamoto Y (2001) *Angew. Chem. Int. Ed.* 40:3206-3208.

Subjects of Revision

Additions to C=O groups. Lewis acid-catalyzed Diels-Alder reactions.

Level 1 – Case 4
Diasteroselective Reductions of β-Ketoesters

Key point: *Chelation versus non-chelation control*

Diastereoselective Reductions of β-Ketoesters

The stereochemical result of the Lewis acid-mediated reductions of α-alkyl-β-ketoesters **1** strongly depends on the Lewis acid employed, the reducing agent and the solvent. For example, the reduction of **1** with $BH_3 \cdot py$ in the presence of $TiCl_4$ and CH_2Cl_2 as solvent, yields the *syn*-isomer **2** with diastereomeric *syn/anti* excesses higher than 95/5, whereas reduction of **1** with lithium triethylborohydride ($LiEt_3BH$) in THF, using $CeCl_3$ as Lewis acid, yields *anti*-isomer **3** with diastereoselectivities higher than 90/10 in the cases studied (Scheme 4.1).

Although the reduction of β-ketoesters **1** with $BH_3 \cdot py/TiCl_4$ was studied in different solvents (CH_2Cl_2, THF, and Et_2O) dichloromethane proved to be the best solvent of choice since the use of THF or Et_2O led to the complete loss of diastereoselectivity.

Discuss the role of the Lewis acid and the solvent polarity in the control of the diastereoselectivity of the reaction. Considering the previous discussion, predict the stereochemistry of the product/s obtained in the reduction of 2,3,3-trimethyl-1-phenyl-1-butanone under the same conditions employed for compound 1.

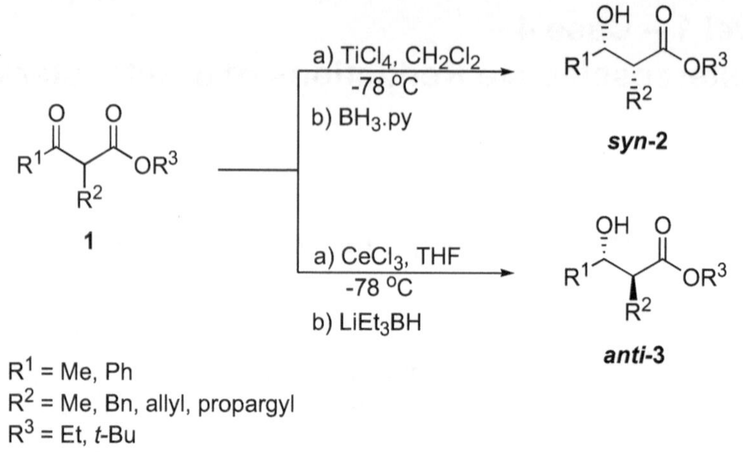

Scheme 4.1

R^1 = Me, Ph
R^2 = Me, Bn, allyl, propargyl
R^3 = Et, *t*-Bu

Experimental Data

Donor numbers *DN* (kcal mol^{-1}): 1,2-dichloroethane (*DN* 0.0); Diethyl ether (Et$_2$O) (*DN* 19.2); Tetrahydrofurane (THF) (*DN* 20.0).

Discussion

The rationalization of the diastereoselectivity observed in the nucleophilic additions to carbonyl groups with an adjacent chiral center has been a subject of interest since the early studies of Cram. Along the years, several models based on conformational analysis have been developed, and among them, probably the *Felkin-Anh's model* is the most widely used. According to the Felkin-Anh's model, the reactive conformation (conformation of lowest energy in the transition state) for the addition of a nucleophile to a carbonyl group has the bonds to the L (large), M (medium) and S (small) substituents staggered relative to the C=O function (Figure 4.1). The nucleophile will attack preferentially on the side of the plane containing the small group, following a non-perpendicular trajectory called *Bürgi-Dunitz trajectory*. That is, the nucleophile approaches at an angle of about 109° with respect to the plane of the carbonyl group.

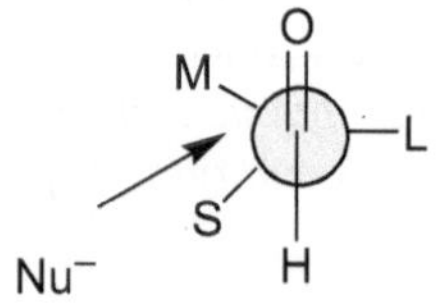

Figure 4.1

The additions of nucleophiles to aldehydes and ketones are promoted by coordination of a Lewis acid to the oxygen atom of the carbonyl group. The coordination with the metal enhances the electrophilicity of the C=O group facilitating the attack of the nucleophile. From a stereochemical point of view, the presence of a Lewis acid is particularly important when a substituent with a heteroatom able to coordinate with the metal is placed next to the carbonyl group. In such cases, the prediction of the stereoselectivity of the reaction requires a chelated reactive conformation as that represented in Figure 4.2. This model is known as *Cram's cyclic model* and again the attack of the nucleophile takes place preferentially from the less-hindered side.

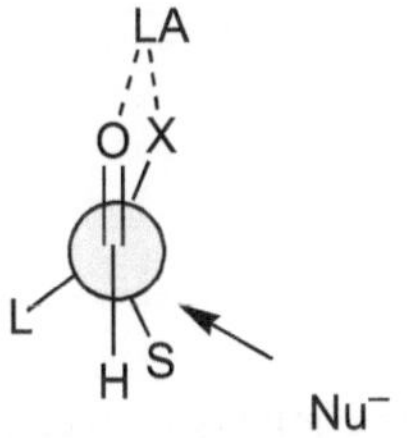

LA = Lewis acid
X = O, N, S

Figure 4.2

The dramatic change of the stereoselectivity with the Lewis acid observed in the reductions of β-ketoesters **1**, could be rationalized on the basis of the different chelating ability of the metals involved in the process. In this case, there are major differences between $TiCl_4$ and $CeCl_3$; *whereas $TiCl_4$ is a strong chelating agent, $CeCl_3$ is not.*

The results obtained in the $TiCl_4$-mediated reductions of β-ketoester **1** could be understood considering the formation of a chelate between the metal atom, the carbonyl function and the β-carbonyl group. Compound **1** has a stereogenic center between the ester and the keto group and we have only represented the model for one of the two possible enantiomers (Scheme 4.2). The $TiCl_4$ complex can be represented as an equilibrium between the conformations **4** and **5**, although the unfavorable steric interaction between the R^2 substituent and the oxygen atom of the C=O group in **5** makes this conformation less stable than **4**. The cyclic intermediate is then attacked by the incoming hydride preferentially from the less hindered side of the most populated conformation **4** leading to the *syn*-alcohol **2** with high diastereoselectivity (*syn* means that the OH and the R^2 groups are on the same side of the molecule on staggered conformation).

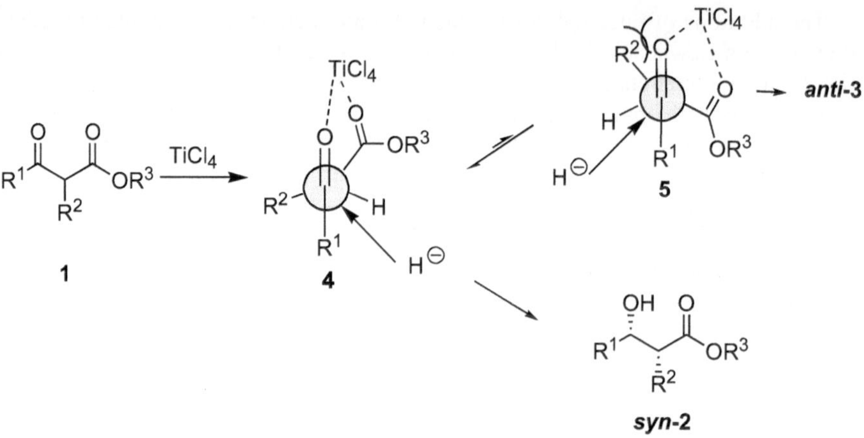

Scheme 4.2

The formation of the chelates between β-ketoesters **1** and TiCl$_4$ is favored in a non-coordinating solvent such as CH$_2$Cl$_2$ but is more difficult when coordinating solvents like THF or Et$_2$O are employed.[1] In such cases the stereochemical control due to the formation of the chelation products with the Lewis acid is lost.

The stereochemistry of the reaction products is reversed when the poor chelating Lewis acid CeCl$_3$ is employed. In this case, an open-chain Felkin-Anh's model is the most adequate to justify the *anti*-selectivity observed during the reduction of β-ketoesters **1**. The first step of the reaction should be the coordination of the CeCl$_3$ with the oxygen atom. As indicated in the Scheme 4.3, the **1**-CeCl$_3$ complex can be represented as equilibrium between conformations **6** and **7**. Invoking the Bürgi-Dunitz trajectory, the attack to the conformation **6** should be preferred by steric effects, leading to the *anti*-alcohol **3** as the major reaction product (*anti* means that the OH and the R^2 group are on opposite sides of the molecule on staggered conformation).

---

[1] The donor numbers (*DN* kcal mol^{-1}) are a measure of the coordinating ability of a solvent. They have been defined as the negative ΔH values for 1:1-adduct formation between antimony pentachloride and an electron pair donor solvent (D) in dilute solution of a non-coordinating solvent (1,2-dichloroethane), according with Eq. 4.1.

$$D: + SbCl_5 \rightleftharpoons \overset{\oplus}{D} - \overset{\ominus}{SbCl_5} \qquad (4.1)$$

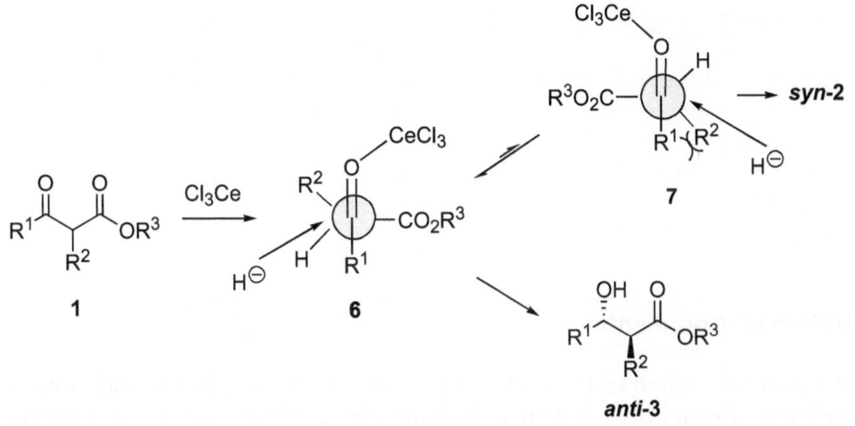

Scheme 4.3

Considering the previous discussion, we can predict the diastereoselectivity in the reduction of 2,3,3-trimethyl-1-phenyl-1-butanone (**8**) either with BH$_3$. py/TiCl$_4$ or with LiEt$_3$BH/CeCl$_3$.

The 2,3,3-trimethyl-1-phenyl-1-butanone is an α-substituted ketone, but the α-substituents have no ability to chelate metals. In that case, the role of the Lewis acid will be just to coordinate the oxygen atom, enhancing its reactivity towards the nucleophilic attack. The same open Felkin-Ahn's model previously commented will be valid for both, TiCl$_4$ and CeCl$_3$-mediated reductions, as it is shown in Scheme 4.4. Again, the first step of the reaction is the coordination of the Lewis acid with the carbonyl group. Although *a priori*, conformations **9** and **10** could be considered, hydride attack to the carbonyl group in **9** should be favored by steric effects leading to the *anti*-alcohol **11** as the main reaction product in both cases.

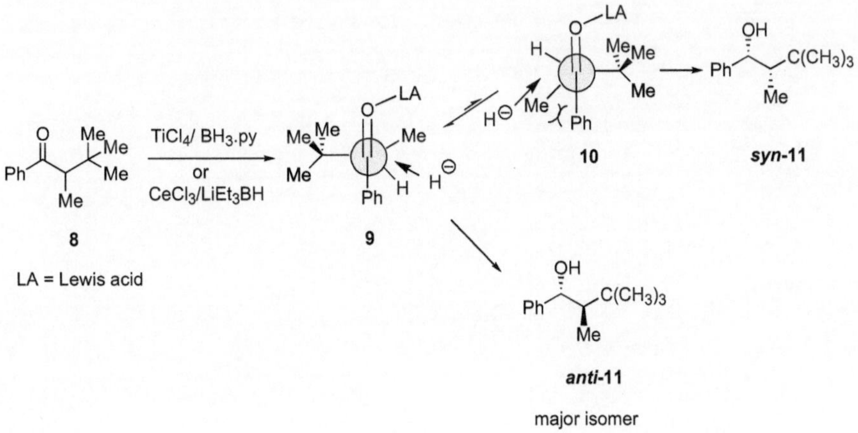

Scheme 4.4

In Summary

The nature of the Lewis acid is determinant in the control of the diastereoselectivity of the hydride reductions of β-ketoesters **1**. Strongly chelating $TiCl_4$ leads mainly to the *syn*-alcohol **2** with $BH_3 \cdot py$ as the reducing agent, while the *anti*-isomer **3** is obtained in preference during the reduction of **1** with $CeCl_3/LiEt_3BH$. The differences in the chelating ability of the metal can explain the reversed stereochemistry observed in the reaction.

Additional Comments

This problem is based on the work by Marcantoni E, Alessandrini S, Malavolta M, Bartoni G, Belluci MC, Sambri L, Dalpozzo R (1999) *J. Org. Chem.* 64:1986-1992.

Subjects of Revision

Models for the diastereoselectivity of additions to C=O groups. Parameters of solvent polarity.

Level 1 – Case 5
Rearrangements from Tetrahydropyran Derivatives

Key point: *Crossover experiments*

Rearrangements from Tetrahydropyran Derivatives

When tetrahydropyranyl ethers **1** and **2** are exposed to excess (3.6 equiv) of SnCl$_4$ at low temperature, an oxygen to carbon rearrangement occurs, resulting in several distinct products that are obtained as diastereomeric mixtures (Scheme 5.1). The formation of these products could be rationalized by considering the initial generation of a carbocationic intermediate (**3** or **4**) from which different final products can be obtained. For example, trapping of cation **3** by chloride ion (already present in the medium) yields the chlorohydrins **5**, whereas aldehydes **6** would be obtained from a 1,2-hydride shift in the intermediate **3**. In other cases, as in **4**, the cation and the oxygen bearing the Lewis acid (still sufficiently nucleophilic) are placed at a distance that allows the cyclization faster than the other competing pathways, leading to bicyclic compounds like **7**.

To account for the formation of carbocations **3** and **4** during the reaction, two alternatives were considered. The first would consist on a concerted process involving simultaneous cleavage of the C-O bond and formation of the C-C bond, as indicated by transition state **8** (path a, Scheme 5.2). The other alternative could be a stepwise process consisting on initial C-O bond cleavage to give fragments **9**

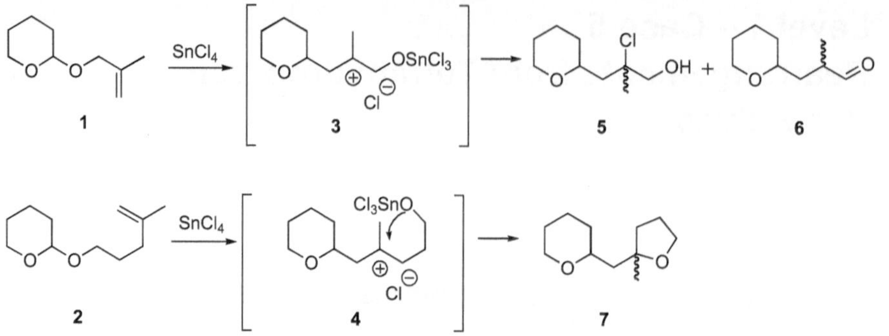

Scheme 5.1

and **10**, which recombine to form the carbocationic intermediate. Fragments **9** and **10** could be *solvent-separated fragments* or a *solvent-caged ion pair* (path b, Scheme 5.2). The investigation of the reaction mechanism was finally done by means of *labeling crossover experiments.*

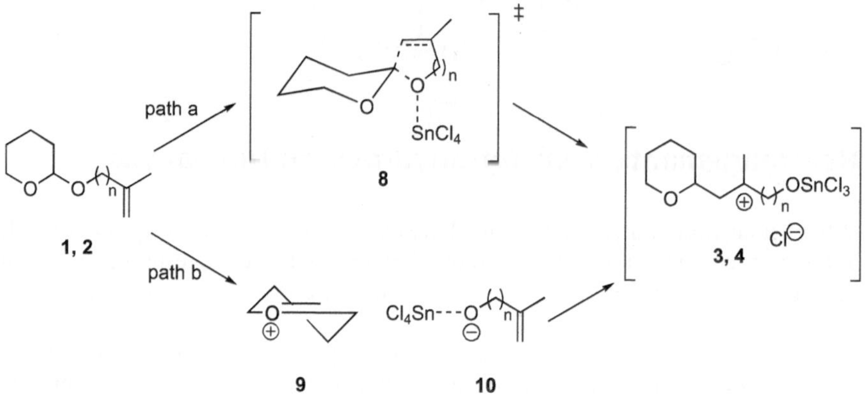

Scheme 5.2

Discuss the following experiments and determine if the rearrangement proceeds via an inter- or an intramolecular pathway

Experimental Data

Experiment 1

Unlabeled *cis*-**11** was treated with SnCl$_4$ in dichloromethane (DCM) at –30°C, yielding exclusively *trans*-**12** as a (3:1) mixture of diastereomers (Scheme 5.3).

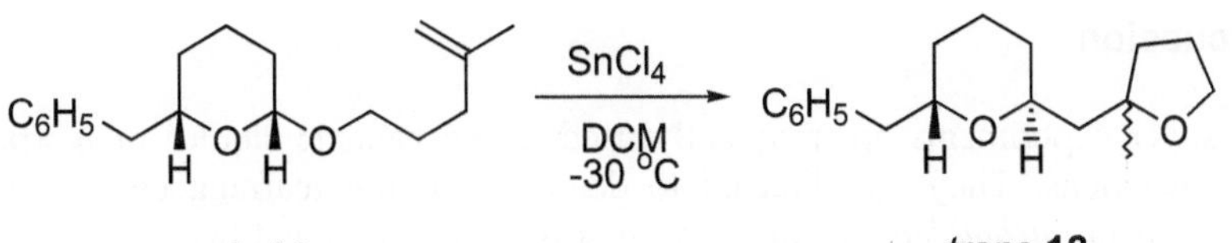

cis-**11** *trans*-**12**

Scheme 5.3

Experiment 2

Labeled *cis*-**13** was treated with $SnCl_4$ under the same conditions used in experiment 1, yielding exclusively *trans*-**14** as a (3:1) mixture of diastereomers (Scheme 5.4). Compounds *cis*-**13** and *cis*-**11** rearranged at comparable rates.

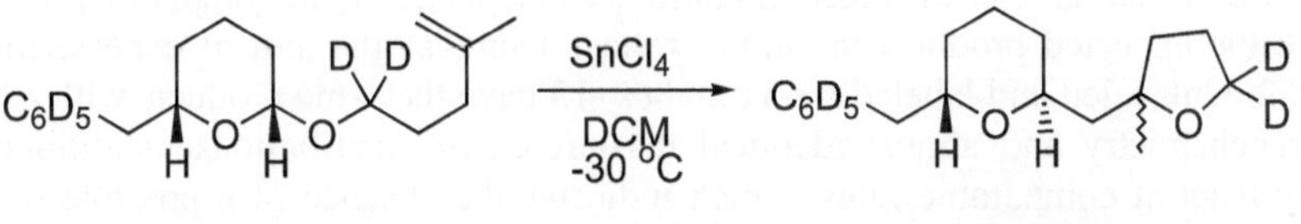

cis-**13** *trans*-**14**

Scheme 5.4

Experiment 3

When equimolar amounts of unlabeled *cis*-**11** and deuterium-labeled *cis*-**13** acetals were treated with $SnCl_4$, a mixture of four compounds **12**, **14**, **15** and **16**, with the expected *trans* stereochemistry in the pyran ring, was obtained in approximately the same yield (determined by mass spectrometric analysis) (Scheme 5.5). The diastereomeric ratio was approximately (3:1) in each case.

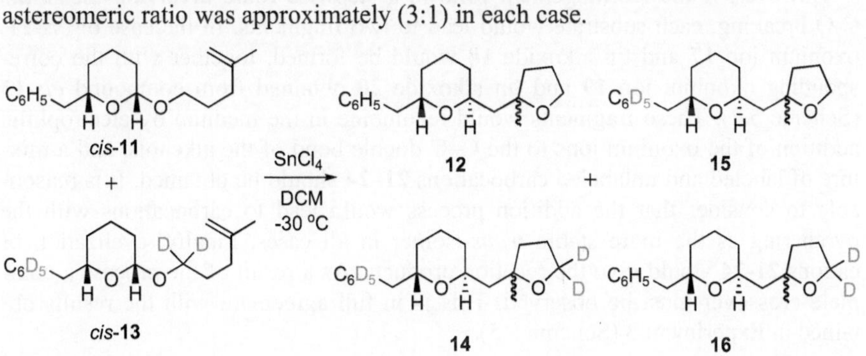

Scheme 5.5

Discussion

Crossover experiments are frequently used in mechanistic studies of rearrangement reactions. They are directed to determine if the rearrangement is *intermolecular* or *intramolecular*. In a typical crossover experiment two substrates, differentiated by substitution or isotopic labeling, are allowed to react together. The absence of cross-products (mixed products) in these experiments is generally interpreted as an argument in favor of an intramolecular rearrangement. Obviously, in this type of experiments it is essential to use substrates that react at comparable rates. That is why the distinguishing substitution should involve very similar groups and should not affect the reactive positions of the molecule. Isotopic labeling is one of the methods most frequently used to distinguish between two substrates in a crossover experiment.

Prior to effecting the crossover experiment, it is necessary to check if the two substrates that have been selected, rearrange independently at comparable rates to give the expected products in similar ratios. This was the aim of experiments 1 and 2. Unlabeled and labeled *cis*-**11** and *cis*-**13** gave the same products with *trans*-stereochemistry and almost identical diastereomeric distributions. Additionally, they react at comparable rates, which indicates the absence of a possible deuterium kinetic isotope effect during the rearrangement of *cis*-**13** and confirms that the doubly deuterated position in this compound has little influence in the reaction course.

Therefore, the results obtained in experiments 1 and 2 confirm that substrates **11** and **13** are a good choice for the crossover experiment.

Next, the crossover experiment (Experiment 3) is conducted. If the reaction occurs through a concerted transition state (like **8** in Scheme 5.2), each substrate should rearrange independently and only a mixture of unlabeled **12** and labeled **14** should be obtained (Scheme 5.6, labeled positions in red).

However, if the rearrangement follows a stepwise route involving the initial C-O breaking, each substrate would lead to two fragments. In the case of *cis*-**11**, oxonium ion **17** and tin alkoxide **18** would be formed, together with the corresponding oxonium ion **19** and tin alkoxide **20** obtained from compound *cis*-**13** (Scheme 5.7). These fragments would recombine in the medium by electrophilic addition of the oxonium ions to the C=C double bond of the alkenols, and a mixture of labeled and unlabeled carbocations **21–24** should be obtained. It is reasonable to consider that the addition process would lead to carbocations with the pyran ring as the more stable *trans*-isomer in all cases. Finally, cyclization of cations **21-24** would give the reaction products. As a result of the process a complete crossover must be observed. This is in full agreement with the results obtained in Experiment 3 (Scheme 5.5).

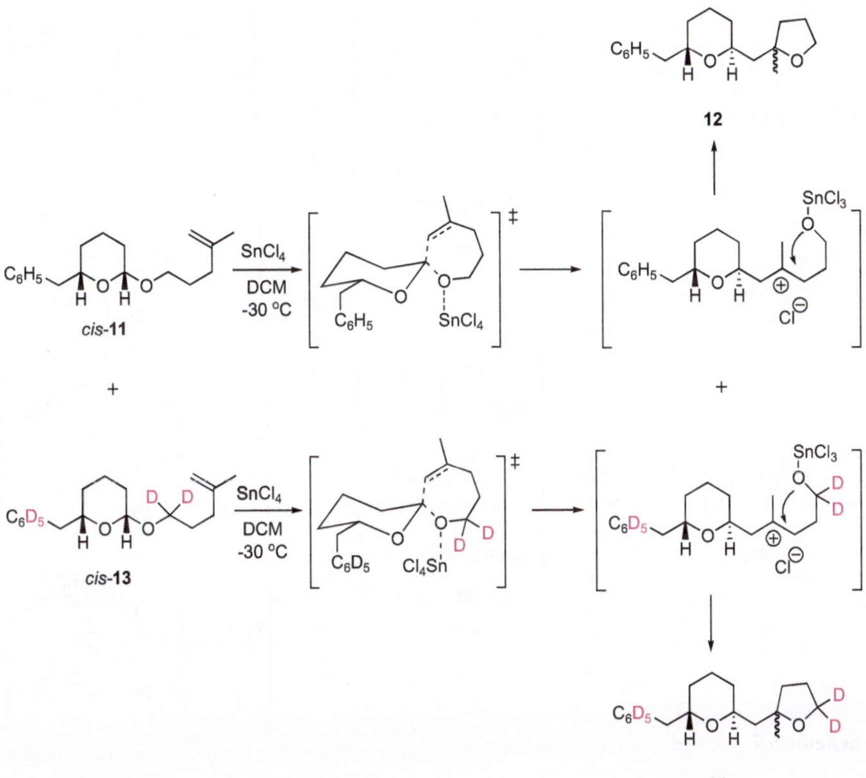

Scheme 5.6

On the basis of the results obtained in the crossover experiments, the rearrangement of the studied tetrahydropyran derivatives is a stepwise process, probably involving a fast fragmentation leading to an oxonium ion and an alkenol, with the oxygen bound as a tin alkoxide. These species are more likely two *solvent-separated fragments* that recombine in a much slower step to form the carbocation intermediate. The possible formation of an *ion pair* seems to be very unlikely in this case, as the recombination of an ion pair within a solvent cage should proceed very rapidly after fragmentation has occurred. Hence, the diffusion of the fragments through the solution does not occur and in consequence, the levels of crossover would be minimal.

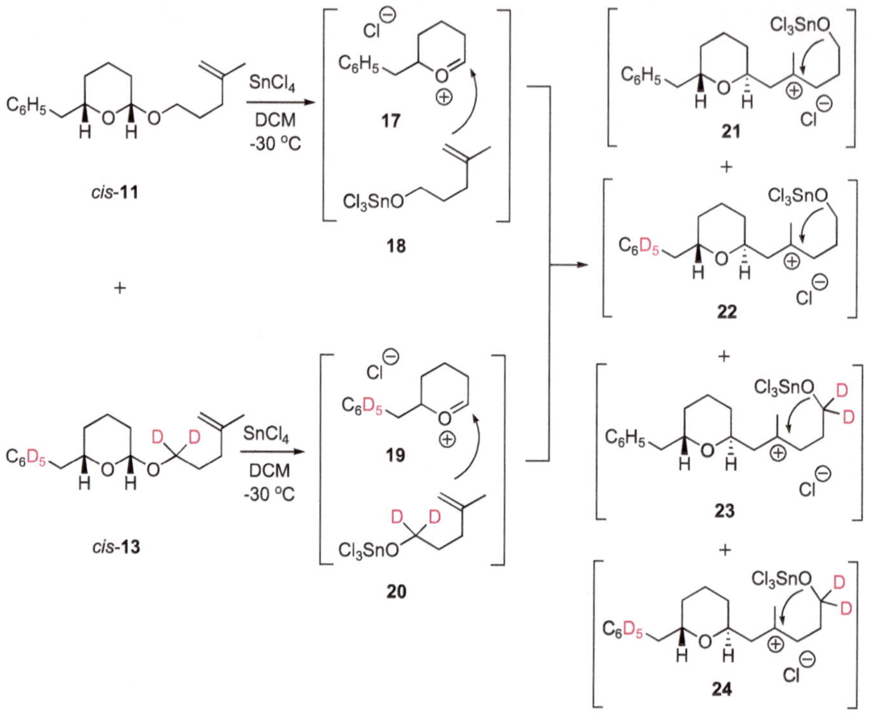

Scheme 5.7

In Summary

By means of crossover experiments it has been demonstrated that the rearrangement of tetrahydropyran derivatives **1** and **2** proceeds by an intermolecular mechanism.

Additional Comments

This problem is based on the work by Buffet MF, Dixon DJ, Edwards GL, Ley SV, Tate EW (2000) *J. Chem. Soc., Perkin Trans 1*, 1815-1827.

Subjects of Revision

Crossover experiments. Carbocations. Ion pairs.

Level 1 – Case 6
Stereospecific Substitution Reactions of Epoxy Sulfides

Key point: *Stereochemistry.*
Neighboring group participation

Stereospecific Substitution Reactions of Epoxy Sulfides

The reaction of epoxy sulfides **1** and **2** with organoaluminium reagents is both *regioselective* and *stereospecific*. Thus, treatment of *cis*-1-(phenylthio)-2,3-epoxyalkanes **1** with excess of Me₃Al yielded after acid quenching, *anti*-hydroxy sulfides **3**, whereas *trans*-1-(phenylthio)-2,3-epoxyalkanes **2** under the same conditions, yielded *syn*-hydroxy sulfides **4**. In both cases the yields of isolated products were greater than 95% (Scheme 6.1).

The reaction rates were very sensitive to a change in the solvent. Whereas dichloromethane was found to accelerate the process, the same reaction in hexane gave very low yields of the open chain products together with and a large amount of unreacted starting materials.

Explain the regio- and stereoselectivity of the reaction and justify the influence of the solvent in the process.

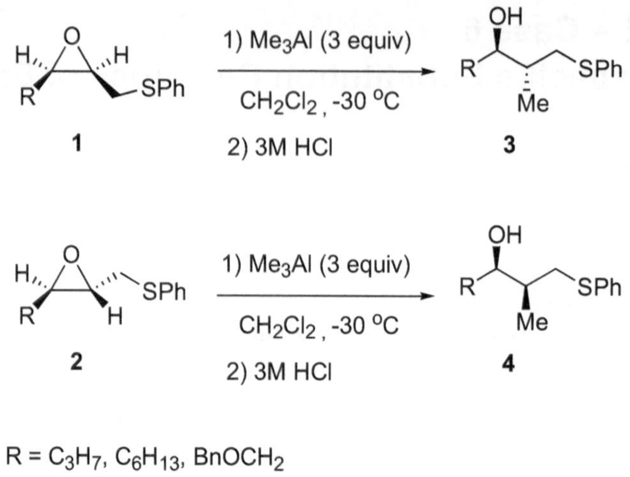

Scheme 6.1

R = C$_3$H$_7$, C$_6$H$_{13}$, BnOCH$_2$

Experimental Data

When 2,3-epoxy amine **5**, structurally related to epoxyalkanes **1** and **2** was treated with trimethylsilyl trifluoromethanesulfonate (TMSOTf) in CDCl$_3$ at − 40°C, and the reaction mixture was allowed to warm to room temperature, the ^{1}H-NMR spectrum of the resulting solution clearly showed the formation of a new compound **6**, stable at room temperature (Scheme 6.2). Compound **6** was very reactive in the presence of nucleophiles yielding, after hydrolysis, products with the structure of aminoalcohols.

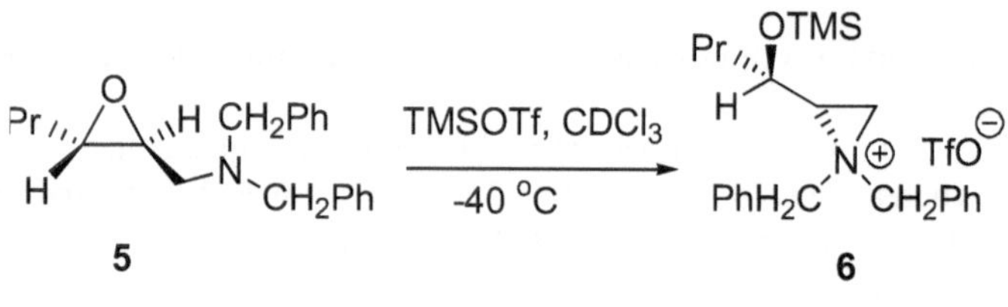

Scheme 6.2

Discussion

It is evident that the reaction of epoxides **1** and **2** with Me$_3$Al is not a simple epoxide ring-opening by nucleophilic attack at the C2 position. As shown in Scheme 6.3, the expected S$_N$2 backside nucleophilic attack at C2 in *cis*-epoxide **1** would yield **7**, which has inverted the configuration at C2 during the process. After quenching, **7** would yield **8**, a compound in which the stereochemistry at the C2 position is exactly the opposite of that observed in the reaction product **3**. In other words, a standard epoxide ring-opening mechanism is not able to justify why during the transformation of **1** into **3** the configuration at C2 is retained.

The same situation could be found in the case of *trans*-epoxide **2**, which by means of a simple backside nucleophilic attack should yield the ring-opening product **9**, again with the wrong stereochemistry at the C2 position if we compare the structures **4** and **10**.

Summarizing: in the reaction of epoxides **1** and **2** with Me₃Al the stereochemistry of the products is exactly the opposite of that expected after a normal S_N2 process. In both cases retention of the configuration at C2 is observed and in addition, the substitution occurs with total regioselectivity at C2.

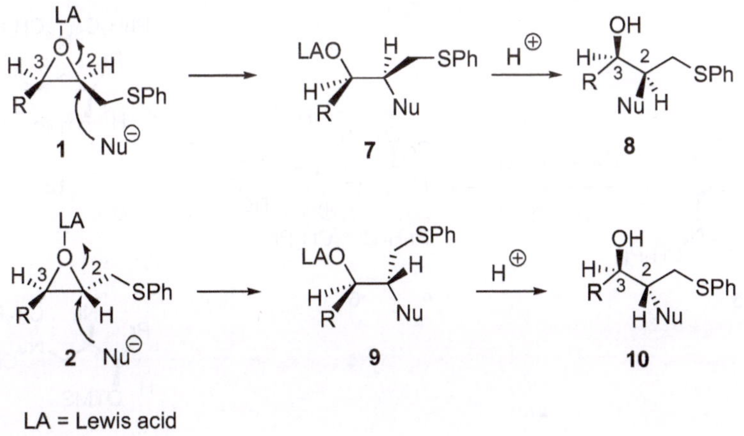

LA = Lewis acid

Scheme 6.3

When we come across a nucleophilic substitution reaction in which the configuration at a chiral carbon is *retained* (and not inverted or racemized), in a substrate in which there is a group with an unshared pair of electrons β to the leaving group, we have to consider that a *neighboring group mechanism* may be operative. In this case, the two requirements are fulfilled, as we have a substitution process, with retention of the configuration at C2 and a sulfur atom nicely placed to promote the intramolecular ring opening of the epoxide. As a result of the sulfur-directed ring opening, an episulfonium (tiiranium) ion intermediate like **11** should be obtained (Scheme 6.4).

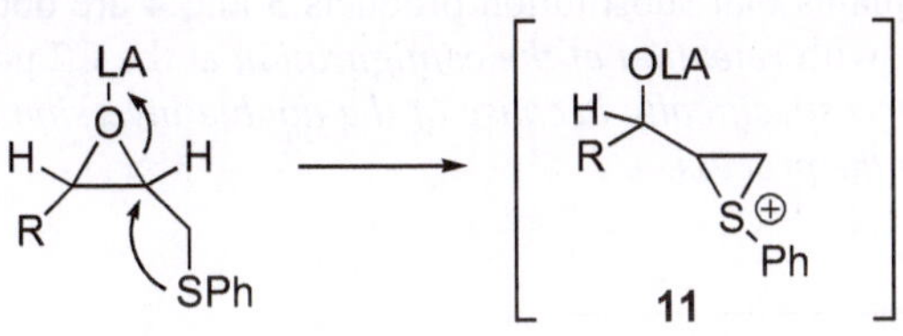

Scheme 6.4

Are these species formed in the reaction of epoxides **1** and **2** with Me₃Al? We are not aware of any experimental data in support of intermediates like **11**. However, in the reaction of 2,3-epoxy amine **5**, (structurally related to **1** and **2**) in acidic medium, an aziridinium intermediate **6** has been isolated and characterized. This result is decisive in the understanding of the reaction course. A neighboring group mechanism could also be proposed to explain the formation of **6**. After coordination of the oxygen with the Lewis acid, the nitrogen atom could promote the ring opening of the oxirane, leading to the obtained aziridinium salt **6** (Scheme 6.5). Aziridinium salts are very reactive electrophiles and they can be opened even in the presence of weak nucleophiles. A nucleophilic attack on **6** would lead to products like **12** or **13**, which after hydrolysis of the TMS group would give amino alcohols.[1]

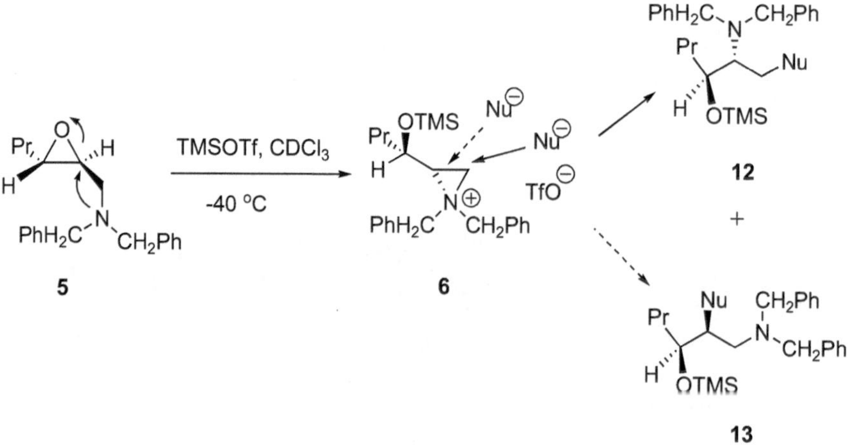

Scheme 6.5

A similar process can be formulated for epoxides **1** and **2** (Scheme 6.6). The first step should be the coordination of the oxygen atom with the Lewis acid to yield complexes **14** and **15**. The coordination enhances the reactivity of the epoxide towards the nucleophilic attack of the sulfur atom and episulfonium ion intermediates **16** and **17** should be formed. Finally, intramolecular migration of the methyl group from the aluminium complex moiety yields the reaction products. The migration occurs exclusively to the former C2 position of the oxirane ring. The formation of episulfonium ion intermediates **16** and **17** by sulfur-directed ring-opening, explains that substitution products **3** and **4** are obtained in a *regioselective* manner, with *retention of the configuration* at the C2 position. *The reaction proceeds stereospecifically because of the double inversion of the configuration at C2 during the process.*

---

[1] The nucleophilic attack in compounds **6** occurs with almost total regioselectivity at the less-hindered position and in fact, products with the structure **13** have not been isolated.

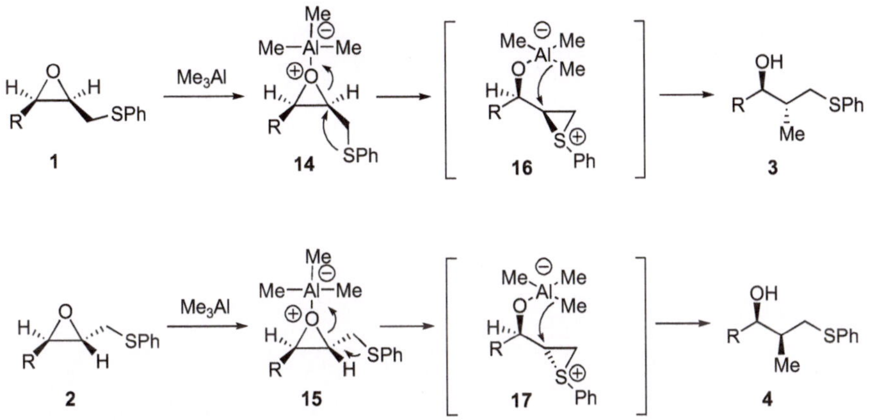

Scheme 6.6

Are the solvent effects in agreement with the proposed mechanism? The reaction rate is increased when more polar dichloromethane is employed instead of less polar hexane. These results are reasonable, as polar intermediates (like episulfonium ions) are better stabilized in polar solvents.

In Summary

Me₃Al acts as a nucleophile in the regioselective and stereospecific ring-opening of 1-(phenylthio)-2,3-epoxialkanes **1** and **2**. The reaction proceeds through a neighboring group mechanism with formation of an episulfonium ion intermediate and leads to the products with retention of the configuration at the C2 position.

Additional Comments

This problem is based on the work by Sasaki M, Tanino K, Miyashita M (2001) *J. Org. Chem.* 66:5388-5394 and on the work by Liu Q, Simms MJ, Boden N, Rayner CM (1994) *J. Chem. Soc. Perkin Trans. 1* 1363-1365.

Subjects of Revision

Neighboring group participation. Epoxides: ring-opening mechanisms. Regioselectivity. Stereospecificity.

Level 1 – Case 7
NaBH₄ Reduction of α,β-Unsaturated Chromium Carbene Complexes

Key point: *Isotopic labeling*

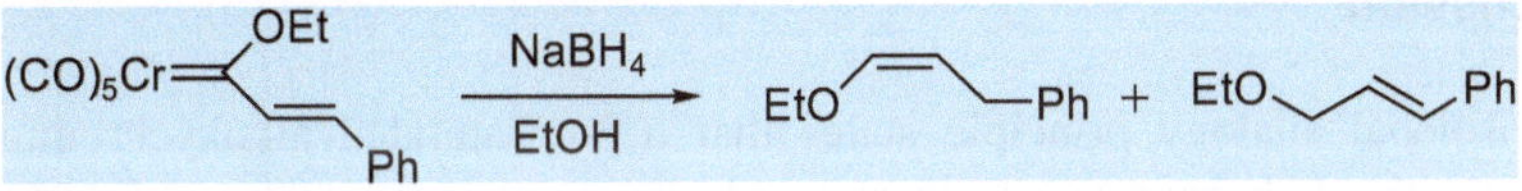

NaBH₄ Reduction of α,β-Unsaturated Chromium Carbene Complexes

Conjugated chromium Fischer alkoxy carbene complex **1** is quantitatively transformed into a mixture of Z-vinyl ether **2** and E-allyl ether **3** by NaBH₄ reduction in EtOH (Scheme 7.1).

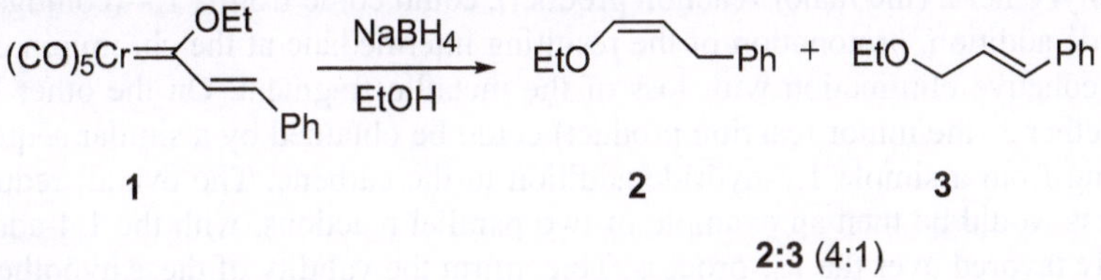

2:3 (4:1)

Scheme 7.1

Propose a reasonable mechanism to account for the observed products and to explain all the following experimental data.

Experimental Data

1. The reduction of alkoxy carbene **1** with NaBD₄ in EtOH yielded the deuterated products **4** and **5** (Scheme 7.2).

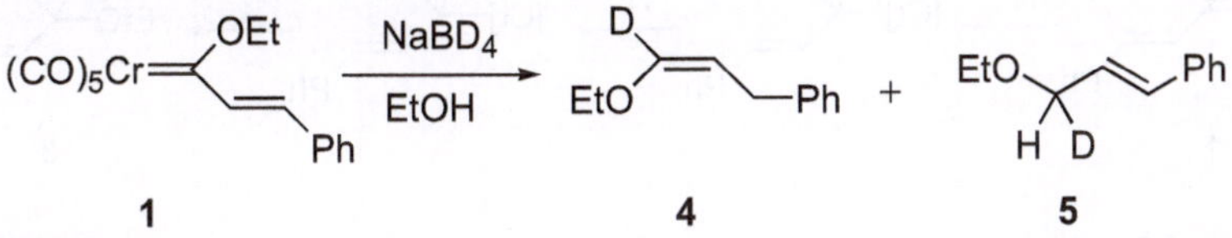

Scheme 7.2

2. The reaction of **1** with NaBH$_4$ in deuteromethanol as solvent yielded the deuteration products **5** and **6** (Scheme 7.3).

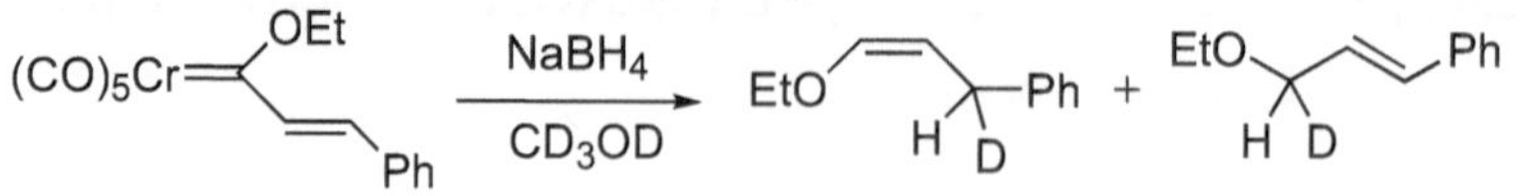

Scheme 7.3

Discussion

The isolobal analogy principle states that α,β-unsaturated alkoxychromium(0) carbenes should behave as α,β-unsaturated esters in processes taking place outside of the coordination sphere of the metal. In consequence, such complexes may be regarded as a kind of super-esters, with their reactivity enhanced due to the presence of a metal in their structures. In this context, the behavior of α,β-unsaturated alkoxychromium(0) carbene complexes and α,β-unsaturated organic esters towards the addition of nucleophiles, should be similar. Hydrides are nucleophiles and, if we consider that complexes **1** should behave like a conjugated system, their reactivity towards NaBH$_4$ could be interpreted considering the two possible alternative 1,2- or 1,4-addition processes (Scheme 7.4a and b).

Vinyl ether **2** (the major reaction product), could come from a 1,4-(conjugated)-hydride addition, protonation of the resulting intermediate at the chromium center and reductive elimination with loss of the metallic fragment. On the other hand, allyl ether **3** (the minor reaction product) could be obtained by a similar sequence, starting from a simple 1,2-hydride addition to the carbene. The overall reduction process would be then an example of two parallel reactions, with the 1,4-addition clearly favored over the 1,2-process. To confirm the validity of these hypotheses it is necessary to check if they are able to justify the results obtained in the deuteration experiments.

In Schemes 7.4a and b we have represented the expected deuterium-labeled positions in the final products. When NaBD$_4$ is employed, the labeled hydrogen atoms are colored red and in the experiment in CD$_3$OD as solvent, they are colored blue.

1,2-Hydride addition

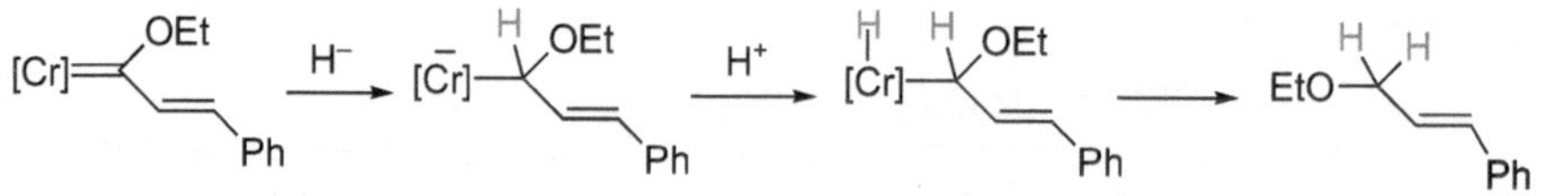

Scheme 7.4a

1,4-Hydride addition

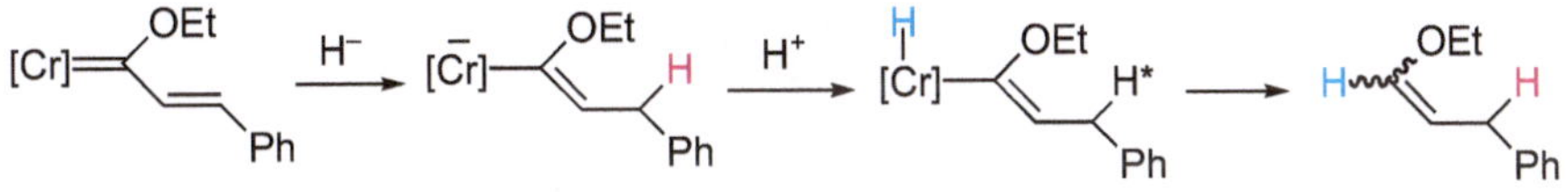

1   **2**

H labeled position with NaBD₄
H labeled position with CD₃OD

Scheme 7.4b

Figure 7.1 shows that in the 1,2-addition pathway, the same labeled position is obtained either with NaBD₄ or with CD₃OD. In consequence, deuterated compound **5** will be obtained in both experiments, which is in agreement with the experimental results.

expected labeled position with NaBD₄ or CD₃OD

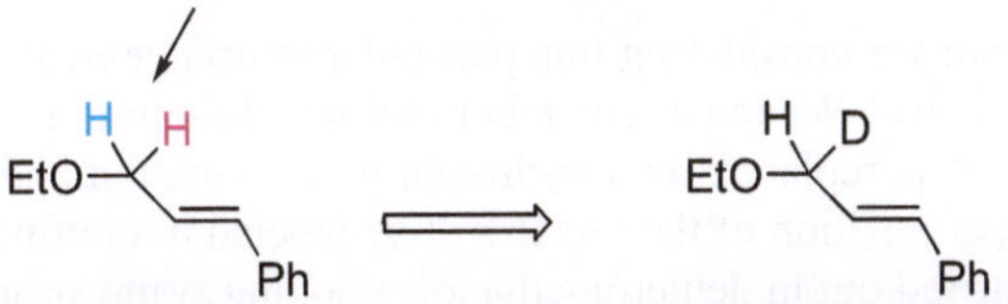

3   **5**

product obtained with NaBD₄ or CD₃OD

Figure 7.1

However, following the 1,4-hydride addition pathway, the labeled positions in the final product **2** are different depending on the reagent/solvent employed. As indicated in Fig. 7.2 the benzylic position should be deuterated in the experiment with NaBD₄, but a deuterated vinylic position should be obtained when the reaction is carried out in deuteromethanol. These predictions do not match with the experimental results, which are exactly the opposite.

It is evident that, although the 1,2-hydride addition mechanism could justify the formation of one of the reaction products (the allyl ether **3**), the alternative 1,4-process is unable to rationalize any of the experimental results.

In consequence, a mechanism based on two parallel 1,2- and 1,4-hydride additions to the carbene complex must be discarded.

Another reasonable alternative would be to suppose that some kind of rearrangement has occurred in the reaction and that the metal could be playing an active role during the process. Considering that at least one of the reaction products (allyl ether **3**) would come from a 1,2-hydride addition to the carbene complex, we could employ this premise as a starting point for the new mechanistic proposal.

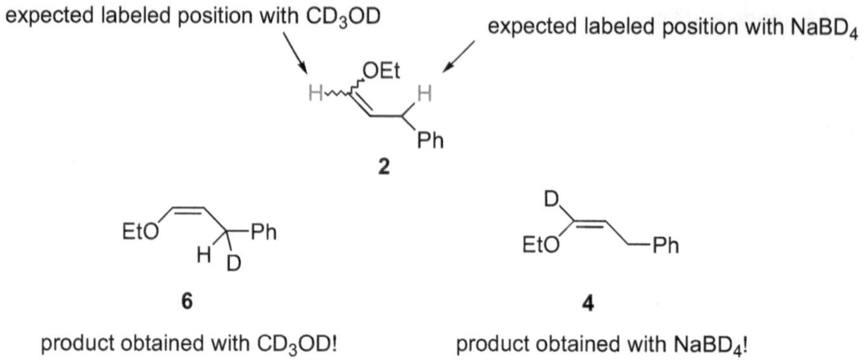

expected labeled position with CD$_3$OD

expected labeled position with NaBD$_4$

2

6

product obtained with CD$_3$OD!

4

product obtained with NaBD$_4$!

Figure 7.2

What else could happen after an initial 1,2-hydride addition to the carbene carbon? We have already mentioned that a rearrangement during the process is possible, e.g., a migration of the metallic fragment Cr(CO)$_5$ to the conjugated position (a 1,3-migration).

The question is: Why are we considering this particular rearrangement and not any other? The reason is simple. We must remember that the chromium is not present in the final products, it is replaced by a hydrogen by protonation and reductive elimination. The former position of the metal will be labeled in the final products in the experiments carried out in deuteromethanol. Looking at the structure of labeled compound **6** in Fig. 7.2, only a 1,3-migration of the metallic fragment could explain the presence of a deuterium atom at the benzylic position.

In Scheme 7.5 we have represented this alternative. The first step would be the addition of the hydride to the Cr=C bond leading to intermediate **7** from which the formation of allyl ether **3** is immediate. Another possible evolution for **7** could be the 1,3-migration of the Cr(CO)$_5$ moiety that would lead to complex **8** from which vinyl ether **2**, the major reaction product, can be easily obtained. To check if both pathways are consistent with the deuteration experiments, we have represented in Scheme 7.5 the expected labeled positions. The incorporation of deuterium from the reagent (H in red) takes place during the hydride addition step, whereas in the CD$_3$OD experiment, the deuterium transfer occurs during the protonation step (H in blue).

The mechanism proposed in Scheme 7.5 explains the results obtained in the deuteration experiments. Only the allylic position is deuterated in the minor reaction product **3** and the vinylic hydrogen of **2** would be replaced by deuterium in the experiment with NaBD$_4$, but the allylic position would be labeled with CD$_3$OD.

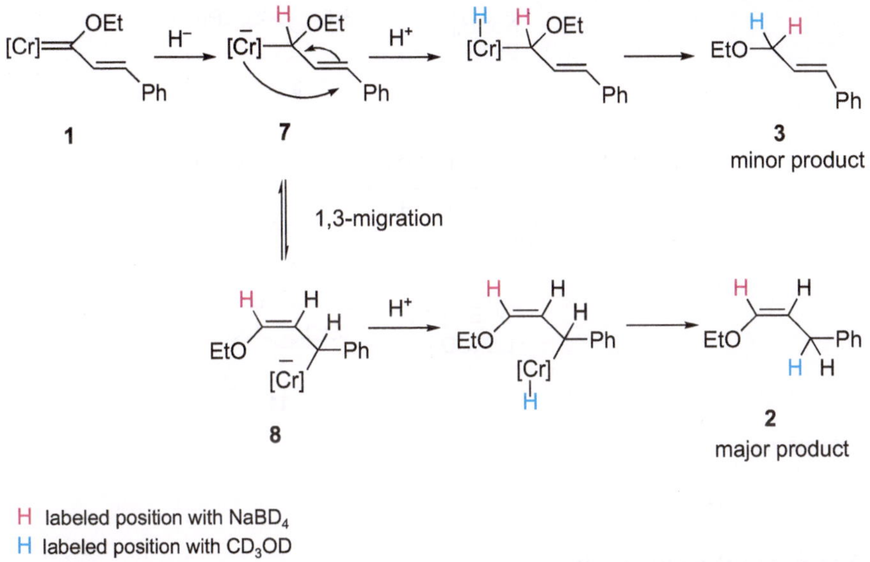

H labeled position with NaBD$_4$
H labeled position with CD$_3$OD

Scheme 7.5

Finally, there is an additional question that we should discuss in light of the proposed mechanism, namely the exquisite *E/Z* selectivity of the newly formed C=C double bond of vinyl ether **2**. The selectivity of the process could be a consequence of the chelation of the basic alkoxy oxygen to the metal center once the 1,3-rearrangement has occurred.

In Summary

The reduction of α,β-unsaturated carbene complexes **1** with NaBH$_4$ takes place with the active participation of the metal in the process. The initial hydride addition to the Cr=C bond is followed by a 1,3-migration of the metallic fragment. This is the key step of the reaction.

Questions

The reduction of alkynyl chromium carbene **9** with NaBH$_4$ in EtOH yields ether **3** as the sole reaction product. With NaBD$_4$/EtOH dideuterated derivative **10** was obtained and finally, when CD$_3$OD was the solvent, the reaction yielded dideuterated compound **11** (Scheme 7.6).

Based on the previous discussion, propose a reasonable mechanism to explain all the experimental results.

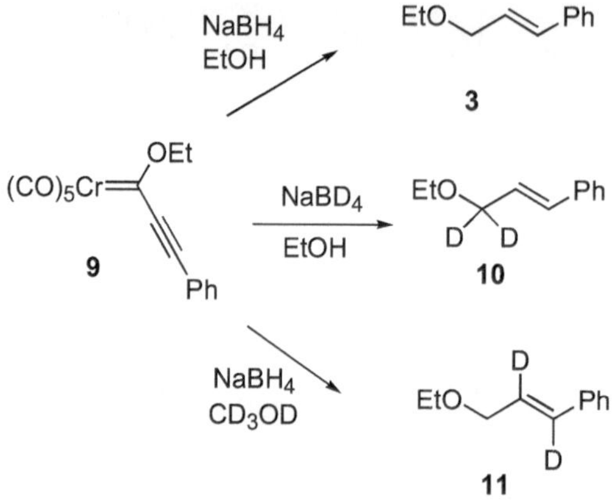

Scheme 7.6

Answer to the Question

Again, it is difficult to understand the results obtained in the reduction of carbene complex **9** without considering the active participation of the metal in the process. The 1,3-migration of the $M(CO)_5$ fragment in the $NaBH_4$ reduction processes seems to be a general behavior for α,β-unsaturated complexes of different structures. The initial 1,2-addition of hydride on the carbene carbon atom of **9** would lead to intermediate **12**. The first incorporation of deuterium occurs at this point when $NaBD_4$ is used as reagent. Alkynyl intermediate **12** may evolve to allenyl complex **13** by a 1,3-propargylic rearrangement. The isolation of ethers **3** from the intermediate **13** requires the addition of a new hydride ion at the end of the allenyl system and explains the incorporation of the second deuterium atom in compound **10** when $NaBD_4$ is employed. The addition of the hydride at the end of the allenyl system is remarkable, since the addition of nucleophiles to allenes usually takes place at the central allenic carbon. Finally, diprotonation of dianion **14** would yield the allyl ethers **3** and explains the formation of dideuterated **11** when CD_3OD was used as solvent (Scheme 7.7).

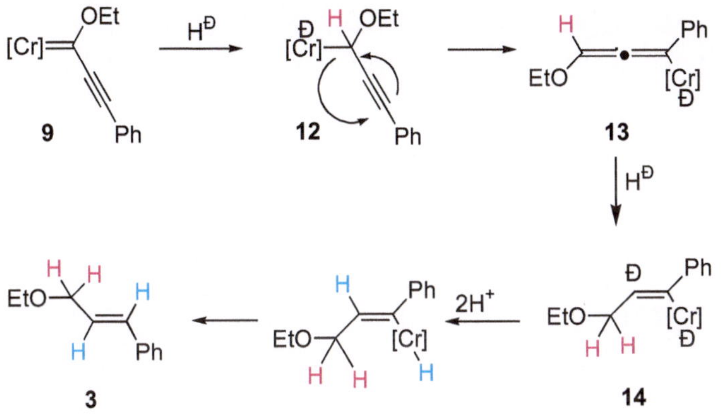

10 (H position deuterated by NaBD$_4$)
11 (H position deuterated by CD$_3$OD)

Scheme 7.7

Additional Comments

An alternative interpretation to the mechanism proposed in Scheme 7.7, could be the protonation of allenyl intermediate **13** before the addition of a new hydride, to form free allene **15**. The subsequent reduction of this compound would also lead to allyl ethers **3** in a pathway fully compatible with the observed deuteration pattern. To exclude this possibility, the authors of the original work prepared allene **15** from 1-ethoxy-3-phenyl-2-propyne and BuLi, (formulate the reaction!) and treated it with NaBH$_4$ in EtOH under the same conditions employed for compound **9**. The allene **15** was recovered unchanged after several hours of reaction (Scheme 7.8).

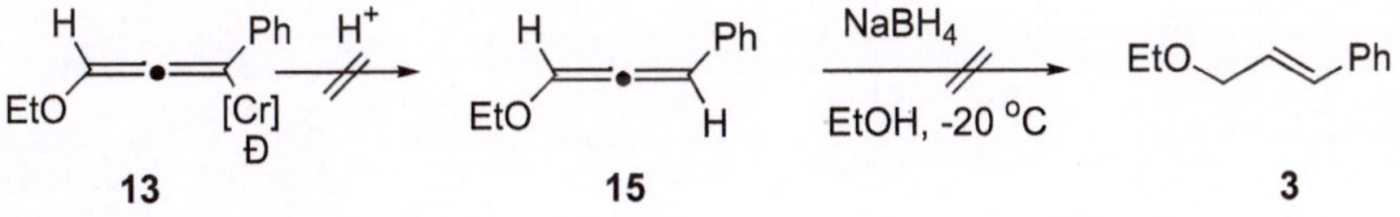

Scheme 7.8

This problem is based on the work by Gómez-Gallego M, Mancheño MJ, Ramírez P, Piñar C, Sierra MA (2000) *Tetrahedron* 56:4893-4905.

Subjects of Revision

Isotopic labeling experiments. Fisher carbenes: structure and reactivity.

Level 1 – Case 8
Addition of Hydroxylamines to α,β-Unsaturated Esters

Key point: *Isotopic labeling.*
Concerted versus polar mechanism

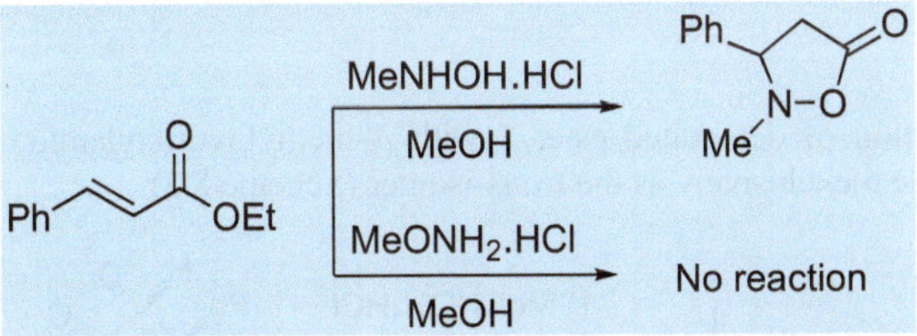

Addition of Hydroxylamines to α,β-Unsaturated Esters

Hydroxylamine derivatives are known to efficiently add to α,β-unsaturated esters and lactones in a conjugated fashion. However, when the reaction was tested with ethyl (*E*)-cinnamate **1**, different results were obtained depending on the reagent employed. Thus, the reaction of **1** with *N*-methylhydroxylamine yielded the isoxazolidinone **2**, whereas no reaction took place (the starting material was recovered unaltered) when *O*-methylhydroxylamine was used (Scheme 8.1).

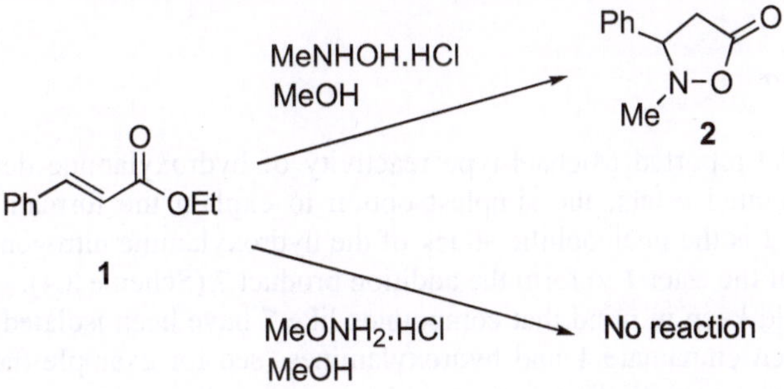

Scheme 8.1

*Propose a reasonable mechanism to explain the formation of the cyclization product **2** and justify the lack of reactivity when O-methylhydroxylamine is used.*

Experimental Data

1. The reaction of deuterated *N*-methylhydroxylamine with **1** gives the ester **3** as a single diastereomer. In the presence of a Lewis acid, **3** cyclized to yield **4**, exclusively as the *cis*-isomer (Scheme 8.2).

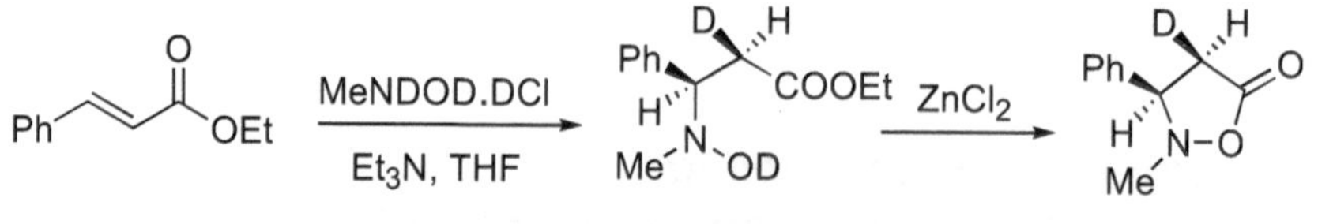

1 **3** **4**

cis-isomer

Scheme 8.2

2. The reaction of deuterated ester **5** with *N*-methylhydroxylamine gave isoxazolidinone **6** exclusively as the *trans*-isomer (Scheme 8.3).

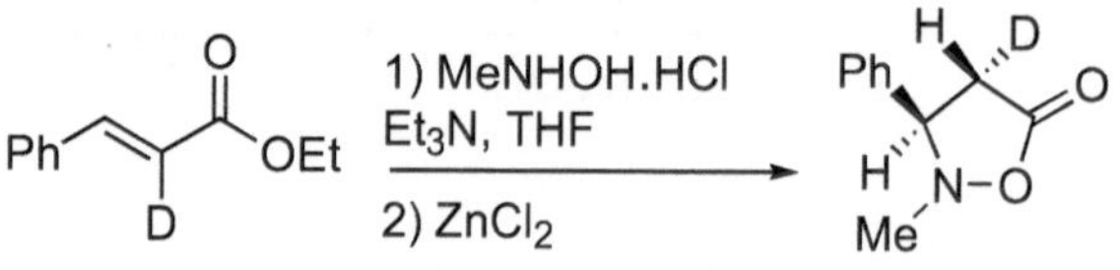

5 **6**

trans-isomer

Scheme 8.3

3. All the reactions between cinnamates **1** and hydroxylamine derivatives were carried out in THF, although no difference was observed, either in stereochemistry or yields, when the solvent employed was EtOH.

Discussion

Based on the reported Michael-type reactivity of hydroxylamine derivatives toward unsaturated esters, the simplest option to explain the formation of isoxazolidinones **2** is the nucleophilic attack of the hydroxylamine nitrogen atom to the C=C bond of the ester **1** to form the addition product **7** (Scheme 8.4).

We should keep in mind that compounds like **7** have been isolated in the reactions between cinnamate **1** and hydroxylamines (see for example the deuterated ester **3** in Scheme 8.2). This fact could be considered an argument in support of the proposed nucleophilic addition mechanism.

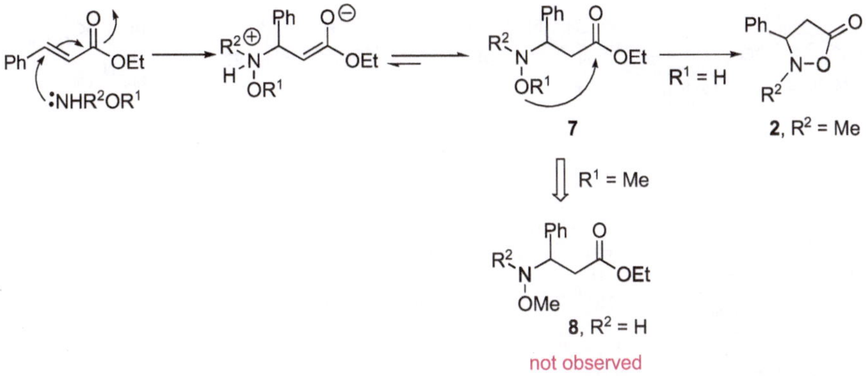

Scheme 8.4

Once **7** has been formed, the reaction products would be different depending on the reagent employed. In the case of *N*-methylhydroxylamine, intermediate **7** (R^1 = H), would be transformed into isoxazolidinone **2** by intramolecular transesterification. However, when the reagent is *O*-methylhydroxylamine the cyclization of **7** (R^1 = Me) by transesterification is not possible and **8** should be obtained.

Nevertheless, this is not the experimental result. Compounds **8** have not been detected and in addition, cinnamate **1** is recovered unaltered after the reaction with *O*-methylhydroxylamine. Clearly, the mechanism depicted in Scheme 8.4 is unable to justify *all of* the experimental results although it could explain *some* of them. Shall it be discarded at this point? The answer is *not yet*.

N-Alkylhydroxylamines and *O*-alkylhydroxylamines are structurally related reagents but in fact they are *different* compounds. The possibility that each reagent could react with cinnamate **1** in a different way should be considered. Hence, the mechanism proposed in Scheme 8.4 still could be useful, at least to interpret the formation of isoxazolidinones **2** in the reactions of **1** and *N*-methylhydroxylamine. To confirm the validity of this option, we would check if the nucleophilic addition mechanism should be able to predict the stereochemistry of the product **3** obtained in the reaction of cinnamate **1** and deuterated *N*-methylhydroxylamine (Scheme 8.5, labeled positions in red).

When we compare the nucleophilic addition mechanism and the experimental results in this experiment, the first oddity is referred to the stereochemistry of compound **3**. The experimental result indicates that deuterated ester **3** is obtained as a single diastereomer, which after cyclization (promoted by $ZnCl_2$) leads to **4** with *cis*-stereochemistry. However, following the nucleophilic addition mechanism previously discussed, product **9** (structurally related to **3**) should be obtained, but as a mixture of diastereomers.

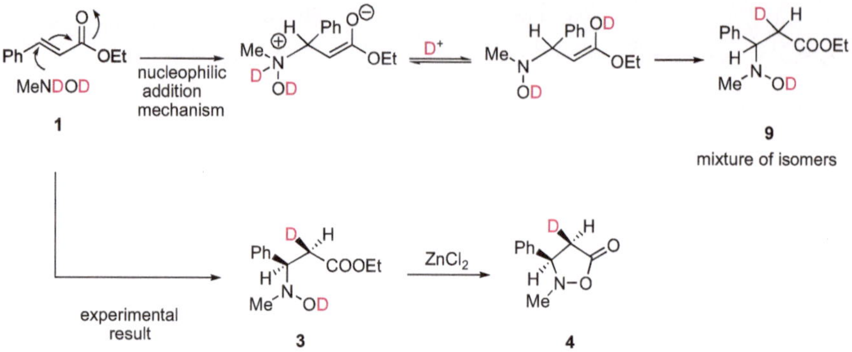

Scheme 8.5

A similar disparity is found if we check the stereochemistry of the reaction products in the reaction between deuterated ester **5** and methylhydroxylamine (in the presence of cyclization promoter Cl_2Zn), to yield labeled *trans*-**6** (Scheme 8.6, labeled positions in red). Again, the nucleophilic addition mechanism would yield mixtures of diastereomers in both the addition product **10** and the cyclization product **11**, which are not experimentally observed (compound **6** is obtained as a single diastereoisomer).

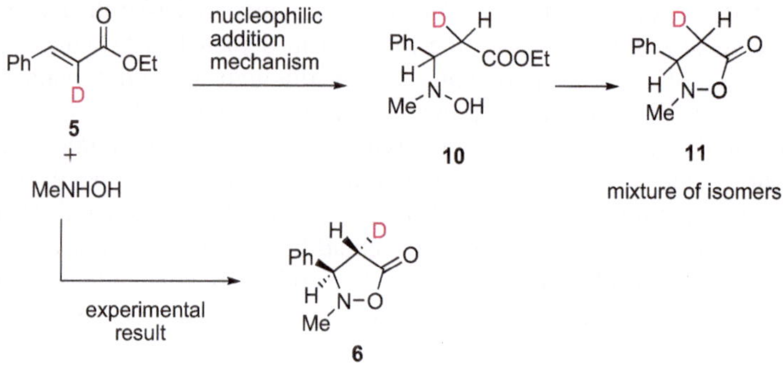

Scheme 8.6

In view of these arguments, the mechanism involving the nucleophilic addition of the hydroxylamine to the conjugated ester must be definitively discarded.

Back to the starting point there is experimental data that has not yet been commented. Until now we have been dealing with a mechanism involving polar transition states that should have been affected to some extent by a change in the polarity of the medium. However, the fact is that the reactions of cinnamate **1** and hydroxylamine derivatives show no difference (either in stereochemistry or yields) when the solvent was changed from THF (E_T^N 0.207) to EtOH (E_T^N 0.654). The lack of sensitivity to a change in the solvent polarity is a characteristic of the reactions involving concerted (isopolar) transition states.

If the reaction between cinnamate **1** and hydroxylamines occurred through a concerted mechanism, the process should take place through a cyclic (five-membered ring) transition state, similar to the ones proposed for [3+2] dipolar cycloadditions or retro-Cope eliminations. The cyclic transition state for the concerted addition of deuterated *N*-methylhydroxylamine to ethyl cinnamate **1** is represented in Scheme 8.7 (atoms involved in the cyclization colored blue). As the nitrogen atom and the deuterium approach the C=C bond from the same side, the addition is *syn,* leading to intermediate **12**. A rapid proton shift would give compound **3** which can be either isolated or cyclized in the presence of a Lewis acid to the isoxazolidinone **4** by intramolecular transesterification. The stereochemistry of the products **3** and **4** is determined during the concerted addition step.

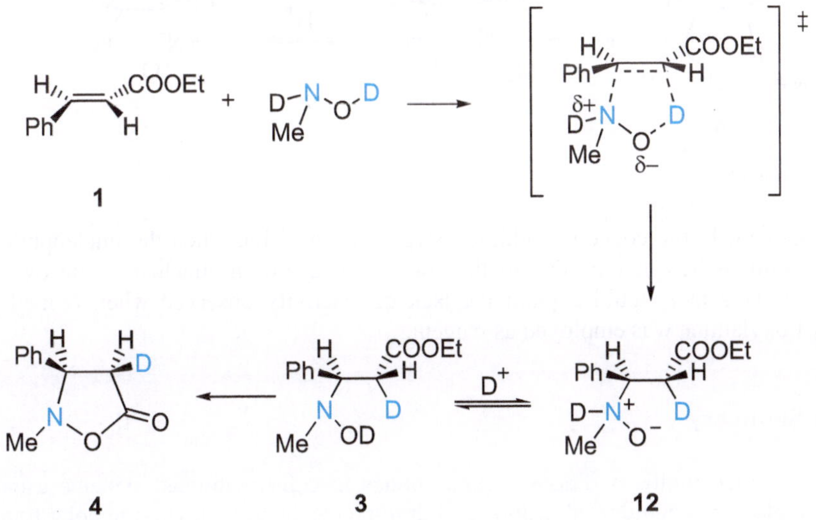

Scheme 8.7

A similar process would explain the *trans*-stereochemistry of **6**, the product obtained in the reaction between hydroxylamine and deuterated cinnamate **5** (Scheme 8.8, labeled atom in red). The nitrogen atom and the proton are incorporated to the C=C bond from the same side, leading to intermediate **13** that after a rapid proton shift and cyclization would give isoxazolidinone **6**. The addition of the reagent is also *syn,* but in this case leading to the cyclic *trans*-isomer.

As we have commented before, the transition state of a concerted process is little influenced by a change in solvent polarity. This fact could fit with the insensitivity of the reaction with the change of the solvent from THF to the more polar EtOH.

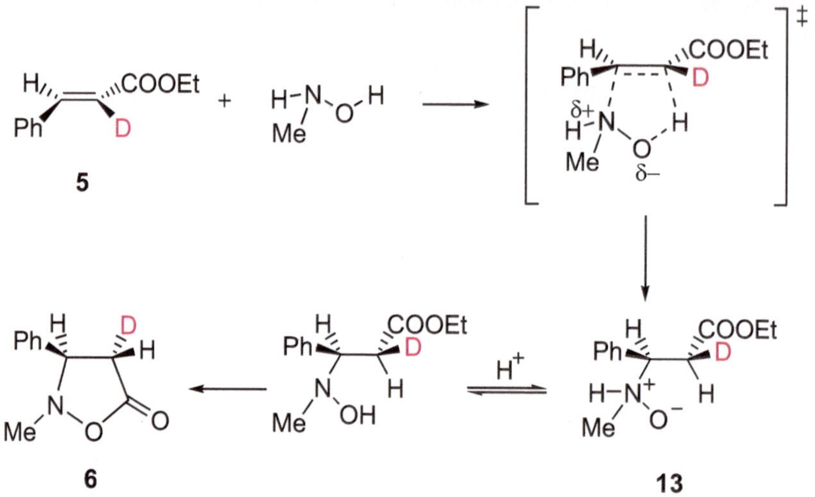

Scheme 8.8

Obviously the concerted addition step is not possible when the nucleophile is *O*-methylhydroxylamine due to the absence of a proton attached to the oxygen atom. This fact would explain the lack of reactivity observed when *O*-methyl-hydroxylamine was employed as reagent.

In Summary

While nucleophilic additions to C=C bonds in α,β-unsaturated systems usually take place in a conjugated fashion and through mechanisms involving polar transition states, in some cases a concerted addition of the nucleophile must be considered. The addition of hydroxylamines to ethyl cinnamate is a clear example of such an unusual type of reaction.

Questions

Based on the concerted mechanism previously discussed, try to predict the stereochemistry of the isoxazolidinones obtained in the reaction between deuterated *N*-methylhydroxylamine and esters **14** and **15**.

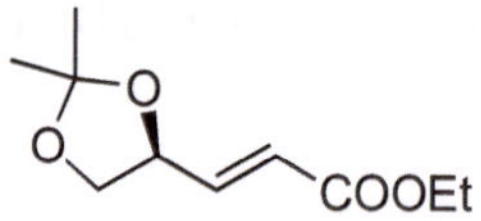

14
E-isomer

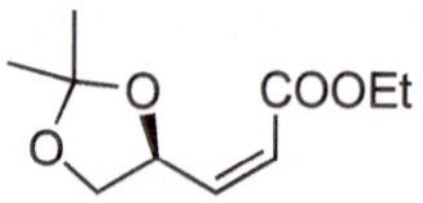

15
Z-isomer

Answer to the Question

Following the mechanism discussed above, the addition of the hydroxylamine to the C=C double bond is *syn* in both cases (Schemes 8.9 and 8.10, labeled positions in red). Transition state **16** would account for the *cis*-stereochemistry of isoxazolidinone **17** obtained from (*E*)-conjugated ester **14**. In a similar manner, transition state **18** would determine the stereochemistry of *trans*-isoxazolidinone **17** obtained from the *Z*-isomer.

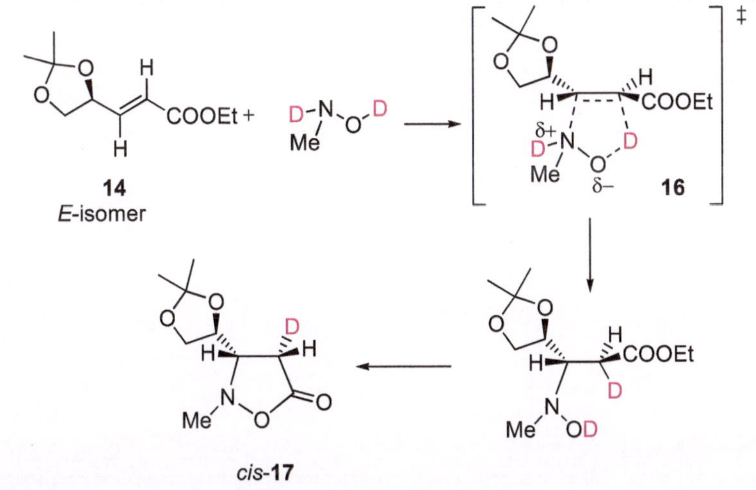

14
E-isomer

cis-**17**

Scheme 8.9

15
Z-isomer

trans-**17**

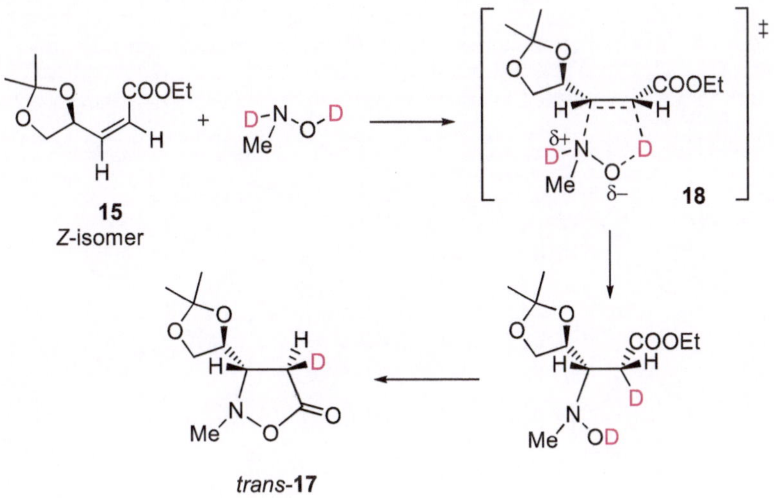

Scheme 8.10

Additional Comments

This problem is based on the work by Niu D, Zhao K (1999) *J. Am. Chem. Soc.* 121:2456-2459.

Subjects of Revision

Isotopic labeling. Nucleophilic conjugated additions. Isopolar reactions.

Level 1 – Case 9
Solvolysis of Electron-Deficient Norbornyl Triflates

Key point: *Reaction Intermediates: Carbocations*

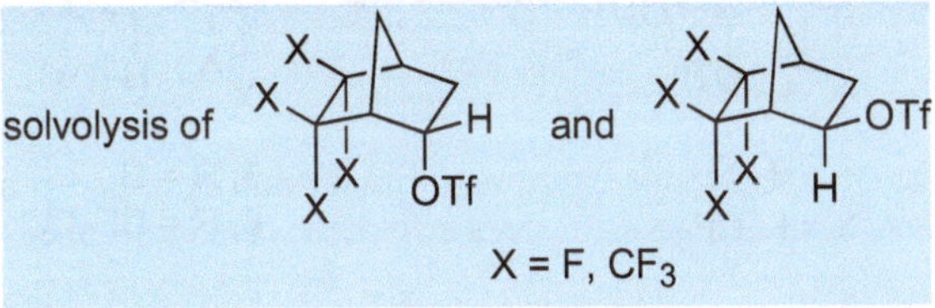

solvolysis of [structure] and [structure]

$X = F, CF_3$

Solvolysis of Electron-Deficient Norbornyl Triflates

The well-known solvolysis of *endo*- and *exo*-2-norbornyl triflates **1** produces exclusively *exo*-derivatives **2**. To establish the influence of electron-withdrawing groups on the mechanism of the reaction, the solvolysis of *endo*- and *exo*-norbornyl triflates **3** and **4** has been studied (Scheme 9.1).

Comment and discuss all the following experimental data, proposing a mechanism that could explain the structure of the products obtained in each case.

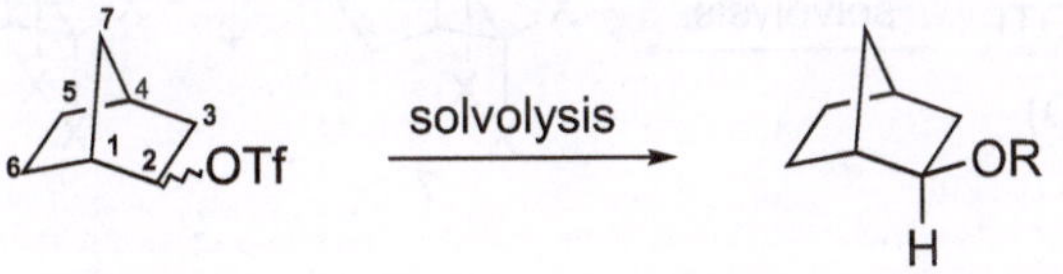

solvolysis

1

2 R = H, alkyl, COalkyl

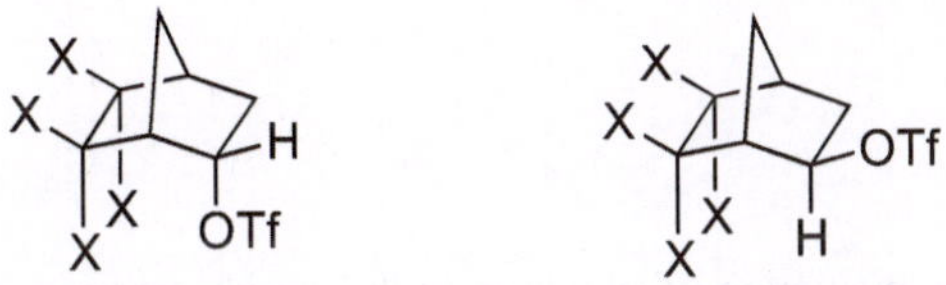

3

4

$X = F, CF_3$

Scheme 9.1

Experimental Data

1. The solvolysis of *exo*-triflates **4** is three to four-times faster than that of *endo*-triflates **3**.
2. In mixtures of dioxane-water, *endo*-triflates **3** lead almost exclusively to *exo*-alcohols **5** (92-99% yield) and similar yields are observed when the solvolysis is carried out in trifluoroacetic acid (TFA) to afford the esters **6**. The solvolysis in TFA of C-2 deuterium-labeled **3** gives deuterated esters **6** *without scrambling of the D-label* (Scheme 9.2).

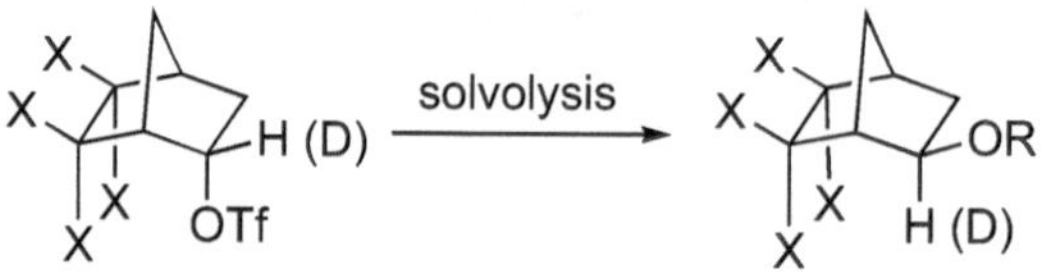

X.
X——[ ]——H (D)   solvolysis →   X.
  X OTf   X——[ ]——OR
  X   X H (D)
                                    X

3   **5**, R = H
X = F, CF$_3$ **6**, R = CF$_3$CO

Scheme 9.2

3. The solvolysis of *exo*-triflates **4** leads to a mixture of norbornenes **7** and substitution products **8** (*exo*-selectivity higher than 90%) in ratios that depend on the reaction conditions (Scheme 9.3). For example, in mixtures of dioxane-water the main reaction products were **7** (60-99%), whereas similar yields of **7** and **8** were obtained when the solvolysis was carried out in TFA. Finally, the solvolysis in TFA of C-2 deuterium-labeled **4**, gave substitution products **8** deuterated in position C-3.

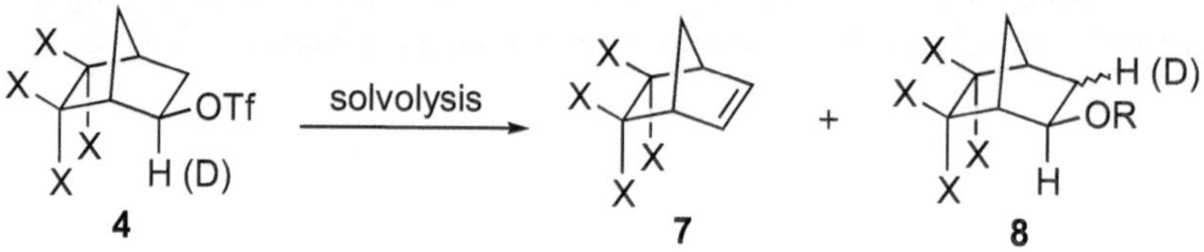

X.
X——[ ]——OTf   solvolysis →   X.   +   X.
  X H (D)   X——[ ]   X——[ ]——H (D)
  X   X   X OR
                                 X   X H

4   **7**   **8**

X = F, CF$_3$ R = H, CF$_3$CO

Scheme 9.3

Discussion

Prior to discussing the influence of electron-withdrawing groups in the solvolysis of 2-norbornyl triflates **3** and **4**, we should remember the different steps of the mechanism that has been proposed for the solvolysis of the structurally related unsubstituted system **1**. The solvolysis of *exo* and *endo*-2-norbornyl derivatives (**1**, OTf or any other good leaving group) has been exhaustively studied, and it is generally accepted that the reaction occurs, in both cases, *via* norbornyl cation **9**, a **nonclassical** (or bridged) **carbocation intermediate** (Figure 9.1). We should remember that in classical carbocations the positive charge is localized on one carbon atom or delocalized by resonance with a conjugated pair of electrons. Nonclassical carbocations however, are a special type of carbocations that have the positive charge delocalized by a double or triple bond that is not in the allylic position, or by a single (σ) bond. This is the case for cation **9** that can be represented by means of the canonical forms **9a** and **9b**.

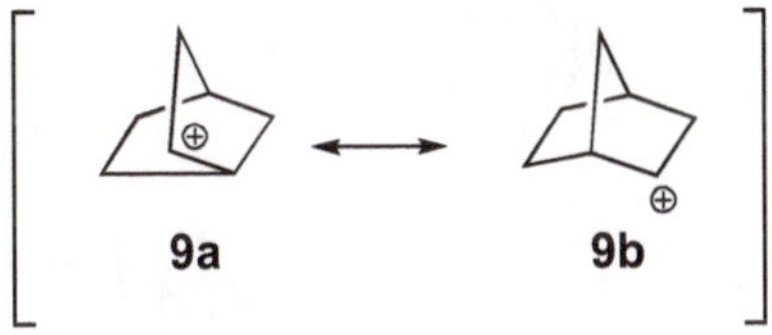

9a   **9b**   **9**

Figure 9.1

The formation of cation **9** from *exo* or *endo*-2-norbornyl derivatives follows a different pathway in each case (Scheme 9.4). For the *exo*-compounds, the departure of the leaving group (X) is assisted by a σ bond (σ neighboring group participation). This does not occur in the *endo*-compounds that form carbocation intermediate **9** by direct ionization. Due to the neighboring group participation of the σ bond, *exo*-2-norbornyl derivatives solvolyze much faster than its *endo*-2-norbornyl isomers.

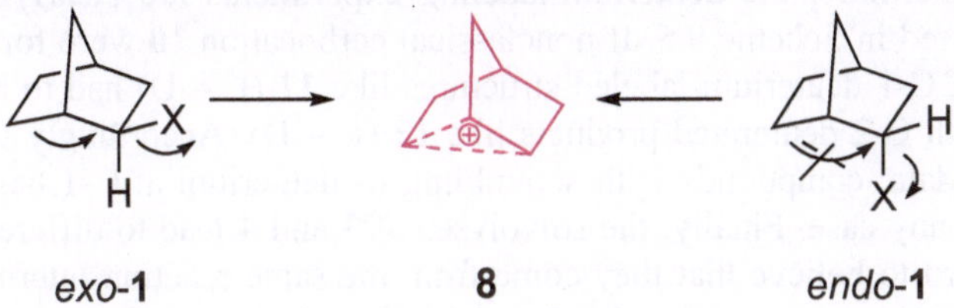

exo-**1**   **8**   *endo*-**1**

Scheme 9.4

Compounds **3** and **4** are structurally related to **1** and it has been observed experimentally that *exo*-triflates **4** react three to four-times more rapidly than *endo*-triflates **3**. These two arguments could be invoked to suggest the involvement of a nonclassical carbocation in the solvolysis process of both compounds, as we have discussed previously for **1**. This time, nonclassical cation **10** should be formed respectively by direct ionization of *endo*-triflates **3** and by assisted departure of the OTf group in the case of *exo*-isomers **4**. The attack of the nucleophile in **10** should finally lead to the solvolysis compounds **11** and **12** (Scheme 9.5).

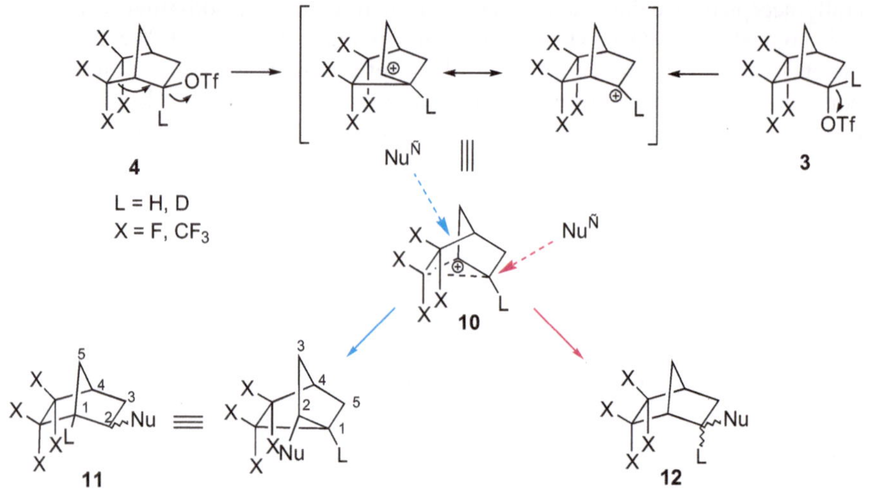

Scheme 9.5

Let us discuss the weak points of this mechanistic pathway. First, the main structural difference between triflates **3**, **4** and **1** is the presence of strong electron withdrawing groups in positions C-5 and C-6 in the two former compounds. If nonclassical carbocation **10** were involved in the solvolysis process, it would be an *electron-deficient unsymmetrically bridged carbocation*, and probably, not very stable. Furthermore, the deuterium-labeling experiments are clearly against the route proposed in Scheme 9.5. If nonclassical carbocation **10** were formed during the process, C-1 deuterium-labeled structures like **11** (L = D) had to be detected, together with C-2 deuterated products like **12** (L = D). Accordingly with the experimental data, compounds with scrambling of deuterium at C-1 have not been obtained in any case. Finally, the solvolyses of **3** and **4** lead to different products and it is hard to believe that they come from the same reaction intermediate. All these arguments are suggesting that **3** and **4** follow different solvolysis pathways.

In consequence, the mechanism proposed in Scheme 9.5 involving nonclassical cation **10** is unable to explain the experimental data and has to be discarded.

Comment

The discussion above illustrates how, when proposing a reaction mechanism, it is much better not to have pre-established ideas in mind. In fact the reactivity of norbornyl derivatives (particularly the solvolysis reactions) is so well known that concepts like **"carbocation generated in a bicyclic system"** *and* **"nonclassical structure"** *are not only frequently associated, but often taken as synonymous. It is important to point out that* **not all** *the reactions involving norbornyl derivatives take place through a nonclassical carbocation intermediate and that* **not always** *the departure of a leaving group in a norbornyl derivative implies a σ neighbor-*

ing-group participation. Unfortunately, this is a common mistake and the readers should be aware of it.

Once we have discarded that a single mechanistic pathway can explain the solvolysis of triflates **3** and **4**, we should propose a reasonable mechanism to interpret the results obtained in each case. We shouldn't forget that a solvolysis is nothing more than a *nucleophilic substitution,* and keeping in mind the different aspects of this class of reactions could be of great help during the analysis of the experimental data.

First, the experimental data indicate that the solvolysis of *endo*-triflates **3** is not solvent-dependent: *exo*-derivatives **5** and **6** are obtained in dioxane-water or TFA respectively, with excellent yields. These data are suggesting that a polar (carbocation) intermediate is not involved in the process. If a direct ionization of the compound (a S_N1 process) had occurred, a carbocation **13** should have been formed in the medium (Scheme 9.6). It is known that carbocations are very sensitive to the solvent nucleophilicity. Thus, in a non-nucleophilic medium (TFA), **13** could live long enough to rearrange (by a H-shift migration, or by a Wagner-Meerwein rearrangement), before being attacked by the solvent. In that case, rearrangement-derived products and scrambling of the deuterium label should have been observed. If the S_N1 is discarded, a direct S_N2 displacement by the nucleophile seems to be much more likely. This is in agreement with the deuteration experiments that show that the reactions occur without scrambling of the 2D-label (Scheme 9.6).

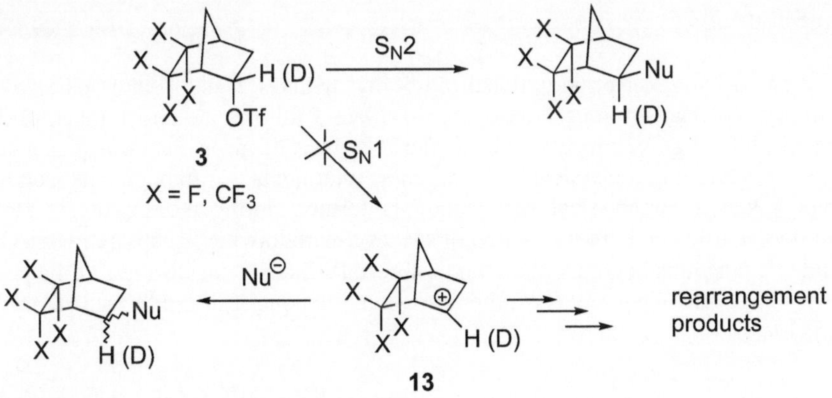

Scheme 9.6

By contrast, the solvolysis of *exo*-triflates **4** is strongly solvent-dependent, as it would be expected if a polar intermediate were involved in the reaction. In dioxane-water the elimination to norbornenes **7** is the main process, whereas in TFA the amount of substitution products **8** increases. Scrambling of deuterium from C-2 to C-3 position in TFA indicates that in this non-nucleophilic solvent a carbocation rearrangement has occurred. A reasonable pathway that could explain all these results is postulated in Scheme 9.7.

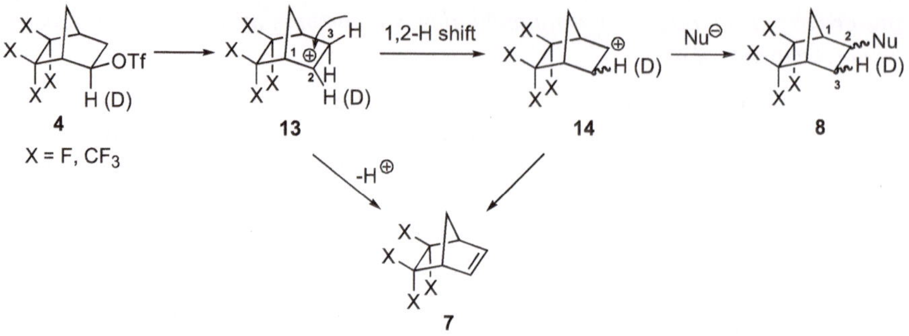

Scheme 9.7

Departure of the triflate substituent would lead to *classical* carbocation **13** precursor of norbornenes **7** by loss of a proton.[1] In more acidic media (less nucleophilic), the lifetime of a carbocation increases and rearrangements could be observed. Thus, 1,2-H shift (from the C3 to the C2 position) in **13** would give **14**, from which substitution products **8** are formed. The H-shift would account for the scrambling of the deuterium that would be placed at C-3 in compounds **8**. Interestingly, no other products derived from Wagner-Meerwein rearrangements of carbocations **13** or **14** have been observed.

In Summary

The solvolysis of electron-deficient norbornyl triflates **3** and **4** takes place by a different mechanism in each case. *endo*-Triflate **3** follows the direct attack of the nucleophile in a S_N2 process whereas the solvolysis of *exo*-triflate **4** is a stepwise reaction involving a **classical** carbocation as intermediate. Although it is well accepted that a **nonclassical** carbocation is formed during the solvolysis of 2-norbornyl triflates **1**, the presence of electron-withdrawing groups (CF_3) at C-5 and C-6 positions in compound **3** and **4** forestalls the σ-delocalization of the positive charge and in consequence, the formation of a nonclassical intermediate in the process.

---

[1] Although it has not been mentioned, partially labeled products **7** should be obtained after the elimination reaction. However, no further information about the mechanism of the reaction can be deduced from a deeper discussion of these labeled products.

Questions

After years of discussion, the nonclassical nature of the 2-norbornyl and 7-norbornenyl cations **9** and **15** seems to be beyond doubt and their structure and charge delocalization is discussed in almost every organic reaction mechanisms textbook (Figure 9.2).

9

norbornyl

nonclassical carbocation

15

7-norbornenyl

nonclassical carbocation

Figure 9.2

Many other cations (generally obtained from solvolysis reactions) have been studied and their classical/non-classical nature has been established. Some of them are shown below in Figure 9.3.

Represent their structures and justify how the positive charge could be stabilized in each case.

15

7-norbornenyl
nonclassical carbocation

16 ⊕

2-bicyclo-[3.2.1]octyl
nonclassical carbocation

17 ⊕

2-phenylnorbornyl
classical carbocation

18

cyclopropylmethyl
nonclassical carbocation

Figure 9.3

Answer to the Question

Classical carbocation:

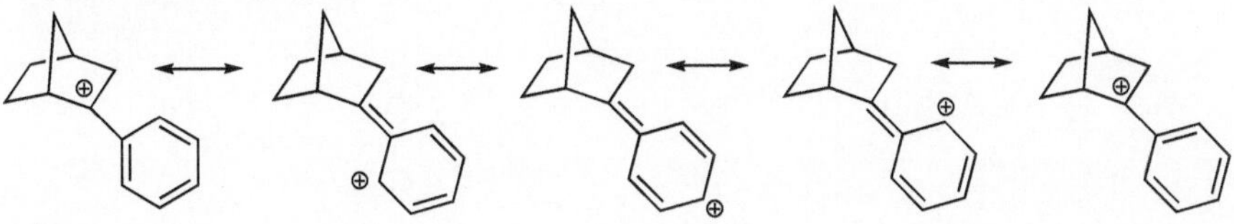

17

Nonclassical carbocations:

15

16

18

Additional Comments

This problem is based on the work by Kirmse W, Mrotzeck U, Siegfried R (1985) *Angew. Chem. Int. Ed. Engl.* 24:55-56.

Subjects of Revision

Reaction intermediates: Classical and nonclassical carbocations. Wagner-Meerwein rearrangements. σ-Neighboring-group participation.

Level 1 – Case 10
Nucleophile Versus Base Catalysis

Key point: *Catalysis*

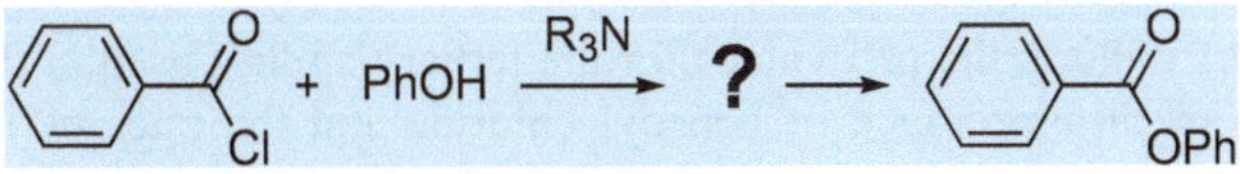

Nucleophile Versus Base Catalysis

To distinguish between nucleophile and base catalysis is not always a simple exercise. The reaction of acid chlorides and alcohols to form esters, in the presence of amines as catalysts, is a nice playground to examine how to distinguish between both processes. This reaction has been exhaustively studied, but most of the work has been done in protic solvents and the mechanisms proposed are based on these reports. The currently accepted mechanism for these reactions suggests a nucleophile catalysis, where the amine attacks the carbonyl group forming a transient tetrahedral intermediate. On displacement of the leaving group, a quaternary acylammonium salt is formed. This ammonium salt is then susceptible to attack by alcohol, water, or other nucleophiles to form the product (Scheme 10.1). However, there is no experimental evidence of the formation of the acylammonium salt. Other proposed mechanisms for amine-catalyzed reactions of carbonyl compounds involve acylium ion and ketene intermediates (the latter can only occur for acid halides containing α-hydrogens), but the actual role of the catalyst is still uncertain.

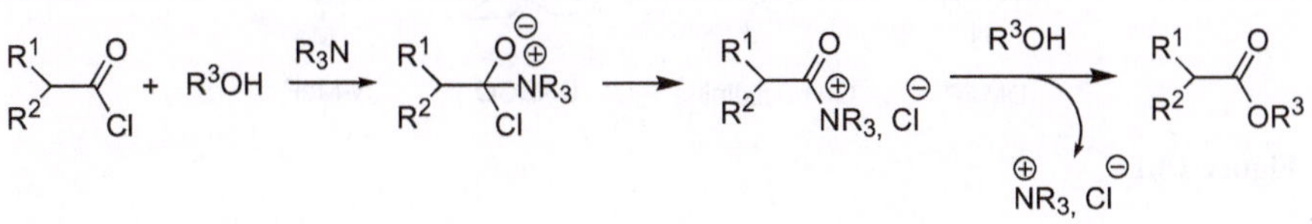

Scheme 10.1

Determine the most probable type of catalysis for these esterification processes, using the reaction of benzoyl chloride and phenol, in dichloromethane (DCM) and in the presence of tertiary amines, as a model.

Experimental Data

1. The reaction between benzoyl chloride and phenol, catalyzed by amines, follows a third-order overall rate, being first order on each of the reactants.
2. The reaction rate does not change noticeably when perdeuterated phenol (C_6D_5OD) was used as reagent.
3. In spite of the fact that both phenol and phenoxide ion have different UV absorptions (λ_{max}(DCM) phenol = 275 nm; λ_{max}(DCM) tetraethylammonium phenoxide = 250 nm), several UV-visible experiments directed towards the detection of the phenoxide ion were fruitless.
4. Table 10.1 collects the pK_a values for the tertiary amines used in this problem, the rates of disappearance of benzoyl chloride and the rates of formation of phenyl benzoate.

Table 10.1

Amine[a]	pK_a	$-d[BC^b]/dt$ (M/s)	$d[PB^c]/dt$ (M/s)
DMAP	9.45	> 0.3	> 0.3
Quinuclidine	10.95	> 0.1	> 0.05
DABCO	8.77	0.00373	0.00277
Et$_2$MeN	10.43	0.00215	0.00223
N-MeP	10.19	0.00128	0.00135
Et$_3$N	10.75	0.00028	0.00025
Pr$_3$N	10.66	0.00011	0.0001
Pyridine	5.21	2.1×10^{-6}	1.7×10^{-6}

[a] See Figure 10.1; [b] BC = benzoyl chloride; [c] PB = phenyl benzoate

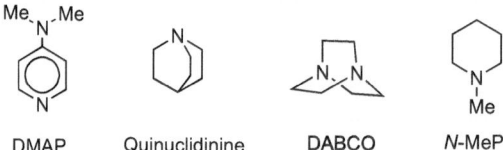

DMAP   Quinuclidinine   DABCO   N-MeP

Figure 10.1

Discussion

Not all the mechanisms proposed for the amine-catalyzed ester formation have to be considered in the reaction between benzoyl chloride and phenol in nonpolar DCM. First, benzoyl chloride has no α-hydrogens and it is evident that any mechanism involving ketene formation should be discarded. Second, the reaction is effected in a relatively non-polar solvent and the presence of acylium ions seems also very unlikely. After these considerations, two catalysis mechanisms can be considered for the model reaction: *nucleophile catalysis* and *base catalysis* (Scheme 10.2). It should be pointed out that both processes are different in origin, since the role of the amine as a nucleophile catalyst is to activate the benzoyl chloride (the electrophile), while in the second case, the amine as base catalyst is activating the phenol (the nucleophile) by forming the more nucleophile phenoxide anion. The mechanisms also differ in the intermediates formed. The nucleophile catalysis involves the intermediacy of an acyl ammonium salt whereas in the base catalysis a phenoxide anion should be formed.

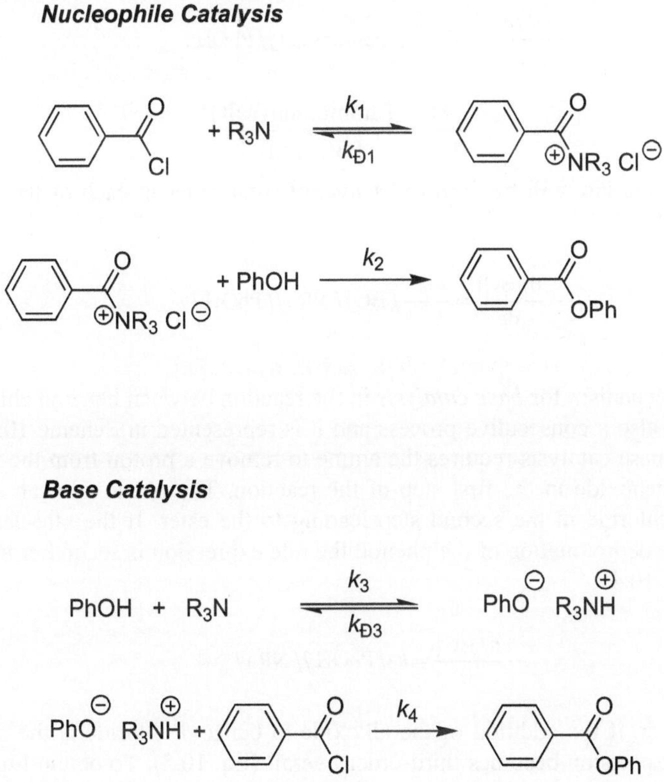

Nucleophile Catalysis

Base Catalysis

Scheme 10.2

Kinetic Considerations

The *nucleophile catalysis* mechanism for benzoate formation depicted in Scheme 10.2 involves a first step producing an intermediate acyl ammonium salt and a second step in which the salt reacts with the phenol to form the ester. As a general rule for the kinetic study of complex reactions, we can make some assumptions to simplify the analysis of this consecutive reaction. Let us suppose that the first step is rate determining. Then k_{-1} is much smaller than $k_2[PhOH]$ and the rate law would be given by Eq. 10.1 (BC = benzoyl chloride):

$$-\frac{d[BC]}{dt} = k_1[BC][NR_3] \tag{10.1}$$

On the contrary, if the second step is rate-determining, the first step will be a fast pre-equilibrium and the rate equation would be given by Eq. 10.2:

$$-\frac{d[BC]}{dt} = k_2[\text{ammonium salt}][PhOH] \tag{10.2}$$

but,

$$\frac{k_1}{k_{-1}} = \frac{[\text{ammonium salt}]}{[BC][NR_3]}$$

then, the rate law will be third-order overall (first order in each of the reactants) (Eq. 10.3):

$$-\frac{d[BC]}{dt} = \frac{k_1 k_2}{k_{-1}}[BC][NR_3][PhOH] \tag{10.3}$$

Equation 10.3 is in agreement with the experimental data.

The mechanism for ***base catalysis*** in the reaction between benzoyl chloride and phenol is also a consecutive process and it is represented in Scheme 10.2. In this case, the base catalysis requires the amine to remove a proton from the phenol to form a phenoxide in the first step of the reaction. The phenoxide can attack the benzoyl chloride in the second step leading to the ester. If the rate-determining step is the deprotonation of the phenol, the rate expression is second order according to Eq. 10.4:

$$-\frac{d[BC]}{dt} = k_3[PhOH][NR_3] \tag{10.4}$$

However, if the addition of the alkoxide to benzoyl chloride is the slow step, the rate expression becomes third-order overall (Eq. 10.5). To obtain Eq. 10.5 we should follow the same reasoning previously discussed for Eq. 10.3.

$$-\frac{d[BC]}{dt} = \frac{k_3 k_4}{k_{-3}}[BC][NR_3][PhOH] \tag{10.5}$$

Equation 10.5 is also in agreement with the experimental data.

Another alternative to obtain a third-order kinetic law would be to consider that the discrete formation of the phenoxide ion is not required. The phenol could just coordinate with the base and form a phenol-amine complex with a partial negative charge on the oxygen. This complex would then react with benzoyl chloride to form the product in the slow step of the reaction (Scheme 10.3).

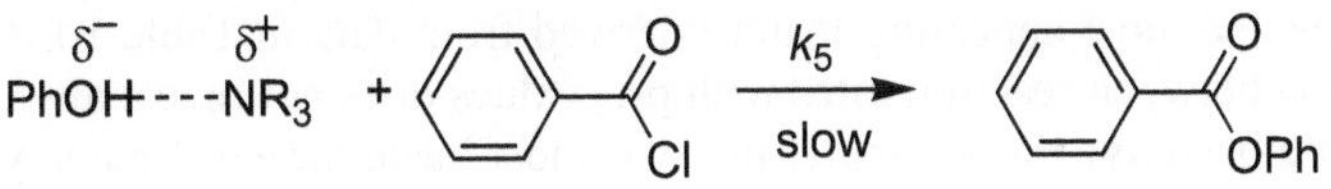

Scheme 10.3

In this case, both the phenol and the amine are present in the rate expression that becomes a third-order rate law (Eq. 10.6).

$$-\frac{d[BC]}{dt} = k_5[BC][NR_3][PhOH] \tag{10.6}$$

By comparison of Eqs. 10.3, 10.5 and 10.6 we conclude that the alternative mechanisms proposed in Schemes 10.2 and 10.3 are kinetically indistinguishable. Provided that in the nucleophile and base catalysis mechanisms the second step would be rate-determining, Eq. 10.3 and Eq. 10.5 are identical, and even a base catalysis by a phenol-amine complex (Eq. 10.6) could not be discarded. Therefore, the discrimination between nucleophile and base catalysis in the model reaction cannot be exclusively made on kinetic grounds.

Isotope Effect

A decrease in reaction rate with phenol-d_6 relative to phenol would indicate O-H (O-D) bond cleavage in the rate-determining step. However, neither the base catalysis nor the nucleophile catalysis require the breakage of the O-H phenolic bond in the slow step of the reaction. Only in the case of the mechanism depicted in Scheme 10.3, a kinetic isotope effect (KIE) could be expected. In consequence, the absence of KIE when perdeuterated phenol is used as substrate is against the mechanism based on the formation of a phenol-amine complex but is not definitive proof to distinguish between the nucleophile and a base catalysis mechanisms postulated in Scheme 10.2.

Effect of the Amine

At this point, we have realized that neither the kinetic data nor the absence of KIE with labeled phenol are conclusive to decide which of the mechanisms proposed in Scheme 10.2 is more likely. The remaining available data are related to the effects of the amine basicity and structure on the rate of the formation of phenyl benzoate (Table 10.1).

For a given amine, almost identical reaction rates are obtained (independent of the kinetic method employed) by monitoring the formation of phenyl benzoate (d[PB]/dt) or the disappearance of benzoyl chloride (−d[BC]/dt). This is consistent with the absence of long-lived intermediates (phenoxide ions) and is in full agreement with the UV-vis experiments that were not able to detect the formation of discrete phenoxide ions under the reaction conditions.

Perhaps the most appealing feature derived from data in Table 10.1 is that the comparison between reaction rates with pK_a values does not reveal any correlation with amine basicity. For example, the two most basic amines,[1] namely quinuclidine and Et_3N react with rates differing by at least three orders of magnitude. *These facts are inconsistent with a base-catalyzed mechanism, which should show a rate increase with increased basicity of the amine employed.*

In Summary

The reaction between benzoyl chloride and phenol may occur through several alternate pathways involving nucleophile catalysis or base catalysis. Such pathways are kinetically indistinguishable and the absence of KIE when phenol-d_6 is used as substrate, is not conclusive to determinate the operative mechanism. Only the absence of a direct relationship between amine basicity and reaction rate and the lack of phenoxide ion in the UV-vis experiments, can rule out a base catalysis mechanism, pointing to a nucleophile catalysis.

Additional Comments

This problem is based on the work by Hubbard P, Brittain WJ (1998) *J. Org. Chem.* 63:677-683.

Subjects of Revision

Nucleophile catalysis. Base catalysis. Spectroscopic detection of reaction intermediates.

---

[1] Using aqueous pK_a values for reactions in aprotic, nonpolar media can be incorrect. In the discussion we have assumed that the relative amine basicity values measured in water are a reasonable measure of their basicity trends in DCM.

Level 1 – Case 11
The Hydrolysis of *p*-Substituted Styrene Oxides

Key point: *Catalysis. Hammett constants*

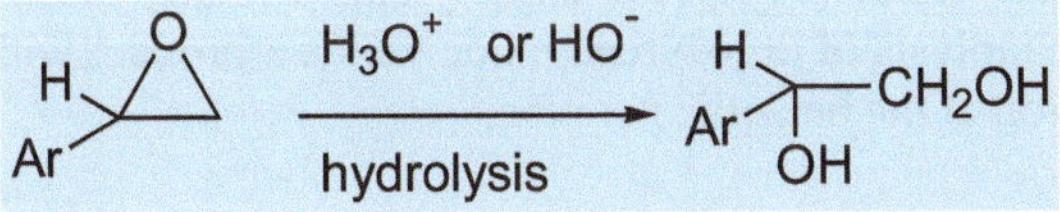

The Hydrolysis of *p*-Substituted Styrene Oxides

Styrene oxides **1** can be hydrolyzed to their glycols **2** in a wide pH range (Scheme 11.1). Interest to understanding the mechanism of this process is related to the fact that epoxide metabolites of polycyclic aromatic hydrocarbons have been claimed to be responsible for the carcinogenic properties of some of these compounds.

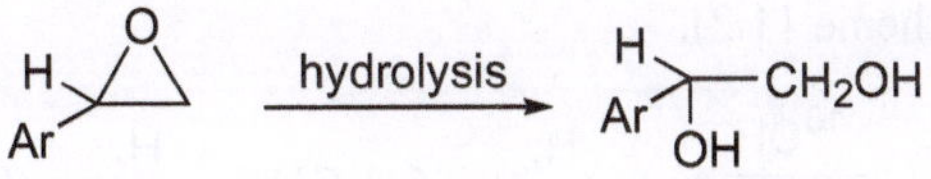

Scheme 11.1

Based on a reasoned discussion of the following experimental data, propose a mechanism for the hydrolysis of substituted styrene oxides in acidic and basic media.

Experimental Data

A study of the hydrolysis of styrene oxide and its *p*-MeO, *p*-Me, *p*-Cl and *p*-NO$_2$ derivatives under acidic (H$_3$O$^+$) and basic (OH$^-$) conditions has been carried out. The experimental results obtained in the study are as follows:

The expression of the rate constant for the hydrolysis of phenyl-substituted styrene oxides **1** to their corresponding glycols **2** is given by Eq. 11.1, where k_0 is the first-order rate constant for the spontaneous (non-catalyzed) reaction, k_H is the second-order rate constant for the hydronium ion-catalyzed reaction and k_{OH} is the second-order rate constant for the hydroxide-catalyzed reaction.

$$k_{obs} = k_0 + k_H[H_3O^+] + k_{OH}[OH^-] \tag{11.1}$$

Hydrolysis in Acidic Medium (H_3O^+)

1. The rate of hydrolysis of p-substituted styrene oxides in acidic medium is strongly affected by the nature of the substituent in the *para* position, varying from $k_H = 1.10 \times 10^4 \ M^{-1} \ s^{-1}$ for the MeO substituent to $k_H = 3.41 \times 10^{-3} \ M^{-1} \ s^{-1}$ in the case of the NO_2 group. A plot of log k_H versus σ^+ for all *para*-substituents gives an excellent Hammett correlation with a slope $\rho^+ = -4.2$.
2. The hydrolysis in $H_2{}^{18}O$ yields glycol products in which all of the ^{18}O is incorporated into the benzyl hydroxyl group.
3. The acid hydrolysis of (+)-styrene oxide yields a totally racemic diol, whereas the acid methanolysis of (+)-styrene oxide yields a product with 89% inversion of configuration at the benzylic position.

Hydrolysis in Basic Medium (OH^-)

4. The rates of hydrolyses in basic medium of p-NO_2, p-Cl, p-H and p-Me substituted styrene oxides are essentially independent of the nature of the substituent (i.e. $k_{OH} = 1.02 \times 10^{-4} \ M^{-1} \ s^{-1}$ in the case of p-NO_2 and $k_{OH} = 1.62 \times 10^{-4} \ M^{-1} \ s^{-1}$ in the case of p-Me).
5. The reactions with $K^{18}OH/H_2{}^{18}O$ show that the addition of $^{18}OH^-$ to styrene oxide occurs equally at both the α-(benzylic) and the β-carbons (51% and 49%, respectively) (Scheme 11.2).

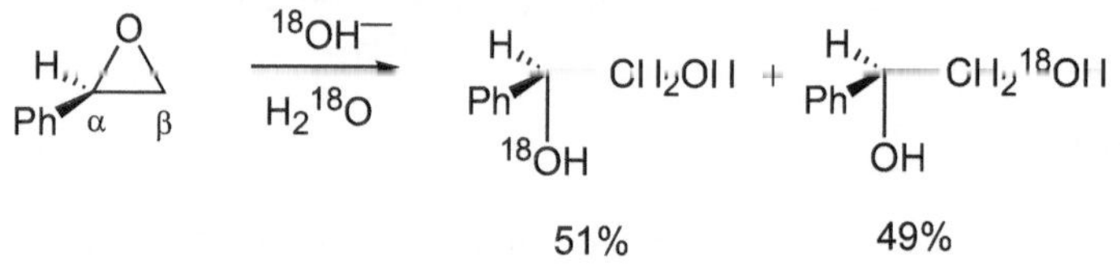

Scheme 11.2

6. The ratio of α-attack/β-attack of the hydroxide ion on styrene oxides was also found to be a function of the *para*-substituent. For example, addition of ^{18}OH to p-methylstyrene oxide occurs mainly at the α-carbon (63%), whereas the addition to p-chlorostyrene occurs mainly at the β-position (only 37% of α-attack). Hammett ρ values for α- and β-additions have been obtained from the corresponding constants k_α and k_β and are calculated to be -0.9 and $+0.2$, respectively.
7. Sodium methoxide reacts with styrene oxides **1** to yield mixtures of compounds derived from α- and β-attack (**3** and **4** respectively in Scheme 11.3) in a ratio that is a function of the electronic nature of the substituent. Furthermore, the disparities in ratios observed for the different substituents are quite similar to that found for the addition of hydroxide ion in water.

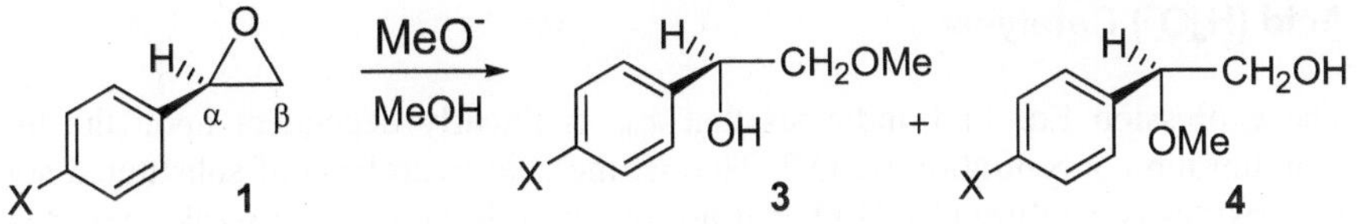

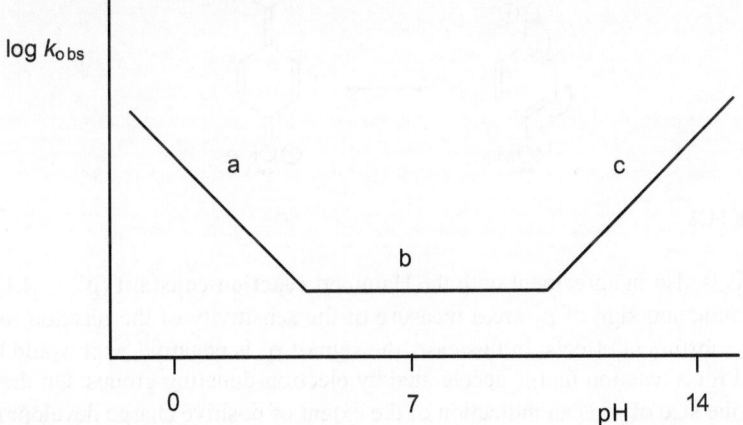

X = MeO (**3:4**, 70:30)
X = Me   (**3:4**, 51:49)
X = H   (**3:4**, 38:62)
X = Cl   (**3:4**, 29:71)
X = NO$_2$ (**3:4**, 15:85)

Scheme 11.3

Discussion

The second and third terms in rate expression Eq. 11.1 ($k_H[H_3O^+]$ and $k_{OH}[OH^-]$, respectively) clearly indicate that the hydrolysis of styrene oxides is subject to both H_3O^+ and OH^- catalysis. Acid-base catalyzed reactions have characteristic profiles when plotting log k_{obs} versus pH (a typical profile is represented in Fig. 11.1). The plots would reflect the contribution of the acid catalysis (a), the spontaneous (uncatalyzed) mode (b) and the basic catalyzed reaction (c).[1]

Next, we will discuss the experimental evidence in order to establish the mechanism of hydrolysis in acid or basic media.

log k_{obs}

a

c

b

0   7   14

pH

Figure 11.1

---

[1] It has been reported that the hydrolysis of styrene oxide proceeds *via* the spontaneous pathway in the range pH 8-12.

Acid (H_3O^+) Catalysis

The expression Eq. 11.1 indicates that k_{obs} is linearly dependent upon the hydronium ion concentration $[H_3O^+]$. That is, the acid hydrolysis of substituted styrene oxides is catalyzed by H_3O^+ but not by any other unionized weak acid (AH) present in the medium. This is distinctive for a **specific acid catalysis reaction**.

$$k_{obs} = k_0 + k_H[H_3O^+] + k_{OH}[OH^-] \tag{11.1}$$

On the other hand, the experimental data indicate that the hydronium ion-catalyzed hydrolysis of styrene oxides is **highly influenced** by the electronic substituent effects in the aromatic ring, being accelerated by p-electron-donating groups. The values of Hammett substituent constants σ measure the electronic interaction between a substituent and the aromatic ring. However, in this case, the best Hammett correlation was obtained when the σ^+ values were employed. The σ^+ (or σ^-) scales modify and improve the σ values by considering that some p-substituents could resonate directly with the reactive center when this is directly linked to the aromatic ring. This effect is known as **through conjugation** and it is associated with the development of charge (positive or negative) at the position next to the ring. The excellent correlation with the σ^+ values observed experimentally is suggesting that a **significant positive charge** is being developed at the benzyl carbon in the transition state. In consequence, p-electron-donating substituents, able to supply π-electron density (as the methoxy group), can stabilize the positive charge at the benzylic position by through conjugation (Fig. 11.2).

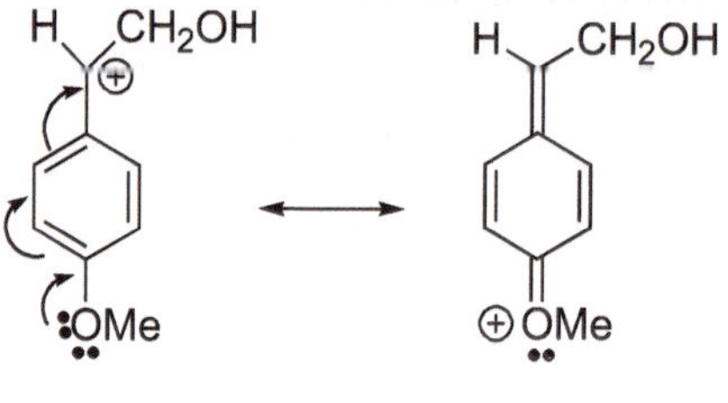

Figure 11.2

This is also in agreement with the Hammett **reaction constant** ($\rho^+ = -4.1$). The magnitude and sign of ρ^+ are a measure of the sensitivity of the reaction to electronic substituent effects. In this case, the sign of ρ^+ is negative, as it would be expected for a reaction that is accelerated by electron-donating groups. On the other hand, the size of ρ^+ is an indication of the extent of positive charge development at the carbon directly linked to the ring, in passing from ground to transition state. In this case ρ^+ is quite large, suggesting that a carbocation must be involved in the rate-determining step of the process.

With the data we have discussed until now we have plenty of information to propose a reasonable mechanism for the H_3O^+-catalyzed hydrolysis of styrene oxides. First, we know that it is a specific acid-catalyzed reaction and second, the intermediacy of a carbocation seems very likely. When writing a reaction mechanism, to deal with a specific acid catalysis process is of great help because generally they all follow the same pattern: *a rapid protonation of the substrate as a pre-equilibrium, followed by a slow step not involving proton transfer* (Scheme 11.4). In this case, the protonation of the oxygen in **1** would lead to **5** that after breaking the C-O bond in the slow step, would be transformed into carbocation **6**. Attack of water in **6** followed by deprotonation of the intermediate alcohol **7** would form the final diols **2**.

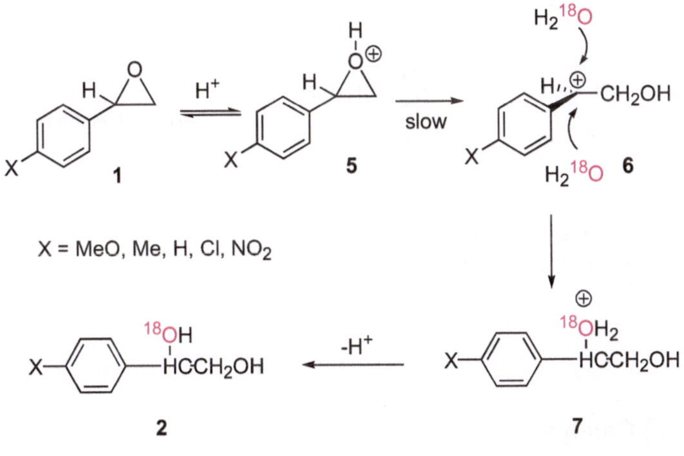

X = MeO, Me, H, Cl, NO$_2$

Scheme 11.4

We have to check now if the formulation of a mechanism involving a carbocation intermediate is supported by the rest of the experimental data. The isotopic labeling experiments carried out with H$_2$^{18}O confirm that the attack of the nucleophile occurs exclusively at the benzylic position, which is in agreement with the mechanism proposed in Scheme 11.4 (labeled atoms in red). Furthermore, the formation of a planar intermediate, like a carbocation, should justify the fact that when chiral (+)-styrene oxide is employed, only the racemic diol is obtained.

Why then does the acid methanolysis of (+)-styrene oxide lead to the diol with 89% inversion of configuration at the benzylic position? Clearly the high percentage inversion of the configuration is incompatible with the formation of a planar intermediate like a carbocation. However, we have made a change in the solvent and methanol is more nucleophilic than water. The inversion of the configuration could be due to a S$_N$2-type attack by the nucleophile at the benzylic position before the carbocation is formed. In consequence, a competition between S$_N$2 displacement by solvent molecules (inversion) and ionization (racemization) may happen. By considering this possibility, in Scheme 11.5 we have indicated the possible mechanisms for the acid solvolysis of substituted styrene oxides. After the protonation of the oxirane ring two alternative pathways could be considered. Electron-donating groups in para position of the aromatic ring and non-nucleophilic solvents will allow for the evolution of protonated oxiranes **5** to carbocations **6** (path a), whereas highly nucleophilic solvents and absence of p-electron-donating groups would favor the direct solvolysis of the oxirane ring (path b).

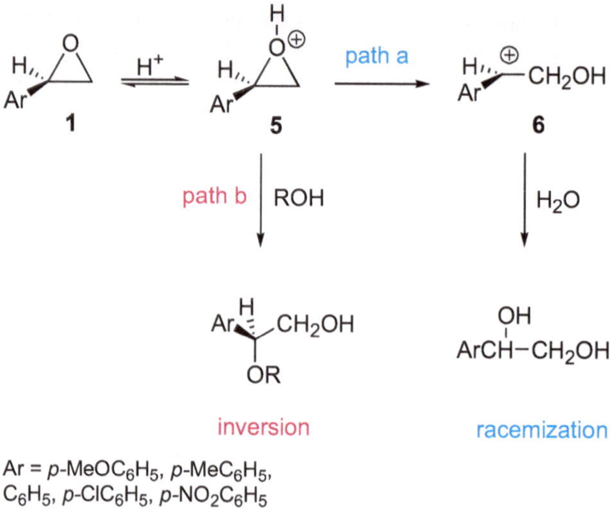

Scheme 11.5

Base (OH⁻) Catalysis

The experimental data obtained in the OH⁻-catalyzed hydrolyses of styrene oxides are completely different from the ones previously commented. First, the reaction products result from the incorporation of the nucleophile (OH⁻) to both, the α and the β positions of the oxirane ring, as it has been demonstrated in the reaction with $K^{18}OH/H_2{}^{18}O$. Second, a change in the nucleophile (solvolysis in MeO⁻/MeOH) does not make any major difference, and again both products resulting from the α- and the β-attack are obtained (Scheme 11.6).

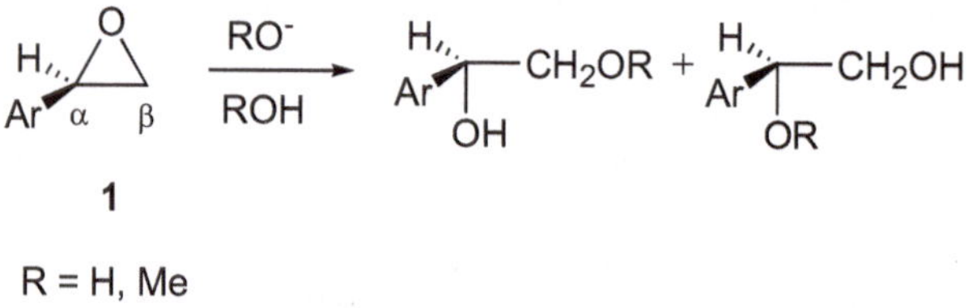

Scheme 11.6

Clearly, the OH⁻ (MeO⁻ in the methanolysis reaction) is acting as a nucleophile (not as a base) in the process. In consequence, the hydrolysis (methanolysis) of styrene oxides in basic medium is a **nucleophile-catalyzed reaction**, not a base-catalyzed process. This is an important point to be considered when proposing a reaction mechanism, because the nucleophile has to be involved in the rate-determining step.

Another essential aspect is the discussion of the electronic effects. The Hammett ρ values for α- and β-addition of OH$^-$ (-0.9 and $+0.2$, respectively) are very low in magnitude and opposite in sign. This results in the observed insensitivity of the reaction rates to the substituents in the aromatic ring (k_{OH} 1.02×10^{-4} M^{-1}s^{-1} in the case of the *p*-NO$_2$-styrene and k_{OH} 1.62×10^{-4} M^{-1} s^{-1} in the case of *p*-Me-styrene). However, despite its low magnitude, ρ_α is indicating some charge development near the ring in the transition state.

Furthermore, the effect of the substituents in position *para* of the ring seems to be determinant in the control of the regiochemistry of the reaction. The experimental results with K^{18}OH/H$_2{}^{18}$O indicate that α-addition (more hindered position) is preferred when electron-donating groups (*p*-MeO) are placed in the aromatic ring. This observation is in agreement with the (small) development of positive charge suggested by the value of the Hammett ρ at the α-position (Fig. 11.3). Electron-withdrawing groups in the styrene ring (*p*-Cl, *p*-NO$_2$) seems to favor the attack at the β-carbon, the less-hindered position. Similar α/β-attack ratios are observed when the reaction is carried out in MeO$^-$/MeOH.

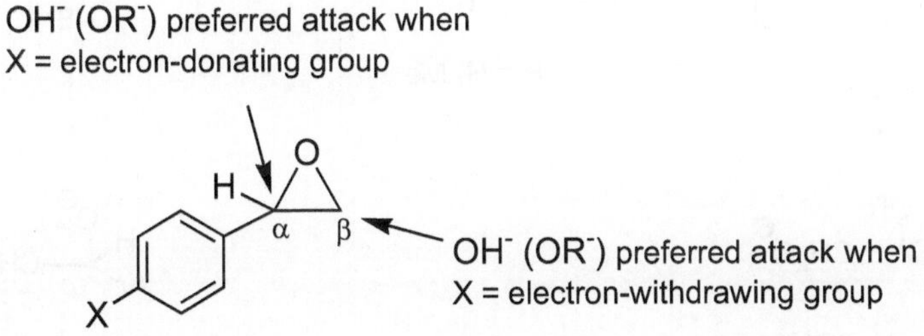

Figure 11.3

To explain this complete set of data we could consider two alternative pathways for the hydrolysis of substituted styrene oxides under basic conditions. In both cases the process should start by the nucleophilic attack of the HO$^-$ (MeO$^-$) ion to the oxirane ring (Scheme 11.7). α-Nucleophilic addition will lead to transition state **8** in which a small positive charge is developed at the benzylic carbon. Electron-donating groups placed at the para position of the aromatic ring should stabilize the positive charge and hence, the formation of **8** is preferred. In absence of electron-donating substituents, the reaction takes place via transition state **9**, resulting from the attack of the nucleophile to the less-hindered β-carbon.

Comment

It is generally assumed that the nucleophilic attack to an oxirane ring occurs preferentially at the less-hindered carbon and the steric hindrance is claimed to be responsible for the regioselectivity of the reaction. In the current example the experimental results indicate that the electronic effects can play an important role in the control of the regioselectivity of the addition.

In Summary

As expected, the hydrolysis (solvolysis) of *p*-substituted styrene oxides follows a different mechanism in acidic or basic conditions. At low pH values the reaction is a **specific acid-catalyzed process** and a carbocation is involved as intermediate. Under basic conditions, the hydrolysis is **nucleophile-catalyzed** and the attack of the nucleophile takes place at both carbons of the oxirane ring. The α-attack (more hindered position) is preferred when strong *p*-electron-donating substituents are placed in the styrene ring. *p*-Electron-withdrawing substituents favor the β-attack (less-hindered position).

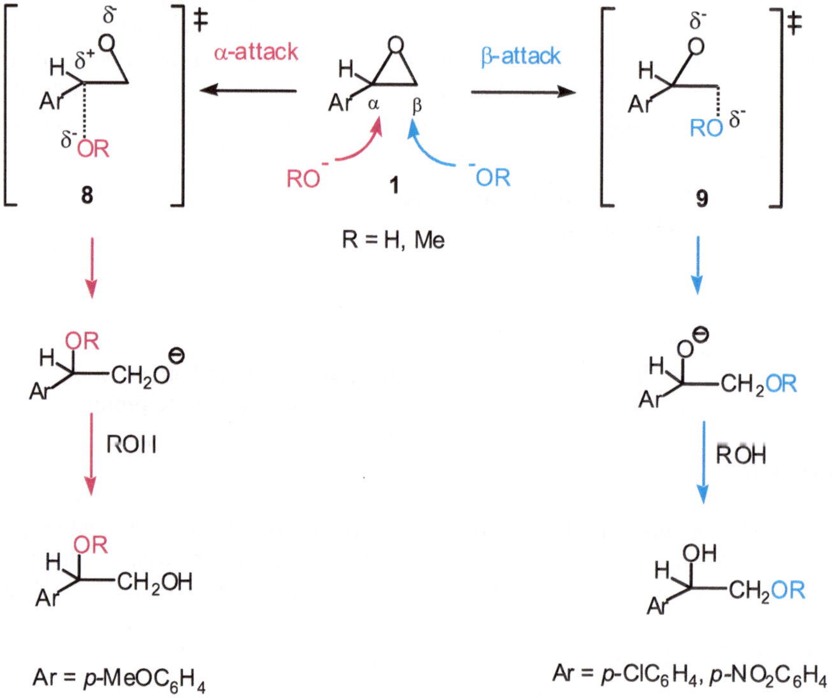

Scheme 11.7

Question

Design an experiment to trap the carbocation intermediate. Discuss the structure of the most appropriate substrate, the type of solvent and propose a suitable trapping agent.

Answer to the Question

Additional evidence in favor of the intermediacy of a carbocation in the acid solvolysis of styrene oxides could be obtained from a trapping experiment. In order to increase the lifetime of the carbocation, it would be convenient to use a *p*-MeO-sustituted styrene oxide as substrate and very little nucleophilic solvent. The trapping agent should be very reactive and much more nucleophile than the solvent, to avoid any undesirable competition reaction between both species. The highly nucleophile azide ion could be suitable to trap the intermediate which has sufficiently long lifetimes in aqueous solutions to be trapped by the reagent (Scheme 11.8).

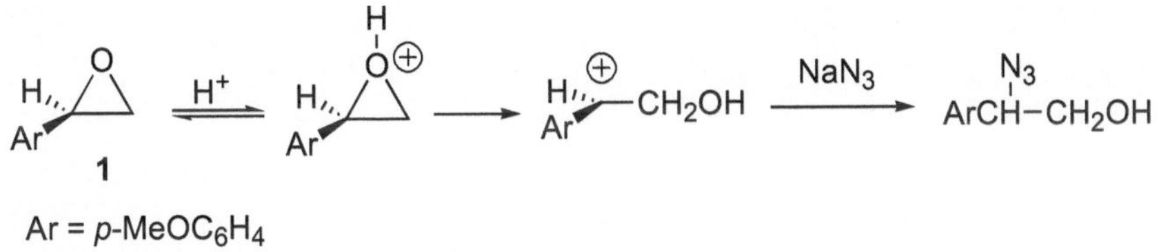

Ar = *p*-MeOC$_6$H$_4$

Scheme 11.8

Additional Comments

This problem is based on the work by Blumenstein JJ, Ukachukwu VC, Mohan RS, Whalen DL (1993) *J. Org. Chem.* 58:924-932.

Subjects of Revision

Catalysis, types and mechanisms. Hammett ρ and σ constants. S$_N$1 versus S$_N$2 reactions. Carbocations.

Level 1 – Case 12
Elimination Reactions of
Benzaldehyde *O*-Benzoyloximes

Key point: *Hammett constants. Isotope effects*

Elimination Reactions of Benzaldehyde *O*-Benzoyloximes

E- and *Z*-Benzaldehyde *O*-benzoyloximes **1** and **2**, give benzonitriles in quantitative yield upon treatment with DBU (1,8-diazabicyclo[5.4.0]undec-7-ene) (Scheme 12.1).

Discuss the mechanism of the elimination reaction and justify the differences in reaction rates, k_H/k_D and Hammett ρ values found for Z- and E-isomers. Interpret the experimental data and propose a reasonable structure for the transition state in both cases.

Experimental Data

1. A complete kinetic study has been carried out to determine the mechanism of the elimination reaction for both isomers and the results obtained are compiled in Table 12.1.
2. The elimination reactions exhibit general-base catalysis in both cases.

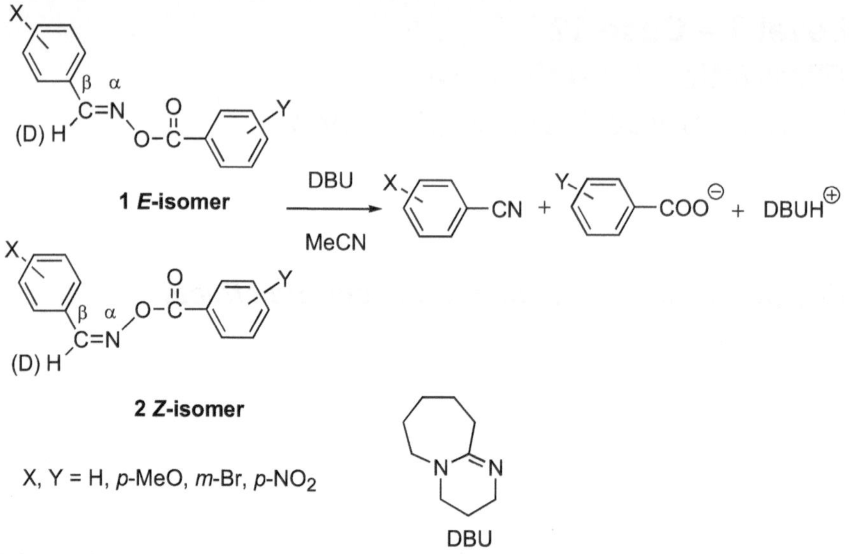

Scheme 12.1

3. The reactions have second-order kinetics (first order in oxime and base).
4. The influence of the β-aryl substituents (X = H, p-MeO, m-Br, p-NO$_2$) upon the elimination rates gave excellent correlations with σ values in all cases.

Table 12.1

	E-isomer	Z-isomer
Relative rate	1	36000
k_H/k_D	3.3 ± 0.2	7.3 ± 0.2
ρ	2.19 ± 0.05	1.21 ± 0.05
β_{lg}	-0.49 ± 0.02	-0.40 ± 0.01

Discussion

Since the base-promoted elimination reactions of E- and Z-benzoyloximes **1** and **2** exhibit second-order kinetics (first order in base and oxime), the kinetic law would have the form of Eq. 12.1:

$$v = k_{obs} \text{ [base][oxime]} \tag{12.1}$$

Among the different types of elimination processes only bimolecular pathways will be in agreement with Eq. 12.1 and hence we will center the discussion on E2 and E1cB mechanisms. Prior to the discussion of the experimental data, it may be necessary to talk about the main features of the E2 and E1cB reactions in detail.

E2 MECHANISM

The E2 elimination is a one-step process and neither a cationic nor a carbanionic intermediate is formed. Proton abstraction by the base and leaving group departure are concerted, although not necessarily synchronous. That is, C_β-H bond breaking may be ahead or behind C_α-X bond breaking (Scheme 12.2).

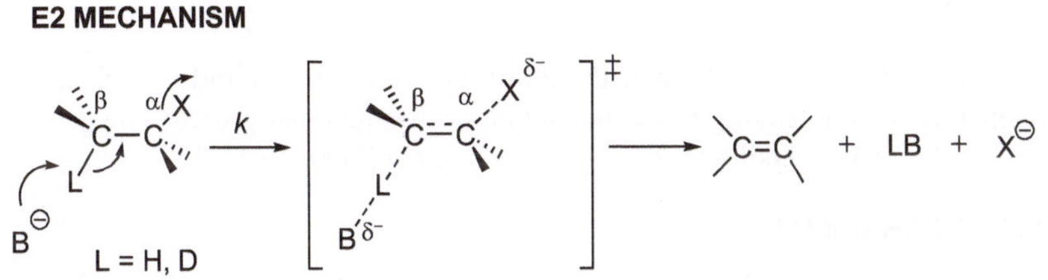

E2 MECHANISM

Scheme 12.2

As C_β-H bond breaking occurs in the rate-determining step of the reaction, the primary deuterium kinetic isotope effect (PDKIE) is clearly observed in the E2 reactions. Even more, the magnitude of k_H/k_D on the β-H is an excellent indicator of the extent of C_β-H bond breaking in the transition state and values of k_H/k_D ranging from 2 to 8 have been described in such reactions. On the other hand, in the E2 mechanism the C_α-X bond cleavage also occurs in the slow step of the reaction. The extent of the C_α-leaving group bond breaking should correlate with the ability of the leaving group to depart (nucleofugacity). Among the different ways to evaluate the nucleofugacity of a group, the β_{lg} values (effect of leaving group basicity) are frequently employed when the leaving group has a basic character, as in the oxime eliminations that we are discussing in this chapter. The absolute values of β_{lg} ($|\beta_{lg}|$) normally range between 0 and 1, from total insensitivity to extreme sensitivity to the leaving group pK_a.

In conclusion: In an E2 mechanism significant k_H/k_D and $|\beta_{lg}|$ values should be expected.

E1cB MECHANISM

In contrast with the previously commented one-step E2 reactions, E1cB eliminations are stepwise processes, consisting in the initial loss of a proton by the action of the base, followed by the departure of the leaving group (Scheme 12.3). The relative magnitudes of the rate constants involved (k_1, k_{-1}, k_2) determine which is the slow step of the reaction. Hence, different types of E1cB mechanisms could be considered.[1]

---

[1] For a detailed study of base-promoted elimination reactions see: Lowry TH, Richardson KS (1987) Mechanism and Theory in Organic Chemistry. 3th edn., Harper and Row, New York, pp 598-616.

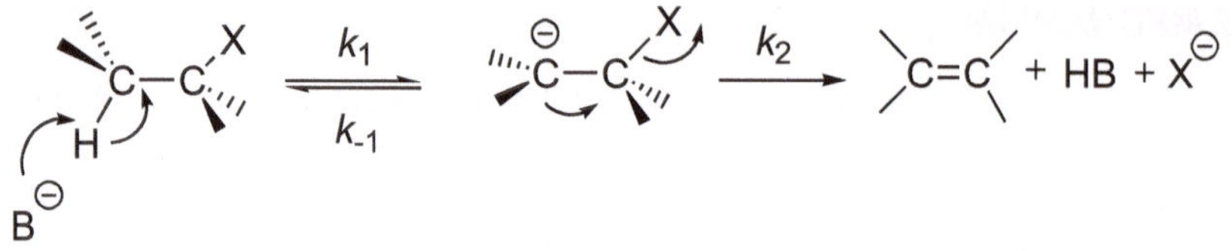

Scheme 12.3

The first situation arises when the anion is formed in a rapid equilibrium and the departure of the leaving group occurs in a subsequent slow step. The process is then called E1cB-reversible or $(E1cB)_R$ mechanism (Scheme 12.4).

$(E1cB)_R$ MECHANISM

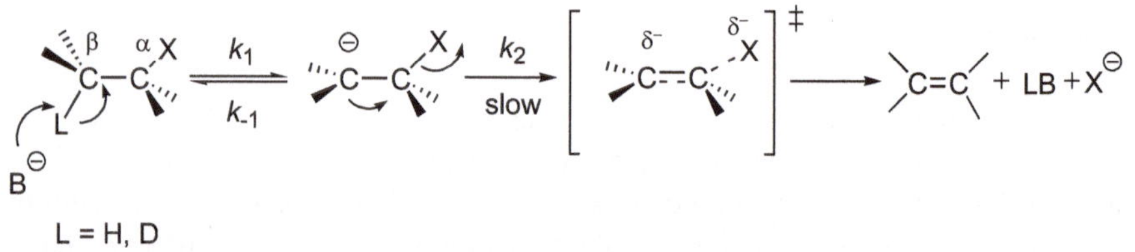

L = H, D

Scheme 12.4

Obviously the C_β-H bond breaking happens before the rate-determining step and no PDKIE should be observed at the H_β position. However, as the C_α-X bond breakage occurs in the slow step, the reaction rate should be sensitive to the nucleofugacity of the leaving group. As the proton abstraction by the base is not directly involved in the rate-determining step, the $(E1cB)_R$ mechanism is not a general-base catalyzed reaction.

In conclusion: In an $(E1cB)_R$ mechanism significant $|\beta_{lg}|$ values and absence of PDKIE should be expected.

Alternatively, the C_β-H proton abstraction could be rate-determining and the departure of the leaving group occur in a further step. The process is called E1cB-irreversible or $(E1cB)_{irr}$ mechanism (Scheme 12.5).

$(E1cB)_{irr}$ MECHANISM

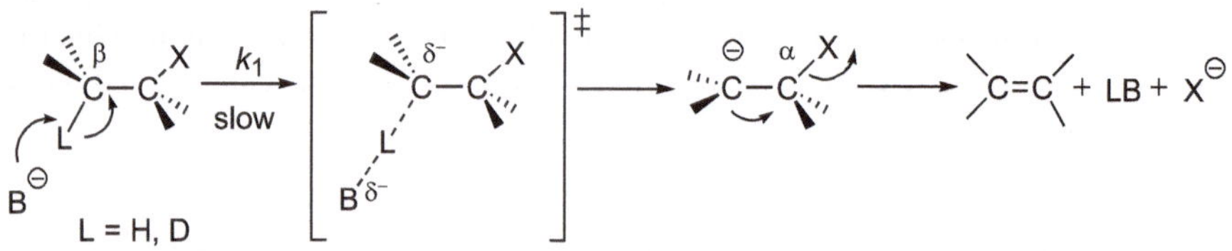

L = H, D

Scheme 12.5

In this case, a high primary kinetic isotope effect (k_H/k_D) and negligible effect of the nucleofugacity of the leaving group should be expected.

Once we have discussed the main features of the E2, (E1cB)$_R$ and (E1cB)$_{irr}$ mechanisms we are ready to determine which one is in better agreement with the experimental data obtained for substrates **1** and **2**.

How can we decide among all these mechanisms?

Frequently, distinguishing among the various elimination mechanisms is not an easy task, but careful analysis of the experimental data will give us the key in this case. First, we know that the DBU-promoted eliminations in *E*- and *Z*-benzaldehyde *O*-benzoyloximes are general-base catalyzed reactions, that exhibit noticeable k_H/k_D values (3.3 and 7.3) and significant extents of leaving group cleavage in the transition state ($|\beta_{lg}|$ 0.49 and 0.40) for *E*- and *Z*-isomers, respectively. The general catalysis and the high k_H/k_D values are clearly incompatible with the E1cB$_R$ mechanism, as we have commented above. On the other hand, the significant $|\beta_{lg}|$ values support a mechanism in which C-leaving group bond breaking occurs in the transition state (E2) but exclude the (E1cB)$_{irr}$ pathway.

In consequence, the bimolecular elimination (E2) seems to be the most appropriate alternative for the DBU-promoted eliminations of oximes **1** and **2**.

Once we have assumed that the E2 mechanism is the most suitable option for both isomers, we have to explain why if compounds **1** and **2** share the same mechanism, they have different rates of elimination, Hammett ρ constants and k_H/k_D values (see Table 12.1).

The first striking fact is the enormous disparity between the reaction rates. Thus, the rate of elimination from *Z*-benzaldoxyme **2** is approximately 36,000-fold faster than from *E*-benzaldoxyme **1**. If we represent the E2 transition states for oximes **1** and **2** we will realize that their structures are determined by the relative arrangement of the H$_\beta$ and the benzoate groups in the starting compounds (Fig. 12.1).

The *Z*-isomer **2** is less stable than the *E*-isomer **1** due to the unfavorable steric interactions between the bulky aryl and benzoate groups. It has been estimated that the difference of stability between both isomers is 3.73 kcal/mol. However, as we have discussed previously, in an E2 elimination process, the H$_\beta$ and the leaving group depart simultaneously in the transition state. In the case of *E*-isomer **1**, transition state **3** reveals that E2 elimination has to be *syn*: H$_\beta$ and the benzoate group have to depart by the same side of the molecule. In the case of the less stable *Z*-isomer **2**, the elimination is *anti*: H$_\beta$ and the benzoate group departure occurs at opposite sides of the molecule, as shown in transition state **4**.

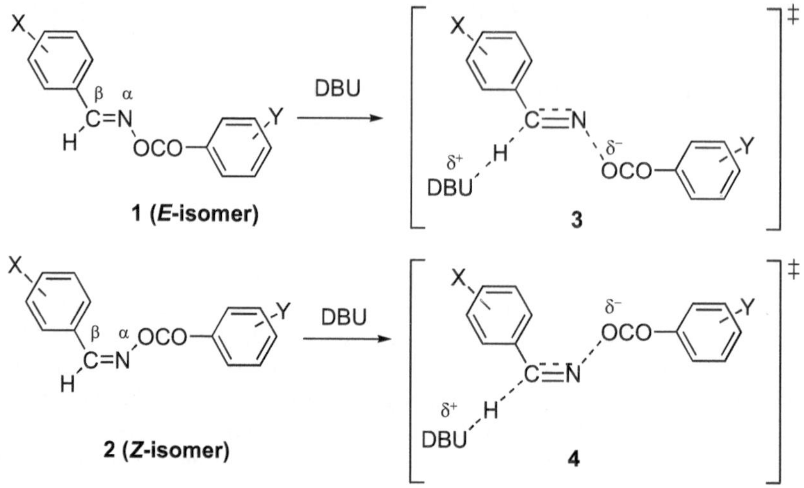

Figure 12.1

Although many examples of both *anti* and *syn* E2 eliminations have been described, it is generally accepted that the *anti*-elimination requires less energy, proceeds *via* a more symmetrical transition state and is usually greatly favored over the *syn* elimination. All these reasons, together with the relief of the steric strain during the process, could explain the much faster rate of *anti*-elimination from Z-isomer **2**.[2]

Other structural features of the transition states **3** and **4**, like the extent of the C_β-H and the *N*-benzoate bond cleavage can be deduced from a careful interpretation of the experimental data in Table 12.1.

Even though E2 reactions are concerted processes, it is rare to find a *pure* (exactly synchronous) E2 case. There is a broad spectrum of E2 mechanisms from those on which the departure of the leaving group is clearly ahead the C-H bond breaking, to those that have undergone more C-H bond cleavage in the transition state. To discuss the degree of synchronicity of the E2 eliminations in oximes **1** and **2** we should consider the $|\beta_{lg}|$ values, the Hammett constant ρ and the k_H/k_D values simultaneously. The $|\beta_{lg}|$ values give an idea about the extent of the N-leaving group bond breaking. As expected for an E2 reaction, the observed values are in the middle of the range (around 0.5), indicating that the reaction is sensitive to the leaving group pK_a. The observed $|\beta_{lg}|$ for the *anti*-elimination in compound **2** ($|\beta_{lg}| = 0.40$) is smaller than the one observed for the *syn*-elimination in compound **1** ($|\beta_{lg}| = 0.49$), indicating a smaller degree of N-leaving group bond cleavage in the E2 *anti* transition state **4**.

---

[2] The much faster rate of *anti*-eliminations has been attributed to the steric strain in the Z-isomer and the favorable overlap between the developing p orbitals at the β-carbon and α-nitrogen atoms in the transition state.

Alternatively, the extent of the C_β-H bond breaking will be determined by the Hammett constant ρ and the k_H/k_D values. The first are related to the extent of charge development at the β-carbon and the second indicate the degree of proton transfer to the base in the transition state. Next we will comment on all these parameters together, in order to find information about the structure of the transition state for both elimination processes.

Z-oxime 2: anti E2 elimination

The Hammett ρ value for *Z*-oxime **2** ($\rho = 1.21$) indicates a small extent of negative charge development at the C_β carbon in the transition state. This small Hammett ρ value is compatible with the observed high H_β deuterium kinetic isotope effect ($k_H/k_D = 7.3$). A very high deuterium PDKIE is expected for an E2 transition state in which the three centers (C, H and DBU) are maintained in a straight line, with the hydrogen positioned *centrally* between the two acceptor centers (Fig. 12.2, L = D). PDKIE, Hammett ρ and $|\beta_{lg}|$ values are consistent with a fairly balanced E2 transition state.

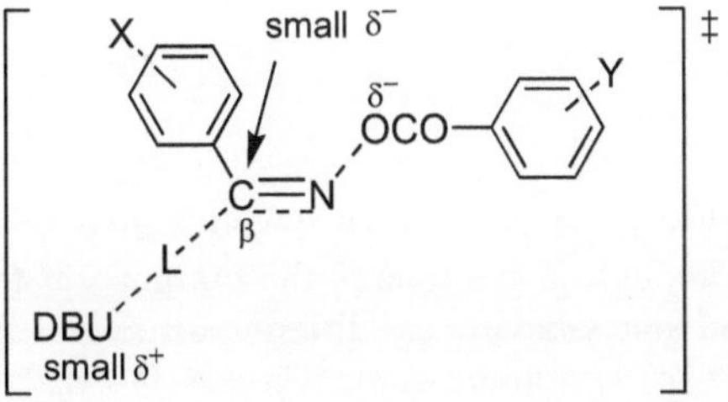

E2 transition state from Z-isomer **2**

Figure 12.2

E-oxime 1: syn E2 elimination

It is known that *syn* E2 eliminations have transition states with a considerable anionic character. This is in agreement with the experimental Hammett ρ value observed for the *syn* elimination of the *E*-oxime **1** ($\rho = 2.19$). This value is significantly higher than the one previously commented and suggests some negative charge development at the C_β carbon in the transition state. Namely, the C_β-H bond breaking must be considerably ahead of the N-leaving group bond breaking. The magnitude of the C-H deuterium kinetic isotope effect ($k_H/k_D = 3.3$) (considerably lower than that of the *Z*-isomer **2**) reflects the fact that the C_β-H bond is broken on a higher extent than the DBU-H bond is formed (Fig. 12.3). We should remember that PDKIE values reflect the degree of hydrogen transfer between the $C\beta$=N carbon and DBU and reach a maximum when the hydrogen is positioned *centrally* between the two centers. Small PDKIES are compatible with unbalanced transition states like the one proposed in Fig. 12.3.

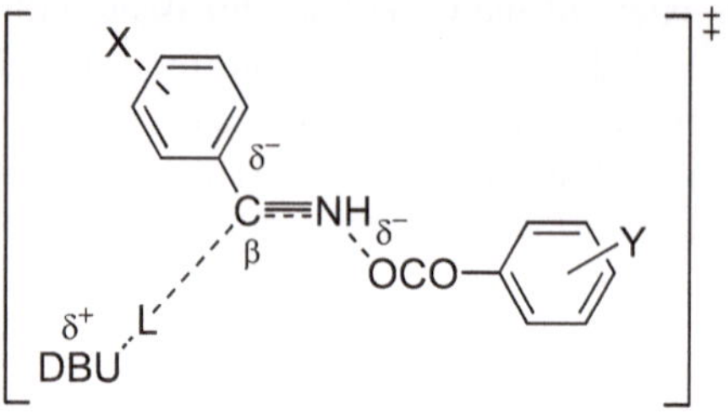

E2 transition state from *E*-isomer **1**

Figure 12.3

With all the data in hand we can conclude that the transition state for the *anti*-eliminations from *Z*-isomer **2** appears to be more symmetrical, with a smaller degree of proton transfer to the base, less negative charge development at the C=N carbon and a smaller extent of leaving group bond cleavage than that for the *syn*-eliminations from *E*-isomer **1**.

In Summary

E- and *Z*-benzaldehyde *O*-benzoyloximes **1** and **2** give benzonitriles upon treatment with DBU. Although in both cases the elimination follows an E2 mechanism, their transition states have very different structures. In the case of the *Z*-isomer **2**, the E2 is an *anti*-elimination, whereas the reaction is *syn* for the *E*-isomer **1**. The *anti*-eliminations from *Z*-isomer **2** are much faster, proceed via a more symmetrical transition state with a smaller degree of proton transfer to the base, less negative charge development at the C=N carbon and have a smaller extent of leaving group bond cleavage than that for *E*-isomer **1**.

Additional Comments

This problem is based on the work by Cho BR, Chung HS, Cho NS (1998) *J. Org. Chem.* 63:4685-4690 and by the work by Cho BR, Cho NS, Lee SK (1997) *J. Org. Chem.* 62:2230-2233.

Subjects of Revision

Elimination reactions. Hammett constants. Primary kinetic isotopic effects.

Level 1 – Case 13
Oxygen Versus Sulfur Stabilization of Carbenium Ions

Key point: *Kinetic Isotope Effects*

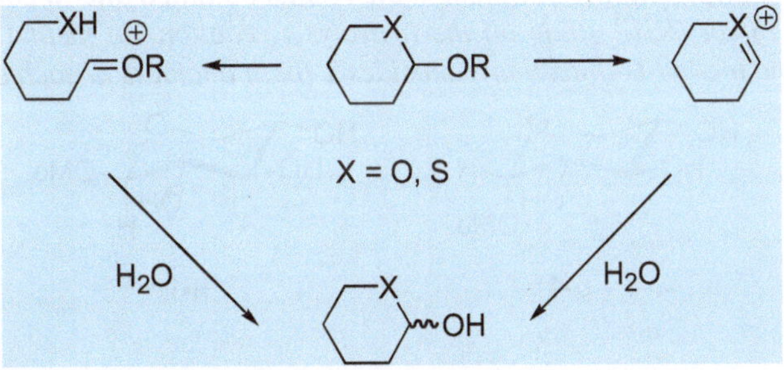

$$X = O, S$$

Oxygen Versus Sulfur Stabilization of Carbenium Ions

The general mechanistic features of the acid-catalyzed hydrolysis of glyco-pyranosides have been recognized for about 30 years. It is generally accepted that the hydrolysis occurs via specific acid catalysis. The rate-determining step involves exocyclic C-O bond cleavage and an oxacarbenium ion is formed (Scheme 13.1).

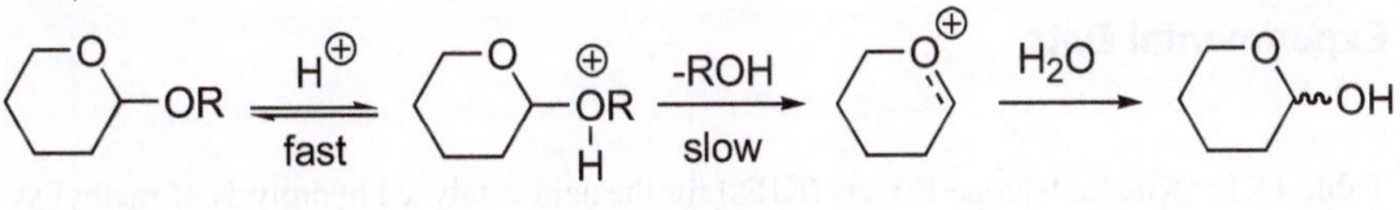

Scheme 13.1

In contrast, little is known about the mechanism of hydrolysis of 5-thioglyco-pyranosides, apart from the fact that they hydrolyze several times faster than the corresponding glycopyranosides. As the relative capabilities of sulfur and oxygen atoms to stabilize an adjacent carbenium ion center has been a controversial subject, an alternative hydrolysis mechanism for both 5-thio- and glycopyranosides has been proposed. This alternative considers that the reaction involves specific acid-catalyzed protonation on the ring heteroatom followed by rate limiting C-X bond cleavage (Scheme 13.2).

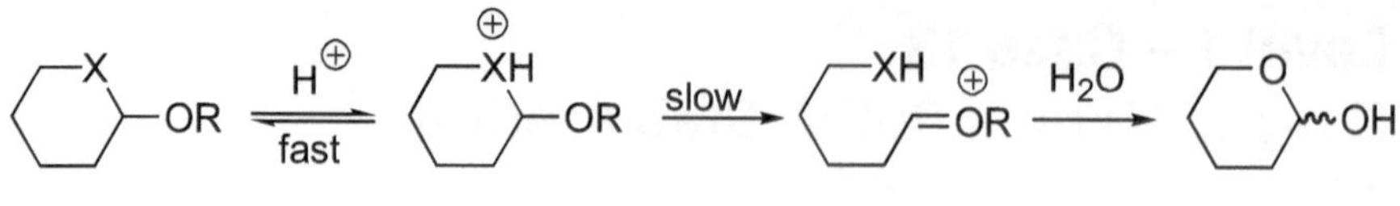

X = O, S

Scheme 13.2

Does the hydrolysis mechanism involve the endocyclic C-X bond cleavage in the slow step of the reaction? Are oxa- or thiacarbenium ions formed in the slow step of the hydrolysis process?

*To answer these questions, comment the results obtained after a complete Kinetic Isotope Effect (KIE) study on the hydrolysis reactions of methyl D-xylopyranosides **1** and methyl D-thioxylopyranosides **2** (both anomers in each case).*

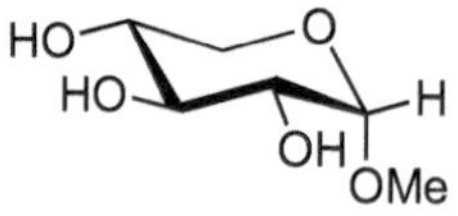

α-**1** 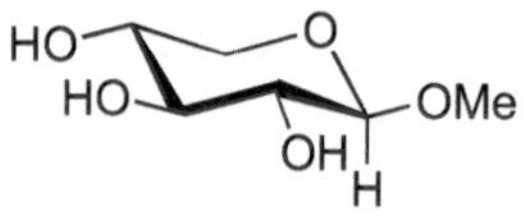   β-**1**

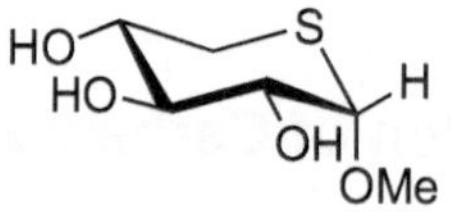

α-**2** 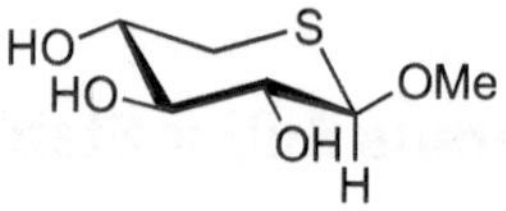   β-**2**

Comment the different types of kinetic isotope data and explain how their values can be used to obtain information about the mechanism of the reaction.

Experimental Data

Table 13.1. Kinetic Isotope Effects (KIEs) for the acid-catalyzed hydrolysis of methyl xylopyranosides **1** at 80°C (average values).

Type of KIE	α-anomer	β-anomer
Secondary α-D	1.128	1.098
Secondary β-D	1.088	1.042
Leaving group ^{18}O	1.023	1.023
Anomeric ^{13}C	1.006	1.006
Ring ^{18}O	0.983	0.978
Solvent (kD_3O^+/kH_3O^+)	2.31	2.24

Table 13.2.   Kinetic Isotope Effects (KIEs) for the acid-catalyzed hydrolysis of methyl 5-thio-xylopyranosides **2** at 80°C (average values).

Type of KIE	α-anomer	β-anomer
Secondary α-D	1.142	1.094
Secondary β-D	1.061	1.018
Leaving group ^{18}O	1.027	1.035
Anomeric ^{13}C	1.031	1.028
Solvent ($k_{D_3O^+}/k_{H_3O^+}$)	2.37	2.63

The calculated isotope effects (HF/4.31G) for the acid-catalyzed formation of an oxacarbenium ion from methanediol as a model compound were: ring $^{18}O = 0.977$ and leaving group $^{18}O = 1.020$

Discussion

As the breakage of endocyclic C-O or exocyclic C-O bonds is the key to distinguishing between the two alternatives of hydrolysis in methyl xylopyranosides **1**, from a mechanistic point of view the ring ^{18}O KIE and the leaving group ^{18}OMe KIE are possibly the most useful probes for determining whether C-X or C-O cleavage occurs during the reaction (Scheme 13.3, X = O). Generally, *normal KIE* values ($k_{16}/k_{18} > 1$) are associated to bond cleavage but *inverse KIE* values ($k_{16}/k_{18} < 1$) are indicative of bond strengthening (bonding to the site of isotopic substitution is increased at the transition state).

Rate limiting exocyclic C-O bond cleavage (path a) should generate a normal leaving group ^{18}O KIE and an inverse ring ^{18}O KIE, whereas the results should be exactly the opposite if endocyclic C-O bond cleavage occurs in the rate-limiting step (path b). The measured leaving group ^{18}O and ring ^{18}O KIEs for α-**1** are 1.023 and 0.983, respectively (1.023 and 0.978 for the β-**1** anomer). These values are definitively supporting the formation of an oxacarbenium ion, as the one proposed in path a (Scheme 13.3).

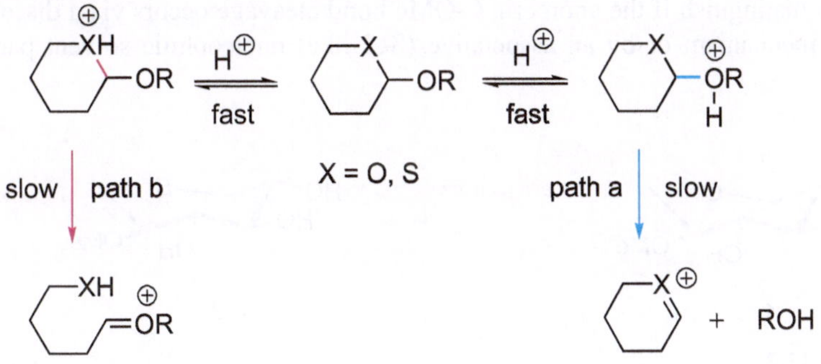

Scheme 13.3

The measured leaving group ^{18}O KIEs for thioxylopyranosides **2** are similar to those observed for their oxa-analogues **1** (Fig. 13.1, labeled oxygens colored red). Therefore, a thiacarbenium ion (X = S, path a, Scheme 13.3) must be formed in these cases, which indicates that a sulfur atom is also able to stabilize an adjacent carbenium ion.

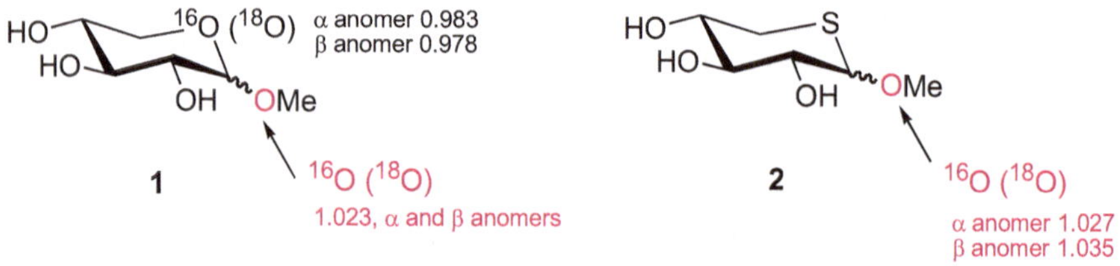

Figure 13.1

The concordance between the leaving group ^{18}O and ring ^{18}O KIEs for compounds **1** and **2** and those calculated for an oxacarbenium ion obtained from a model compound, (leaving group $^{18}O = 1.020$ and ring $^{18}O = 0.977$) is additional support for the exocyclic C-O bond cleavage and endocyclic C-O bond strengthening in the transition state of the reaction.

Therefore, based on the leaving group ^{18}O and ring ^{18}O KIEs, the hydrolysis of xylopyranosides **1** and thioxylopyranosides **2** occurs via rate-determining exocyclic C-O bond cleavage, with formation of an oxa- or thiacarbenium ion as intermediate. To confirm this hypothesis, we have a full set of additional kinetic isotopic data that must be thoroughly interpreted, checking if they are in full agreement with the proposed pathway.

Anomeric ^{13}C KIEs

The formation of an oxa- or thiacarbenium ion requires the cleavage of the anomeric C-OMe bond in the rate-determining step of the reaction. Hence, a noticeable *heavy atom KIE* is expected at this position, (Fig. 13.2, labeled position in red). The magnitude of the anomeric ^{13}C KIEs (k_{12}/k_{13}) can be used as a diagnostic tool to distinguish if the anomeric C-OMe bond cleavage occurs via a dissociative (S_N1) mechanism or by an associative (S_N2-like) nucleophilic solvent participation.

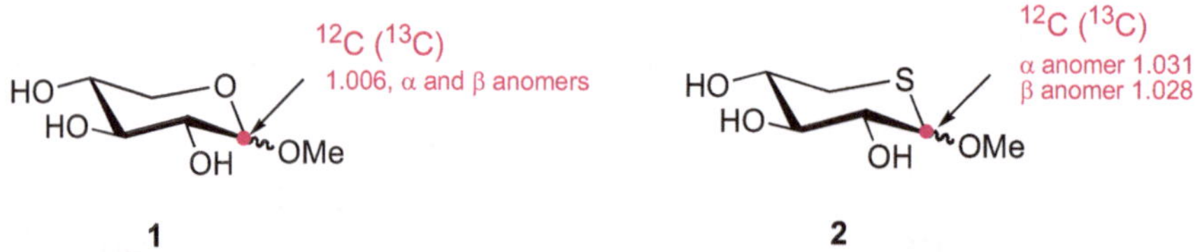

Figure 13.2

^{13}C KIEs (k_{12}/k_{13}) for associative processes in glycopyranosides are estimated to be in the range (1.03-1.08) whereas for dissociative substitution mechanisms are about (1.00-1.01). The measured values for the ^{13}C KIEs in xylopyranosides **1** (1.006) suggest a S_N1 mechanism for the departure of the OMe group whereas for 5-thioxylopiranosides **2** (1.031 and 1.028 for α and β anomers, respectively) could point to a solvent-assisted process.

Secondary DKIES

The secondary deuterium KIEs (SDKIEs) are observed when the labeled position is placed next to the reactive center. Although the bonds involving the labeled atom are not broken in the rate-determining step, it is affected to some extent by the changes that occur in the transition state. Depending on the distance from the deuterium atom to the bond that is actually been broken, the SKIEs can be classified as α-SKIEs, β-SKIEs and more rarely γ-SKIEs.

α-Secondary KIES

Considering the mechanism proposed for the hydrolysis of pyranosides **1** and **2**, a noticeable α-SKIE should be observed in the anomeric C-H(D) bond (colored red in Fig. 13.3). This position is placed next (α) to the bond that is being broken in the transition state (this bond is colored blue in Fig. 13.3). In fact, the origin of the α-SKIEs is considered to arise mainly from the weakening of an out-of-plane Cα-H(D) bending vibration as hybridization at the anomeric carbon changes from sp^3 to sp^2 in the rate-determining step. The estimated α-SKIE values for this change of hybridization are seldom larger than 1.2. The observed data for the α- and β-anomers of compounds **1** and **2** are in the range of the expected α-KIEs for the proposed hydrolysis mechanism (Fig. 13.3).

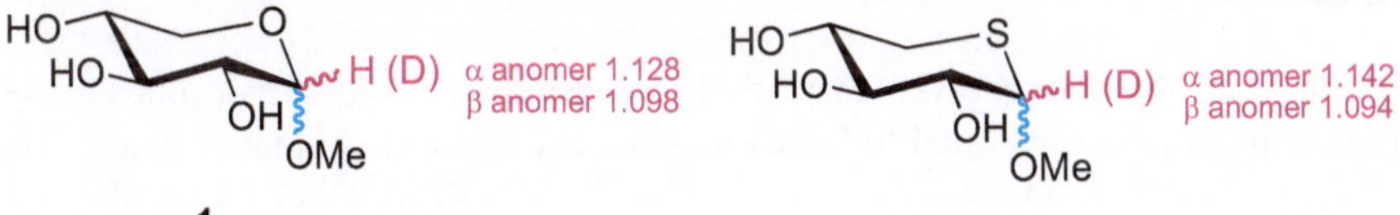

Figure 13.3

β-Secondary KIES

In a β-SKIE the deuterated position is placed β to the bond that is being broken in the transition state. In Fig. 13.4 the anomeric C-OMe bond, which is being broken in the transition state is colored blue and the β-H(D) atom is colored red. β-Secondary KIEs are only of interest when the β-C-H(D) bond can overlap by hyperconjugation with a p vacant orbital and are of particular importance in the study of reactions involving carbocations. The magnitude of a β-SDKIE is influenced by transition state geometry and it reaches its maximum when the β-CH(D) bond and the nascent p orbital are parallel in the transition state. The observation of SDKIEs in the hydrolysis of xylopyranoses **1** and **2** supports the proposed formation of an oxa- or thiacarbenium ion intermediate in the slow step of the reaction.

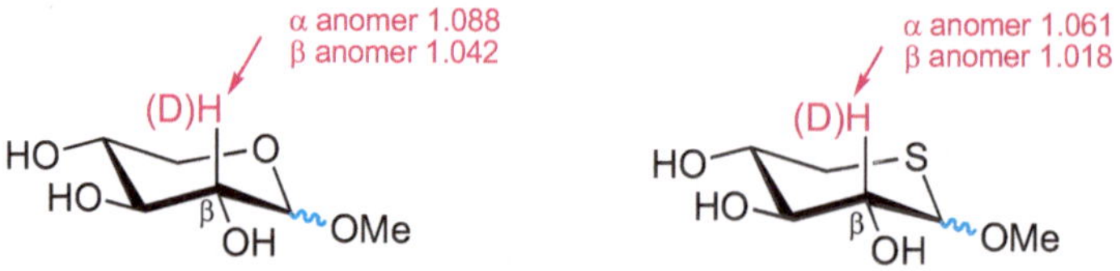

Figure 13.4

Solvent KIEs

The measured solvent deuterium KIES ($kD_3O^+/kH_3O^+ > 2$), indicate that the rate of hydrolysis is strongly affected by a change in the solvent, which is not surprising if a carbonium ion is involved as intermediate. The measured values are in fact the average of different factors: the ability of the solvent to protonate the substrate (it is known that D_3O^+ in D_2O is a stronger acid than is H_3O^+ in H_2O), the exchange of the acid protons of the substrate with the solvent and the formation of hydrogen bonds. The high values of the solvent KIEs for compound **1** and **2** are consistent with specific acid-catalyzed processes in which pre-equilibrium protonation occurs on the exocyclic oxygen atom, followed by rate-limiting C-O cleavage.

In Summary

The acid-catalyzed hydrolysis of methyl 5-thioxylopyranosides **2** occurs via fast protonation of the C-O bond followed by slow exocyclic C-O bond cleavage. The study of the different types of kinetic isotope effects indicates that a thiacarbenium ion is formed in the process. A similar study of the hydrolysis of xylopyranosides **1** has confirmed the mechanism generally accepted for the reaction and has demonstrated the intermediacy of oxacarbenium ions as intermediates.

Questions

α-Glucopyranosyl fluoride **3** reacts with azide ion to give the substitution product **4** exclusively as the β-anomer. However, α-5-thioglucopyranosyl fluoride **5** yields a mixture of α- and β-azides **6** together with the hydrolysis product **7** (Scheme 13.4).

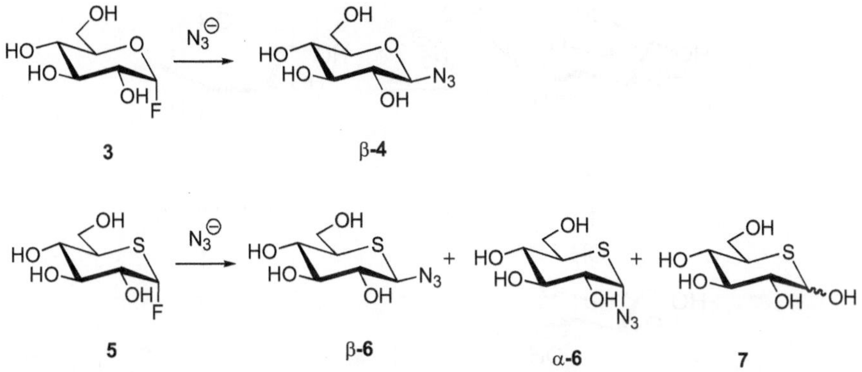

Scheme 13.4

In this latter case, both the yield and the product ratio β-6/α-6 change with azide ion concentration (Table 13.3). *Propose a mechanism to explain the different stereochemical outcome of the reaction in each case.*

Table 13.3

[N₃⁻] (M)	7 (%)	β-6 and α-6 (%)	β-6 / α-6
0.27	58	32	1.52
0.47	43	57	1.73
0.67	33	67	1.98
1.00	24	76	2.14

Data obtained in D_2O

Answer to the Question

The experimental evidence obtained for the azide nucleophilic substitution in fluorides **3** and **5** suggests that the reaction follows a different mechanism in each case. α-Glucopyranosyl fluoride **3** yields the substitution product **4** with *inversion* of the configuration at the anomeric carbon, which is the expected result for a S_N2-type reaction (Scheme 13.5).

*Should we have considered an oxacarbenium ion **8** as intermediate in this case?* Clearly, if **8** were formed, the nucleophilic attack of the azide ion of this species should yield a mixture of α- and β-anomers that is not observed. On the other hand, cation **8** is short living (estimated life around 3×10^{-12} s) and the azide ion is very nucleophile. Obviously, the S_N2 attack occurs before the cation is formed and solvent-equilibrated in the reaction medium.

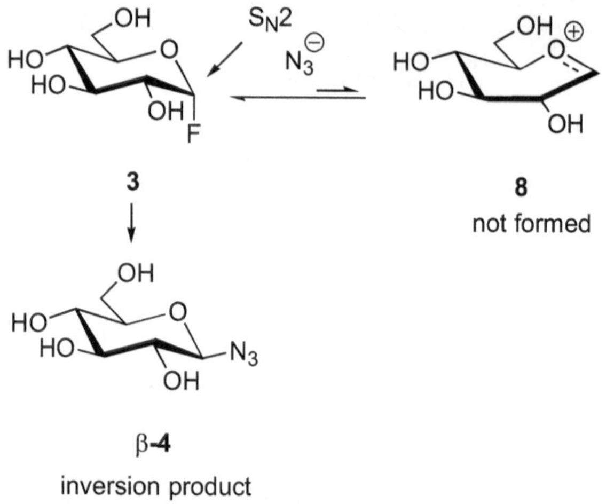

Scheme 13.5

The reaction has a different outcome with 5-thioglucopyranosyl fluoride **5**. In this case, a mixture of *inversion and retention* substitution products was obtained, together with the hydrolysis product **7**. Obviously, these data cannot be explained by a direct S_N2 attack of the azide ion to the substrate. In fact, they are pointing to a stepwise S_N1 mechanism, in which the nucleophilic attack occurs after the departure of the fluoride ion. In this case, a 5-thioglucopyranosyl cation **9** must be formed as intermediate (Scheme 13.6).

Undoubtedly, these experimental results indicate that the replacement of the ring oxygen in glucopyranosyl cation **8** by a sulfur atom in **9** increases its stability. The estimated lifetime for the 5-thioglucopyranosyl cation **9** in water is 1.1×10^{-9} s, a lifetime that is sufficient for the intermediate to become solvent-equilibrated. As the solvent is also a nucleophile, products **7** resulting from the nucleophilic attack of water in **9** would be formed, competing with those resulting from the attack of the azide ion **6**.

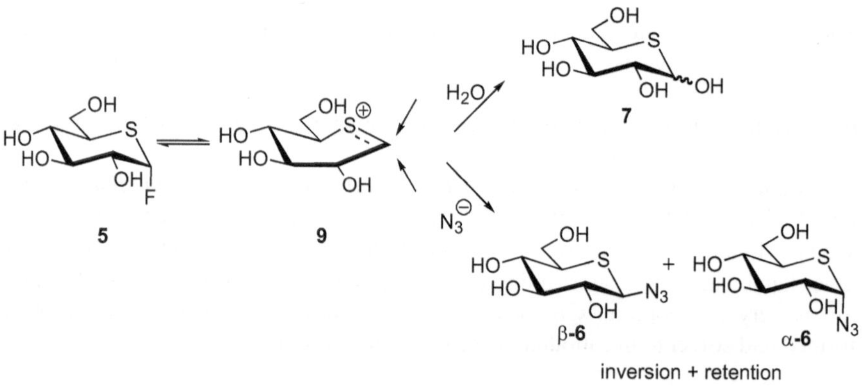

Scheme 13.6

We discussed previously that the slow step of the hydrolysis mechanism of pyranosides **1** and **2** was the formation of a carbenium ion. In the case of **5** we can suggest that the departure of the fluoride ion is rate-limiting and hence the capture of the cation **9** is the fast step of the reaction. An increase in the azide concentration results in an increase of the yield of the substitution products, in detriment of the hydrolysis product **7** obtained by direct attack of the water. As expected, the more stable anomer β-**6** is obtained with preference to the α- anomer.

Additional Comments

This problem is based on the work by Indurugalla D, Bennet AJ (2001) *J. Am. Chem. Soc.* 123:10889-10898 and on the work by Johnston BD, Indurugalla D, Pinto BM, Bennet AJ (2001) *J. Am. Chem. Soc.* 123:12698-12699.

Subjects of Revision

Kinetic isotope effects: primary and secondary deuterium kinetic isotope effects. Heavy atom isotope effects. Solvent isotope effects. S_N1 and S_N2 mechanisms.

Level 1 – Case 14
Cyclization of 2,3-Dibenzylidenesuccinates

Key point: *Electrocyclizations.*
Sigmatropic Rearrangements

Cyclization of 2,3-Dibenzylidenesuccinates

The cyclization of *E,E*-dibenzylidenesuccinate esters to dihydronaphthalenes has been employed as the key step in the synthesis of lignans, a class of natural products found in plants. The stereochemistry of the cyclization products depends on the reaction conditions. For example, irradiation of succinate **1** in EtOH yields almost exclusively *cis*-dihydronaphthalene **2**, whereas heating of **1** with trifluoromethanesulfonic acid (CF_3SO_3H, triflic acid) in CH_2Cl_2 yields *trans*-dihydronaphthalene **3** together with a small amount of the dearylized product **4**. Prolonged heating or the use of excess triflic acid decreases the yield of **3**, making compound **4** the major reaction product (Scheme 14.1).

*Propose a reasonable mechanism for the cyclization processes and justify the stereochemistry of dihydronaphthalenes **2** and **3**.*

Discussion

A photochemical ring closure may point to an **electrocyclic reaction**. In this case six conjugated π-electrons of succinate **1** could be involved in the reaction (four of the diene and two of one of the aromatic rings). Following the Woodward-Hoffmann rules, a photochemical six-electron ring closure must proceed through a

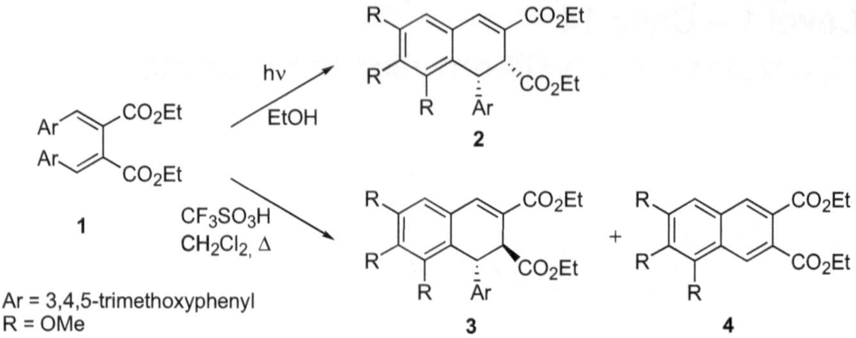

Scheme 14.1

conrotatory transition state as indicated in Fig. 14.1. The ring closure requires that the overlapping between the termini lobes of the π system in the HOMO of the hexatriene (ψ_4^* under photochemical conditions) should move in the same direction.

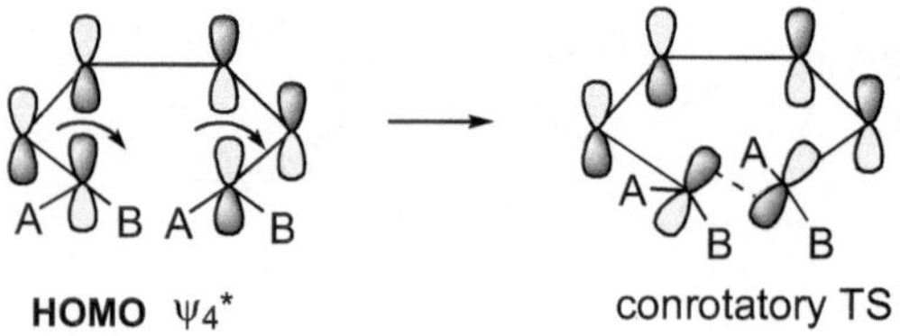

HOMO ψ_4^* conrotatory TS

Figure 14.1

A photochemical conrotatory six-electron ring closure in **1** would lead to intermediate **5**. The stereochemistry of **5** will be determined by the conrotatory electrocyclization mode and the aryl group in C1 and the hydrogen in C8a will be *trans* to each other (Scheme 14.2).

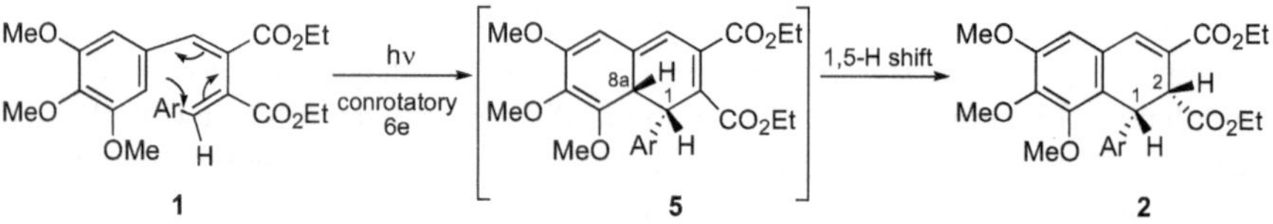

Scheme 14.2

Intermediate **5** is not isolated. It is transformed into the more stable dihydronaphthalene **2** by means of a thermal [1,5] sigmatropic hydrogen shift. This type of sigmatropic rearrangement can be understood in terms of the frontier molecular orbital theory considering the interaction between the H(1s) orbital and the LUMO (ψ_3^*) of the diene component in the transition state (Fig. 14.2). A positive overlap between the orbitals where bond breaking and bond making takes place is produced when the H atom slides across the top face of the planar transition state. This kind of shift is called **suprafacial**.

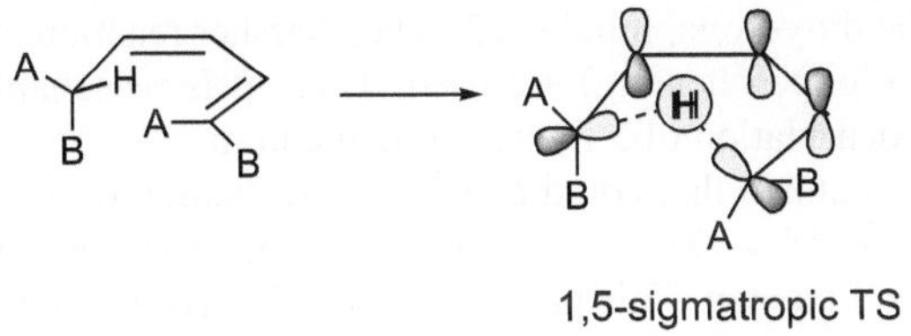

1,5-sigmatropic TS

Figure 14.2

As the [1,5]-sigmatropic hydrogen shift is suprafacial, the hydrogen would end at the same face of the molecule and hence, the substituents on positions C1 and C2 will have a *cis* relationship in the final product **2** (Scheme 14.2).

Considering the Woodward-Hoffmann rules, if the conrotatory six-electron ring closure is allowed under photochemical conditions, it must be forbidden under thermal conditions. In consequence, the cyclization of **1** to *trans*-dihydronaphthalene **3** in the presence of triflic acid should not be a concerted process. Therefore, in this case, it is more reasonable to think of a stepwise acid-promoted mechanism for the cyclization, possibly involving carbocations as intermediates.

Mechanisms under acidic conditions usually start by a protonation step. In this case, protonation of the conjugated double bond would lead to carbocation **6** in equilibrium with the more stable tautomer **7**. Cyclization onto the aromatic ring would lead to conjugated cation **8**, which after aromatization yields the final product **3**. The experimental fact is that the compound having the *trans* arrangement between the substituents on C1 and C2 positions is exclusively obtained. That is, the formation of the thermodynamically more stable product seems to be favored (Scheme 14.3).

This mechanism is very reasonable and could justify the structure and stereochemistry of **3**. However, it does not explain the formation of the aromatic byproduct **4**. Otherwise, it is also unlikely that naphthalene **4** could be formed straight from succinate **1**. In fact, we already know that the yield of **4** increases at

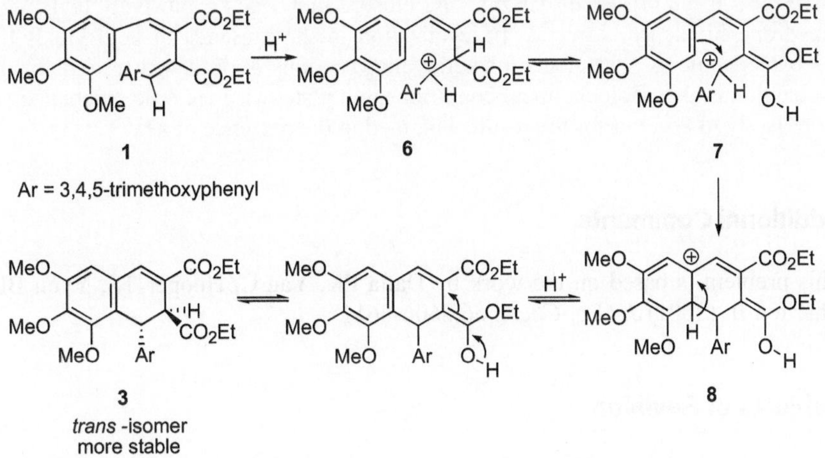

1

Ar = 3,4,5-trimethoxyphenyl

6

7

3
trans -isomer
more stable

8

Scheme 14.3

the expense of the dihydronaphthalene **3**, when harsher reaction conditions (excess of triflic acid, prolonged heating) are used. Then, it is reasonable to think that **4** was obtained by dearylation of **3** in the acidic medium.

A possible mechanism that could explain the formation of dearylated product **4** is shown in Scheme 14.4. Dearylation most likely involves protonation of **3** at the *ipso* carbon as the first step. This protonation would give the highly stabilized carbocation **9** (the positive charge is delocalized by conjugation through the ring). Next, β-elimination on **9**, induced by the TfO⁻ anion, would lead to **4** and 1,2,3-trimethoxybenzene as reaction products. We should remark that the protonation of a benzene ring requires the loss of aromaticity and hence is rarely observed. In this case however, the key is the formation of a very stable intermediate like **9**. Since naphthalene **4** is obtained from dihydronaphthalene **3**, longer reaction times and higher amounts of acid would account for the formation of the aromatic compound **4** as the main reaction product.

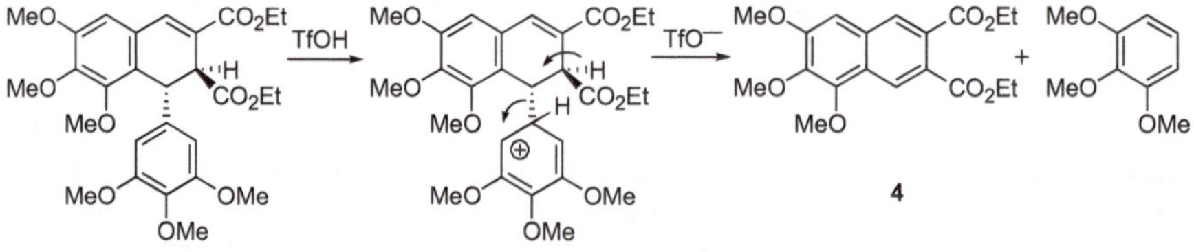

Scheme 14.4

In Summary

1,2-Dihydronaphthalenes can be obtained by cyclization of *E,E*-dibenzylidene succinates **1**. The stereochemistry of the products depends on the cyclization process. Thus, in the presence of light, ring closure in **1** leads exclusively to 1,2-*cis*-dihydronaphthalenes, whereas the cyclization in the presence of acid yields the 1,2-*trans* isomers. A sequence of pericyclic reactions is proposed to explain the cyclization under photochemical conditions, but a stepwise cationic mechanism is more likely to account for the results obtained in the presence of acid.

Additional Comments

This problem is based on the work by Datta PK, Yau C, Hooper TS, Yvon BL, Charlton JL (2001) *J. Org. Chem.* 66:8606-8611.

Subjects of Revision

Electrocyclic reactions. Sigmatropic rearrangements. Reactions of carbocations.

Level 1 – Case 15
Oxazoline *N*-Oxides as Dipoles in [3+2] Cycloadditions

Key point: *Cycloadditions*

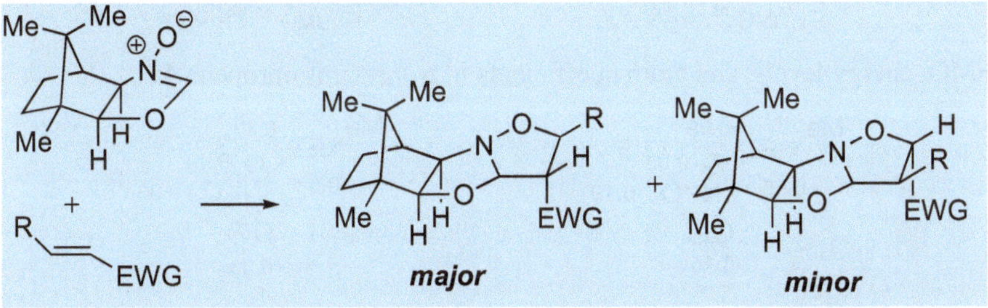

Oxazoline *N*-Oxides as Dipoles in [3+2] Cycloadditions

Camphor-derived oxazoline *N*-oxides **1** have been employed as dipoles for [3+2] cycloadditions with α,β-unsaturated esters or nitroalkenes. The reactions proceed in good yields and are highly *regio-* and *endo*-stereoselective. Thus, regioisomers **2** are obtained preferentially to **3** (the **2:3** ratios are about 95:5), with almost complete *endo*-selectivity (Scheme 15.1).

*Justify the structure and stereochemistry of the products that will be obtained in the reaction of **1** with trans-1-nitropropene (**4**) and trans-2-methyl butenoate (**5**) Indicate which alkene is the most reactive as dipolarophile towards dipole **1**.*

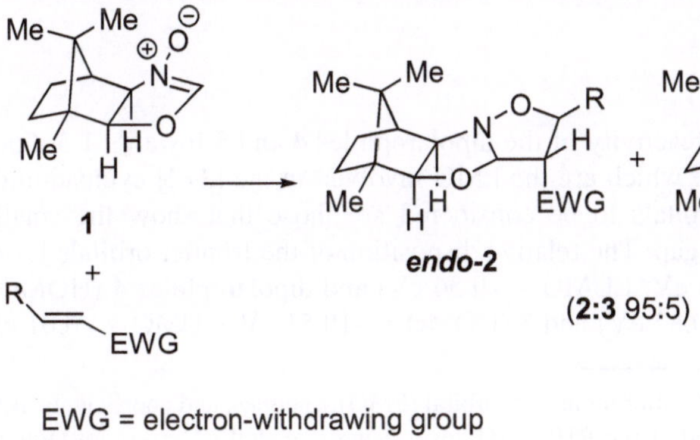

EWG = electron-withdrawing group

Scheme 15.1

Experimental Data[1]

Frontier molecular orbitals (FMO), energy levels, and atom coefficients in oxazoline N-oxide **1**:

$$E_{HOMO} = -8.58 \text{ eV} \qquad\qquad E_{LUMO} = +0.56 \text{ eV}$$

FMO, energy levels, and atom coefficients in *trans*-1-nitropropene **4**:

$$E_{HOMO} = -11.27 \text{ eV} \qquad\qquad E_{LUMO} = -0.87 \text{ eV}$$

FMO, energy levels, and atom coefficients in *trans*-2-methyl butenoate **5**:

$$E_{HOMO} = -10.51 \text{ eV} \qquad\qquad E_{LUMO} = -0.01 \text{ eV}$$

Discussion

Reactivity

Prior to discussing the reactivity of the dipolarophiles **4** and **5** towards 1,3-dipole **1**, we have to determine which are the FMO involved in the [3+2] cycloaddition. The pair of frontier orbitals to be considered are those that show the smaller HOMO-LUMO energy gap. The relative disposition of the frontier orbitals for dipole **1** (HOMO = –8.58 eV; LUMO = +0.56 eV) and dipolarophiles **4** (HOMO = –11.27 eV; LUMO = –0.87 eV) and **5** (HOMO = –10.51 eV; LUMO = –0.01 eV)

[1] The calculations of the frontier molecular orbital (FMO), energies, and coefficients of the reactants were performed at the RHF/AM1 level using the MOPAC program (Version 5.0).

suggests that in both cases the narrower HOMO-LUMO gap involves the HOMO of **1** (dipole) and the LUMO of the alkene (dipolarophile). Therefore, the reactions are HOMO dipole-controlled.

To decide which alkene (**4** or **5**) is more reactive towards dipole **1** we have to calculate the HOMO dipole-LUMO alkene energy difference in each case. That is, the smaller difference in energy will lead to a faster reaction. The HOMO-LUMO gaps are 7.71 eV in the case of nitroalkene **4** and 8.57 eV in the case of ester **5**. Hence, the cycloaddition between oxazoline *N*-oxide **1** and nitroalkene **4** is faster than the reaction with ester **5**.

Regioselectivity

To explain the regioselectivity of the reaction we will consider the coefficients of the atomic orbitals in the FMO involved in the [3+2] process. We should remember that overlapping between two lobes implies that they have wave functions of the same sign and that the interaction is more favored if the lobes have similar sizes. In this case, the oxygen adjacent to the nitrogen atom and the carbon of the C=N bond in **1**-HOMO would combine with the C=C carbons of the alkene-LUMO as it is shown in Scheme 15.2. Considering the signs of the coefficients involved, there is only one possible orientation either with nitroalkene **4** or with ester **5**. The reaction is regioselective leading to adducts **6** and **7**, respectively.

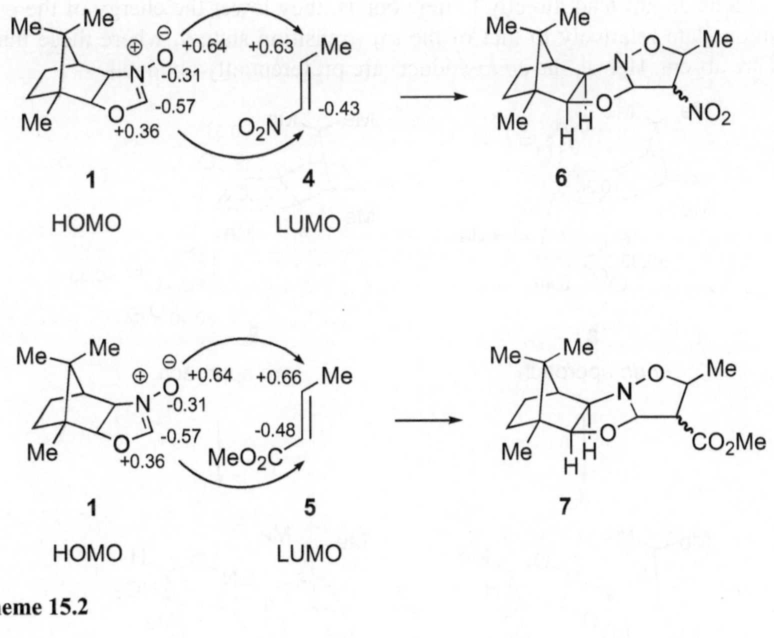

Scheme 15.2

Caution

Frequently when discussing the regioselectivity of a cycloaddition in terms of the FMO approach, we are too aware of the size of the coefficients of the atomic orbitals involved in the reaction. It is true that the greatest bonding interaction is achieved by pairing the two centers with the large coefficients rather than pairing large with small positions (principle of maximum overlap). However, **the first condition for two lobes to overlap is to have the same sign.** Two lobes with opposite signs could never overlap, even if they have similar sizes. Therefore, size is always a secondary factor to consider after sign.

Stereoselectivity

The experimental data indicate that the reaction between dipole **1** and dipolarophiles **4** and **5** is not only regioselective but highly *endo*-selective (*endo:exo* ratio higher than 95:5). Frequently, to account for the stereoselectivity in cycloaddition reactions, other interactions concerning parts of the frontier orbitals, **which are not directly involved in forming new bonds**, are invoked. These interactions are known as **secondary effects** and in this case, are clearly stabilizing the *endo*-transition state in preference to the *exo*-transtition state.

In Fig. 15.1 we have represented the FMO in the transition state for the reaction between dipole **1** and nitroalkene **4**. The preferred *endo* approach indicated by **8** is due to the stabilizing secondary bonding interactions between the p_z orbitals of the central nitrogen (–0.31) and endocyclic oxygen (+0.36) in the dipole, with the NO_2 group of the dipolarophile (N –0.40; O +0.33, respectively). Although these interactions do not lead directly to new bonds, they lower the energy of the *endo*-transition state relatively to that of the *exo*-transition state **9**, where these inteactions are absent. Hence, the *endo*-adducts are preferentially obtained.

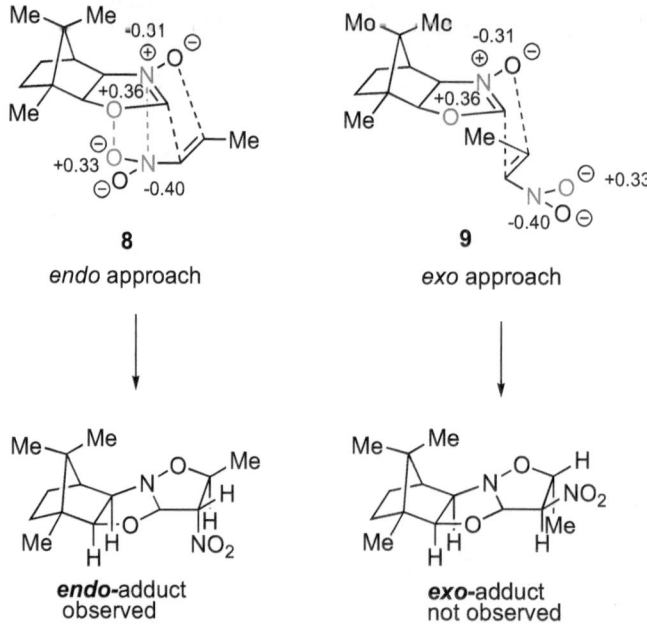

Figure 15.1

Exercise: Represent the *endo*- and *exo*-transition states for the reaction between dipole **1** and ester **5**. In this case, the *endo*-transition state is favored by the secondary bonding interactions between the nitrogen and endocyclic oxygen in the dipole and the CO_2 group of the dipolarophile.

In Summary

The reactivity, regioselectivity and stereoselectivity of [3+2] cycloadditions of oxazoline *N*-oxides and α,β-unsaturated esters or nitroalkenes can be rationalized in terms of the FMO theory. The reactions are HOMO-dipole controlled and the preferred *endo*-selectivity in those cycloadditions can be rationalized by stabilizing secondary orbital interactions only present in the *endo*-approach.

Questions

It is a known fact that oxazoline *N*-oxides show better reactivities and selectivities than the corresponding nitrones in the [3+2] cycloaddition reactions. For example, 2,4,4-trimethyl oxazoline *N*-oxide **10** is more reactive as a dipole than the nitrone (2,2,5-trimethylpyrrolidine *N*-oxide) **11**.

Justify this experimental observation considering the FMO energies in each case.

10	**11**
$E_{HOMO} = -8.41$ eV	$E_{HOMO} = -8.51$ eV
$E_{LUMO} = +0.59$ eV	$E_{LUMO} = +0.69$ eV

Figure 15.2

Answer to the Question

The enhanced reactivity of oxazoline *N*-oxide **10** compared to nitrone **11** could be explained in terms of FMO theory. The data of HOMO energies indicate that **10** has a higher HOMO and a lower LUMO than **11**. In reactions controlled by the HOMO dipole-LUMO dipolarophile interaction, as in Figure 15.3, as well as in processes controlled by the LUMO dipole-HOMO dipolarophile interaction (Fig. 15.4), oxazoline *N*-oxide **10** will provide a smaller energy gap than nitrone **11** when reacting with a dipolarophile.

The presence of the endocyclic oxygen atom in oxazoline *N*-oxide **10** is probably responsible for shifting the HOMO to higher energy and the LUMO to lower

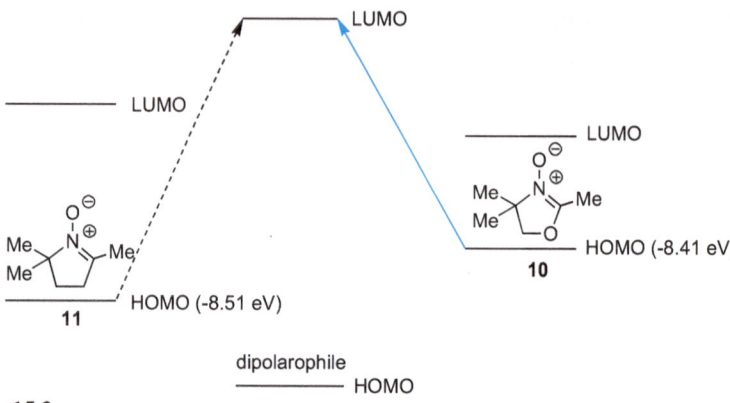

Figure 15.3

energy compared to nitrone **11**. The effect of the substituents on the FMO energies has been established[2] and it is known that adding an extra conjugation to the double bond generally produces a reduction of the HOMO-LUMO energy gap.

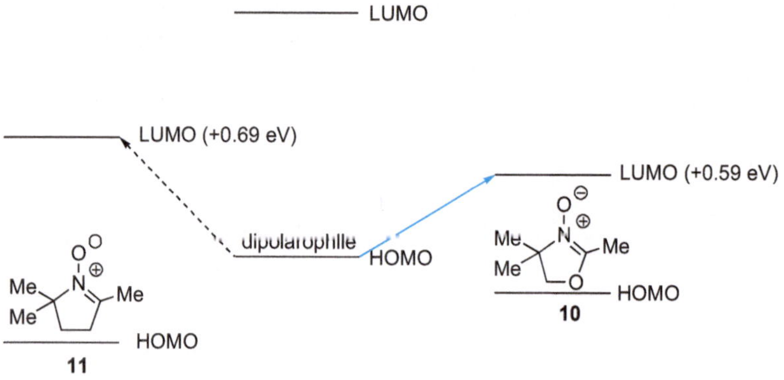

Figure 15.4

Additional Comments

This problem is based on the work by Kouklovsky C, Dirat O, Berranger T, Langlois Y, Tran-Huu-Dau ME, Riche C (1998) *J. Org. Chem.* 63:5123-5128.

Subjects of Revision

FMO theory. Cycloaddition processes: Reactivity, regioselectivity and stereo-selectivity.

---

[2] For a study of the effect of substituents on the energies of the frontier orbitals of dienes and dienophiles see: I. Fleming, *Frontier Orbitals and Organic Chemical Reactions* Wiley, Chichester, **1976**, p. 114.

Level 1 – Case 16
Light-Induced Cycloadditions of
N-Phthaloyl α-Amino Acids

Key point: *Cycloadditions. SET*

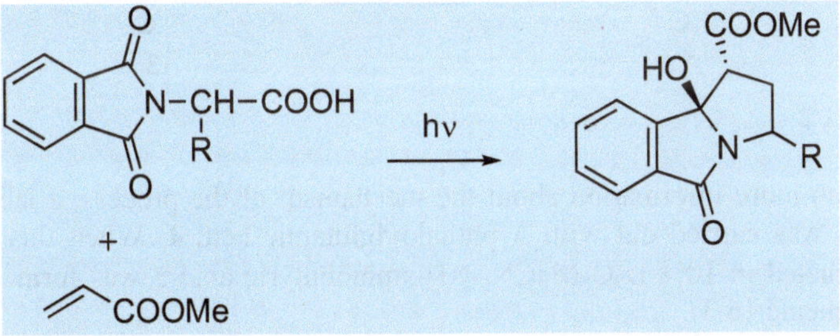

Light-Induced Cycloadditions of *N*-Phthaloyl α-Amino Acids

The irradiation of *N*-phthaloyl α-amino acids **1** in the presence of electron-deficient olefins, such as methyl acrylate, gives benzopyrrolizidines **2** in good yields (Scheme 16.1).

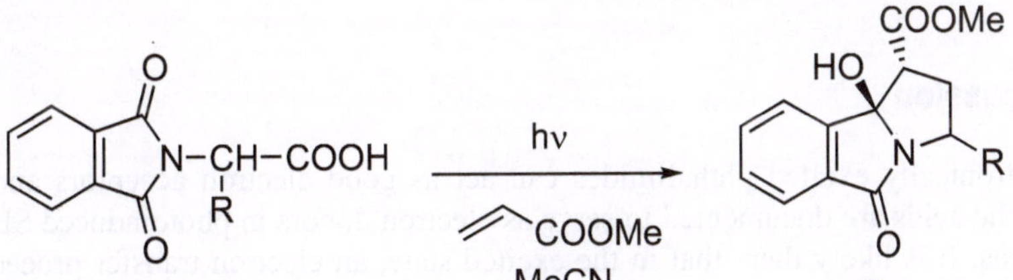

R = H, Me, Ph

Scheme 16.1

Propose a reasonable mechanism to explain how benzopyrrolizidines **2** *are formed.*

Experimental Data

N-Phthaloyl α-amino acids **1** undergo photodecarboxylation to generate the corresponding *N*-alkylphthalimides **3**. It is believed that the photodecarboxylation occurs by photoinduced single electron transfer (SET) involving the electronically excited phthalimide as electron acceptor and the carboxylic acid as electron donor (Scheme 16.2).

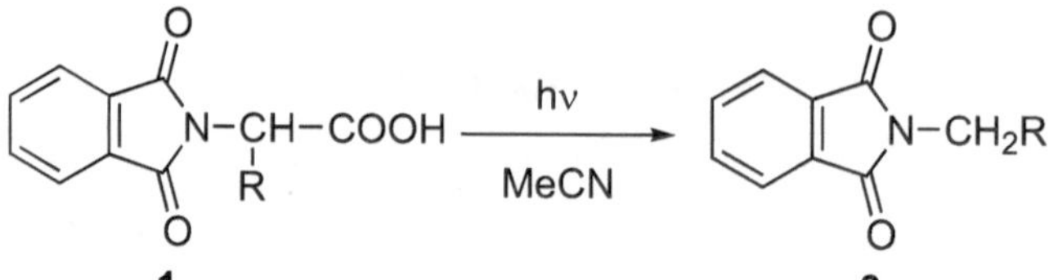

Scheme 16.2

To gain more information about the mechanism of the process, a labeling experiment was carried out with *N*-phthaloylglutamic acid **4**. When this substrate was irradiated in 10% D_2O-MeCN, γ-*D*-aminobutyric acid **5** was formed in high yield (Scheme 16.3).

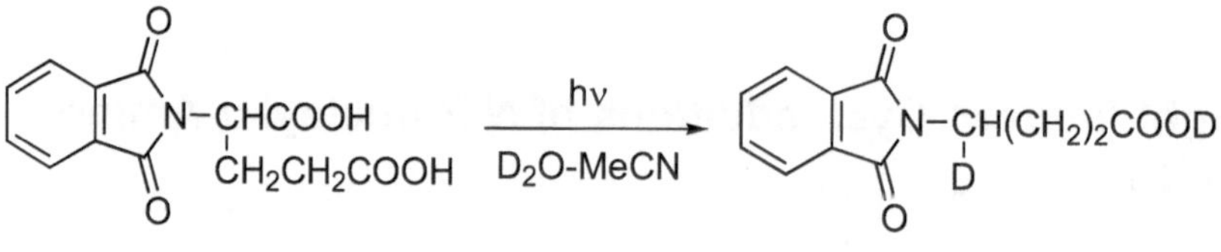

Scheme 16.3

Discussion

Electronically excited phthalimides can act as good electron acceptors and carboxylic acids are documented to serve as electron donors in photoinduced SET reactions. It is likely then, that in the excited state, an electron transfer process between the phthalimide system and the carboxylic acid would occur, leading to charge-separated diradical **6**. Proton transfer from the carboxylic acid function (H^+ is an electrofugal group)[1] would form a carboxy radical **7**, which could undergo rapid decarboxylation to azomethyne ylide **8**. This reactive species is transformed into the final decarboxylated product **3** by intramolecular proton transfer (Scheme 16.4).

---

[1] A leaving group that carries away an electron pair is called *nucleofugal*. If it comes away without the electron pair it is called *electrofugal*.

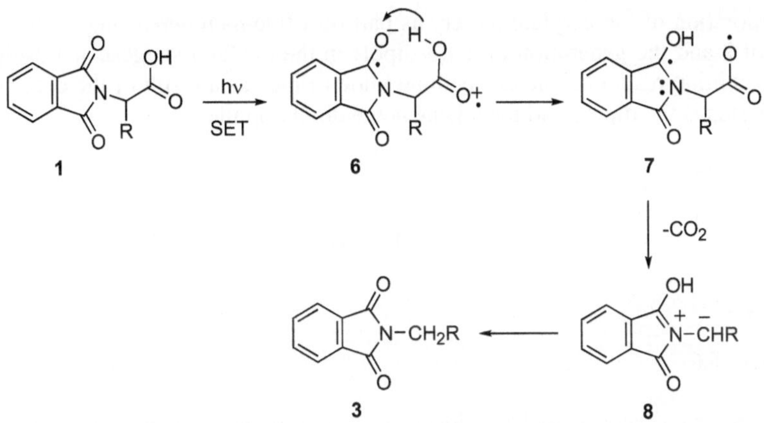

Scheme 16.4

The results obtained in the irradiation of **4** with D_2O-MeCN support the SET-based mechanism proposed in Scheme 16.4. In this case, the first step in the deuterated medium should be the exchange of the acidic protons with the solvent. After the SET process, charged-separated dirradical **9** should be formed. Subsequent deuterium transfer in **9** followed by loss of CO_2 would give deuterated azomethine ylide **10** that after deuterium transfer finally yields the observed γ-labeled *N*-phthalimidobutyric acid **5** (Scheme 16.5).

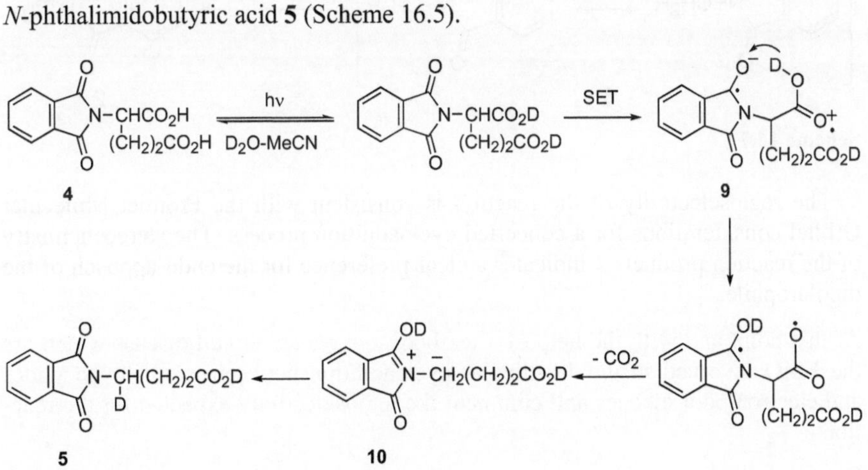

Scheme 16.5

The formation of an azomethyne ylide during the photodecarboxylation of *N*-phthaloyl α-amino acid **1** makes easier the interpretation of the results obtained in the presence of methyl acrylate. If we examine the structure of the reaction product **2**, we would easily recognize the fragments corresponding to the acrylate (colored red), the *N*-phthaloyl α-amino acid **1** that has lost the carboxylic acid function (colored blue) and the new bond formed (in black) (Scheme 16.6). The incorporation of the acrylate moiety as part of a five-membered ring in the structure of **2** and the generation of a 1,3-dipole in the medium suggests a 1,3-dipolar cycloaddition reaction. The azomethine ylide formed as a result of the decarboxylation step is the dipole and the alkene is the dipolarophile.

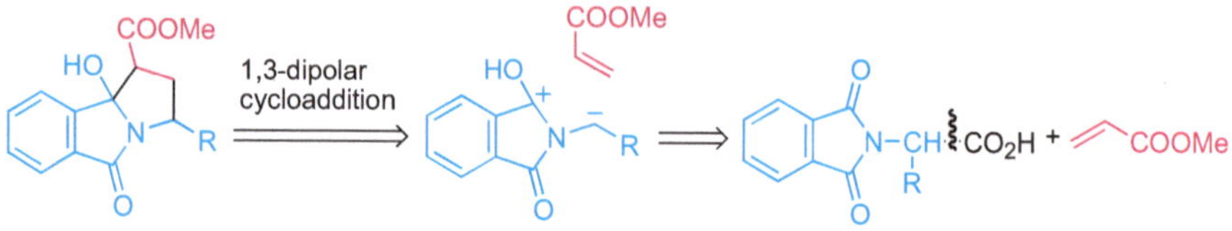

Scheme 16.6

As we have commented previously (Scheme 16.4), azomethyne ylides **8** are short-lived intermediates that in the absence of any suitable trapping agent regenerate the phthalimide function by intramolecular proton transfer leading to the decarboxylation product. However, when a dipolarophile like methyl acrylate is present in the reaction medium, they can be trapped to form the corresponding cycloadducts **2** (Scheme 16.7).

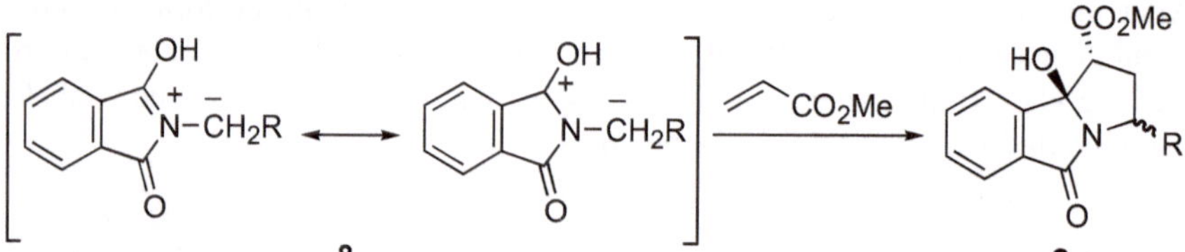

Scheme 16.7

The regioselectivity of the reaction is consistent with the Frontier Molecular Orbital considerations for a concerted cycloaddition process. The stereochemistry of the reaction products **2** indicates a clear preference for the endo-appoach of the dipolarophile.

At this point and with the help of a textbook the reader should discuss which are the FMO involved in the 1,3-dipolar cycloaddition between azomethyne ylides and electron poor alkenes and comment the regioselectivity expected for the reaction.

In Summary

Azomethyne ylides can be obtained by irradiation of *N*-phthaloyl α-amino acids. These intermediates can be trapped with electron poor alkenes to yield benzopyrrolizidines in good yields.

Questions

Irradiation of *N*-(trimethylsilyl)methylphthalimide **11** in MeCN with methyl acrylate produces benzopyrrolizidine **12** exclusively (Scheme 16.8). Explain this experimental result.

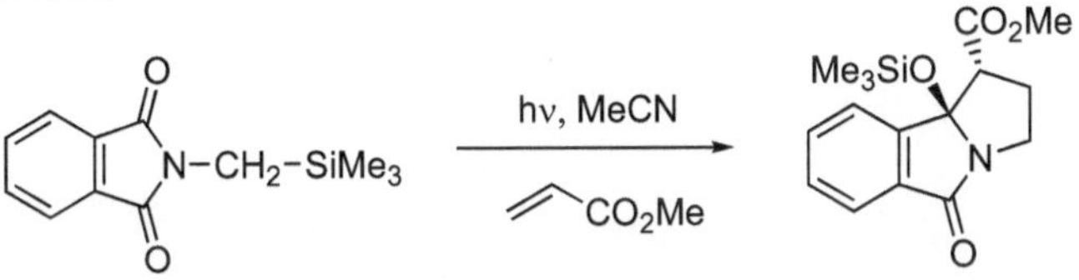

Scheme 16.8

Answer to the Question

To justify the experimental result we could follow the stepwise mechanism discussed in the previous section. The phthalimide system is a good electron acceptor in the excited state and the alkylsilane can be an electron donor, so a photoinduced electron transfer process could lead to the charged separated diradical **13**. Transfer of the electrofugal TMS group to the oxyanion carbonyl oxygen would form azomethyne ylide **14** that can be trapped by the methyl acrylate to yield the *endo*-reaction product **12** (Scheme 16.9).

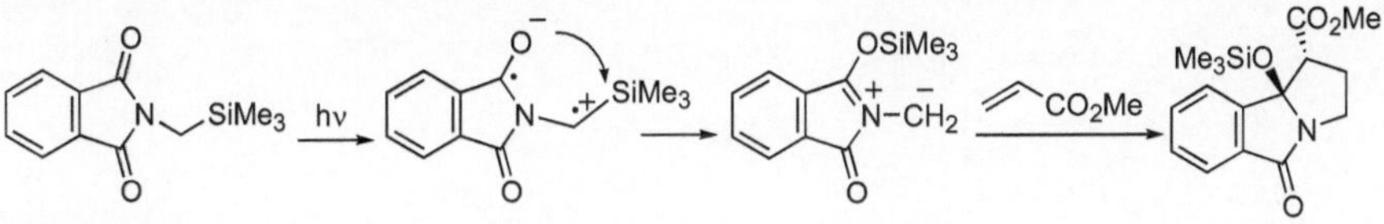

Scheme 16.9

Additional Comments

The formation of azomethyne ylides as intermediates in the photoreactions of *N*-(silylmethyl)phthalimides and *N*-phthaloyl derivatives of α-amino acids has been confirmed by laser flash photolysis and fluorescence spectroscopy.

This problem is based on the work by Takahashi Y, Miyashi T, Yoon UC, Oh SW, Mancheño M, Su Z, Falvey DF, Mariano PS (1999) *J. Am. Chem. Soc.* 121:3926-3932.

Subjects of Revision

SET processes. Cycloadditions

Level 2 – Case 17
Change in Rate-Determining Step in an E1cB Mechanism: Aminolysis of Sulfamate Esters

Key point: *Activation Parameters. Brønsted Plots*

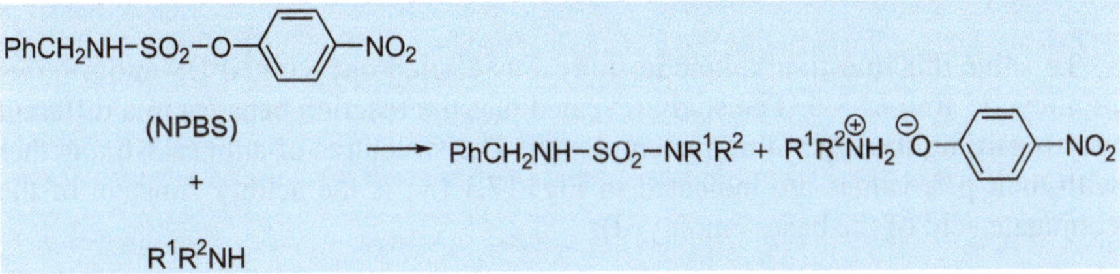

PhCH$_2$NH–SO$_2$–O–⟨benzene ring⟩–NO$_2$

    (NPBS)

      +   ⟶   PhCH$_2$NH–SO$_2$–NR^{1}R^2 + R^{1}R$^2\overset{\oplus}{N}$H$_2$ $\overset{\ominus}{O}$–⟨benzene ring⟩–NO$_2$

    R^{1}R^{2}NH

Change in Rate-Determining Step in an E1cB Mechanism: Aminolysis of Sulfamate Esters

The aminolysis of 4-nitrophenyl *N*-benzylsulfamate (NPBS) **1** yields sulfamides **2** and 4-nitrophenol as ammonium salt (Scheme 17.1).

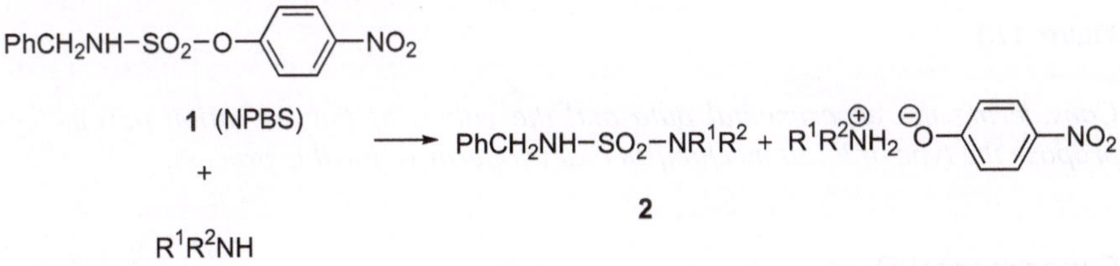

PhCH$_2$NH–SO$_2$–O–⟨benzene ring⟩–NO$_2$

    1 (NPBS)

      +   ⟶   PhCH$_2$NH–SO$_2$–NR^{1}R^2 + R^{1}R$^2\overset{\oplus}{N}$H$_2$ $\overset{\ominus}{O}$–⟨benzene ring⟩–NO$_2$

    R^{1}R^{2}NH   **2**

Scheme 17.1

The reaction is an elimination process that has been interpreted through an E1cB mechanism, as described in Scheme 17.2. The only question left is to determine whether the first step (k_1, removal of the proton) or the second step (k_2, elimination of the leaving group) is rate determining. In other words, to distinguish between an (E1cB)$_{irr}$ (E1cB irreversible) and an (E1cB)$_R$ (E1cB reversible) mechanism.

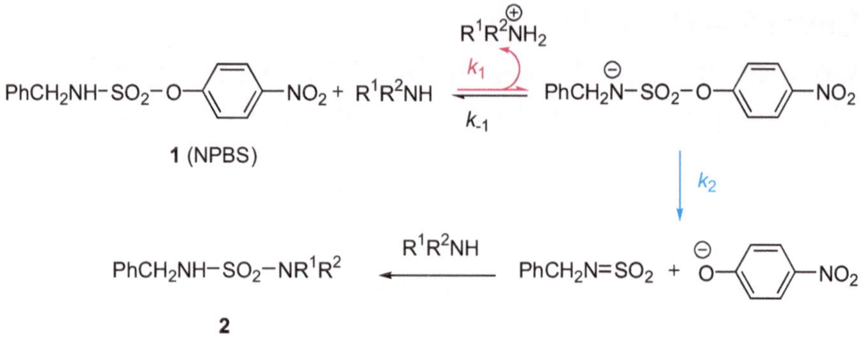

Scheme 17.2

To solve this question, a kinetic study was carried out with NPBS and a series of alicyclic amines **3-6**. The study revealed that the reaction behaves in a different way regarding the type of amine employed. The structures of amines **3-6** together with their pK_a values are indicated in Fig. 17.1 (K_a is the acidity constant of the conjugate acid of the bases employed).

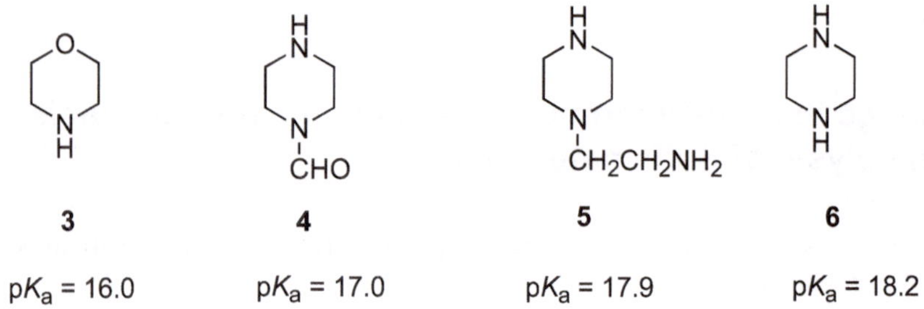

Figure 17.1

Considering the experimental data and the values of the activation parameters propose the type of E1cB mechanism that is operative for this process.

Experimental Data

1. The general rate law followed for the aminolysis reactions studied in this work has the form:

$$v = k_{obs}[NPBS] \tag{17.1}$$

From plots of k_{obs} vs [amine] straight lines were obtained as indicated by Eq. 17.2. The slopes of these plots are k' (second-order rate constants) and the k_0 values are negligible in most cases.

$$k_{obs} = k_0 + k'[amine] \tag{17.2}$$

Considering Eq. 17.2, the overall rate law has the form:

$$v = (k_0 + k')[\text{amine}][\text{NPBS}] \qquad (17.3)$$

2. Brønsted type plot of log k' against amine pK_a has been performed for the aminolysis of NPBS with amines 3-6 (Fig. 17.2). The most notable feature is that the Brønsted plot is biphasic and two β values have to be considered. The β_1 value for the lower part of the plot is 0.7 and β_2 is ~ 0. The change occurs at approximately the pK_a of NPBS (17.68).

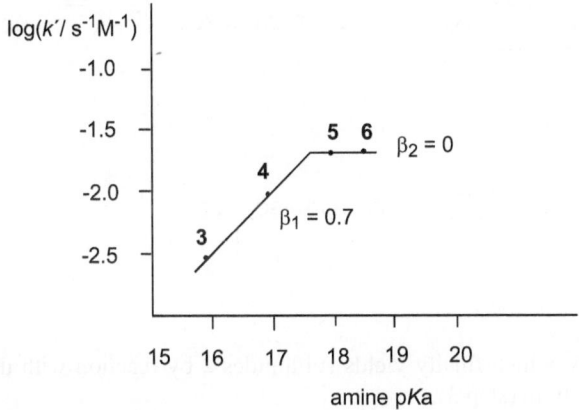

Figure 17.2

3. The activation parameters for the aminolysis of NPBS in acetonitrile are collected in Table 17.1.

Table 17.1

Amine	$\Delta H^{\ddagger}$, kJ mol^{-1}	$\Delta S^{\ddagger}$, J mol^{-1} K^{-1}
3	46 ± 5	-140 ± 15
4	43 ± 4	-143 ± 15
5	77 ± 8	-10 ± 1
6	75 ± 7	-15 ± 1

4. Typical E2, (E1cB)$_{irr}$ and S$_N$2 processes have entropies of activation in the range of −55 to about −170 J mol^{-1} K^{-1}. Less negative values, in the range of −40 to +150 J mol^{-1} K^{-1}, have been associated with the (E1cB)$_R$ mechanism.

Discussion

The E1cB mechanism consists on a stepwise elimination – the different steps are represented in Scheme 17.3. The first step of the reaction is the removal of a proton by the base to yield a negatively charged intermediate **7**. The second step consists of the departure of the leaving group (nitrophenol) in **7** and the formation of

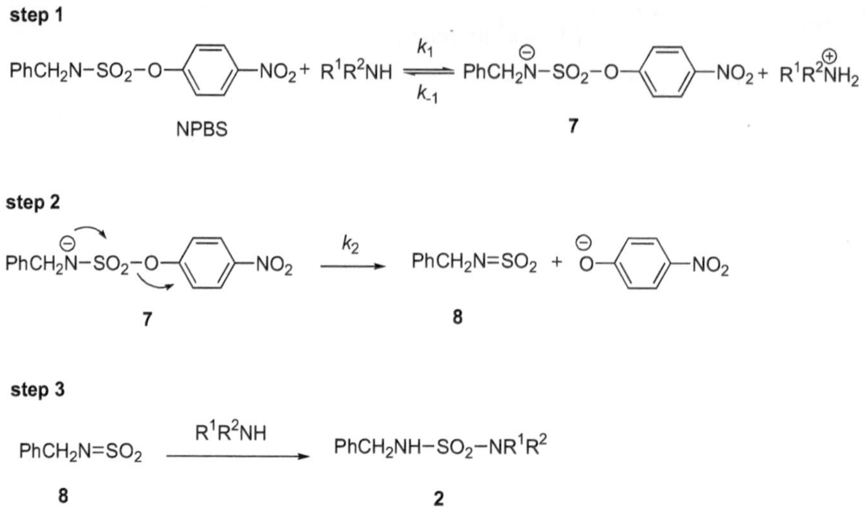

step 1

$$PhCH_2N-SO_2-O-\!\!\!\left\langle\!\!\!\bigcirc\!\!\!\right\rangle\!\!-NO_2 + R^1R^2NH \underset{k_{-1}}{\overset{k_1}{\rightleftharpoons}} PhCH_2\overset{\ominus}{N}-SO_2-O-\!\!\!\left\langle\!\!\!\bigcirc\!\!\!\right\rangle\!\!-NO_2 + R^1R^2\overset{\oplus}{N}H_2$$

NPBS   7

step 2

$$PhCH_2\overset{\ominus}{N}-SO_2-O-\!\!\!\left\langle\!\!\!\bigcirc\!\!\!\right\rangle\!\!-NO_2 \overset{k_2}{\longrightarrow} PhCH_2N=SO_2 + \overset{\ominus}{O}-\!\!\!\left\langle\!\!\!\bigcirc\!\!\!\right\rangle\!\!-NO_2$$

7   8

step 3

$$PhCH_2N=SO_2 \xrightarrow{R^1R^2NH} PhCH_2NH-SO_2-NR^1R^2$$

8   2

Scheme 17.3

intermediate **8**, which finally yields sulfamides **2** by reaction with the amine present in the medium (step 3).

Depending on the rate-limiting step, there are different types of E1cB processes. The kinetic law of Eq.17.3 indicates a second-order base-catalyzed reaction, but it is unable to distinguish whether step 1 or step 2 is rate-determining. However, although the base is essential in both steps, they are not equally affected by the base-catalyst employed. In an $(E1cB)_{irr}$ mechanism (step 1 is the slow step), the base is directly involved in the rate-determining step but in an $(E1cB)_R$ mechanism, the removal of the acidic proton by the base is fast, and the departure of the leaving group (step 2) is the slow step of the reaction.

The Brønsted Catalysis Law establishes that the effectiveness of a catalyst is related to its acid or base strength. Brønsted correlations are indicating that a **general catalysis** process is occurring. That is, the use of a stronger acid or base catalyst leads to a higher catalytic rate constant.

In this case, the aminolysis of NPBS is a base-catalyzed reaction and the expression for the Brønsted catalysis law will be as follows:

$$\log k' = \beta \, pK_a + constant \tag{17.4}$$

where k' is the second-order catalytic constant for the reaction catalyzed by the base, β is the Brønsted parameter associated with the particular reaction being catalyzed by base, and K_a is the acidity constant of the conjugate acid of the base considered.

Values for β usually range between 0 and 1. A reaction with $\beta \sim 1$ is very sensitive to the base strength of the catalyst. If different bases are present at comparable concentrations, the strongest will be most effective. In aqueous solution, the strongest possible base is OH^-, so the rate law of a reaction with $\beta \sim 1$ in alkaline solution will indicate specific catalysis. Values of $\beta = 0$, on the other hand, imply complete indifference to the nature of the catalyst. That is, bases of widely different strengths are almost comparable in catalytic effectiveness.

Figure 17.2 represents Eq. 17.4 for the aminolysis of NPBS. The lower part of the Brønsted plot represented in Fig. 17.2 (β_1) suggests that the aminolysis of NPBS is a general catalysis reaction, very sensitive to the catalyst employed ($\beta_1 = 0.7$, lower pK_a amines 3 and 4). However, there is a pK_a value from which a sudden discontinuity in the plot is observed and the slope becomes zero (higher pK_a amines 5 and 6). At this point there must be a change in the reaction that is no longer being general base catalyzed. This type of discontinuities in a Brønsted plot could indicate a change in the rate-determining step of the reaction. Considering the three steps of the elimination mechanism, β_1 could correspond to a reaction in which step 1 is rate-determining $(E1cB)_{irr}$ whereas β_2 fits better with an elimination in which step 2 is the slow step of the process $(E1cB)_R$.

A look at the mechanism depicted in Scheme 17.3 shows that the base is only involved during the first step of the reaction (k_1, removal of the proton), whereas there is no base involvement in the departure of the leaving group (k_2 second step). The differences between a *bimolecular* and a *unimolecular* rate-determining step should be reflected in the values of the entropy of activation, $\Delta S^\ddagger$. The entropy of activation is related to the loss or gain of degrees of freedom in the activated complex compared to the reactants. The negative sign of $\Delta S^\ddagger$ indicates that the activated complex is *more ordered* than the starting reagents. For the amines shown in Table 17.1, the entropy of activation is changing from approximately -140 J mol^{-1} K^{-1} (compounds 3 and 4, lower pK_a values, $\beta_1 = 0.7$ region), to approximately -10 J mol^{-1} K^{-1} (compounds 5 and 6, higher pK_a values, $\beta_2 = 0$ region). The more negative values observed for amines 3 and 4 are in agreement with a process in which two species (NPBS and base) are associated in the activated complex during the rate-determining step of the reaction. These arguments fit well with a $(E1cB)_{irr}$ mechanism. In addition, the observed values for amines 3 and 4 are in the range of those previously reported for other processes involving bimolecular activated complexes, that can be as high as -170 J mol^{-1} K^{-1} (see experimental data). The differences of the $\Delta S^\ddagger$ values in the case of the more basic amines 5 and 6 (about -10 J mol^{-1} K^{-1}) are remarkable. These values are more in agreement with a quite balanced process in regard to the loss or gain of degrees of freedom when passing from the reagents to the activation complex. This could fit with an $(E1cB)_R$ mechanism consisting on a fast bimolecular first step (removal of the proton by the base) and a slow unimolecular second step (loss of the nitrophenol fragment and simultaneous formation of a π bond). The observed entropy of activation will reflect the contribution of both processes. In fact, less negative and even positive values ranging from -40 to $+150$ J mol^{-1} K^{-1} have been associated with $(E1cB)_R$ mechanisms.[1]

---

[1] If step 2 is rate-determining and step 1 is a pre-equilibrium, the observed entropy of activation ($\Delta S^\ddagger_{obs}$) is also related to the change in the entropy during the first step (Eq. 17.5).

$$\Delta S^\ddagger_{obs} = \Delta S \text{ step } 1 + \Delta S^\ddagger \text{ step } 2 \qquad (17.5)$$

The values of the enthalpy of activation ($\Delta H^{\ddagger}$) complement the trends marked by the entropy of activation and reflect the energy changes between reactants and activated complex ascribable to the breaking and reforming of bonds. Breaking old bonds requires energy and forming new bonds releases energy. Low values of $\Delta H^{\ddagger}$ (lower than 100 kJ mol^{-1}) are estimated for reactions in which two species are involved (bimolecular processes). All $\Delta H^{\ddagger}$ values in Table 17.1 are lower than 100 kJ mol^{-1}, as expected for an overall second-order reaction, and are particularly low in the case of the less basic amines **3** and **4** (46 and 43 kJ mol^{-1}, respectively) which supports the arguments in favor of an $(E1cB)_{irr}$ mechanism in these cases.

In Summary

The aminolysis of 4-nitrophenyl-*N*-benzylsulfamate (NPBS) follows a base-catalyzed E1cB mechanism. However, the rate-determining step of the reaction changes depending on the pK_a of the amine employed. At lower amine pK_a values, an $(E1cB)_{irr}$ reaction takes place and the removal of the proton by the base is the rate-determining step of the reaction. At higher pK_a values, the removal of the proton is fast and the departure of 4-nitrophenol is the slow step of the reaction $(E1cB_R$ mechanism). The change in the rate-determining step occurs at approximately the point where the substrate pK_a is equal to the pK_a of the catalytic amine.

Additional Comments

This problem is based on the work by Spillane WJ, McGrath P, Brack C, O'Byrne AB (2001) *J. Org. Chem.* 66:6313-6316.

Subjects of Revision

Brønsted catalysis law. Brønsted coefficients. Activation parameters: physical significance.

Level 2 – Case 18
Unusual Diels-Alder Reactivity of Acyclic 2-Azadienes

Key point: *Cycloadditions*

Unusual Diels-Alder Reactivity of Acyclic 2-Azadienes

Azadiene **1**, bearing an electron-withdrawing substituent (CO_2Me), is expected to be electron-deficient and, therefore, more likely to react with electron-rich dienophiles in an *inverse electron demand* Diels-Alder reaction. However, for a diene, **1** shows a very unusual reactivity, since it participates in cycloadditions with both electron-rich and electron-deficient dienophiles. As an example, reaction of azadiene **1** with diethyl fumarate yields pyridine **2** whereas in the presence of *N*-cyclopenten-1-ylpyrrolidine, tetrahydropyridine **3** is obtained (Scheme 18.1).

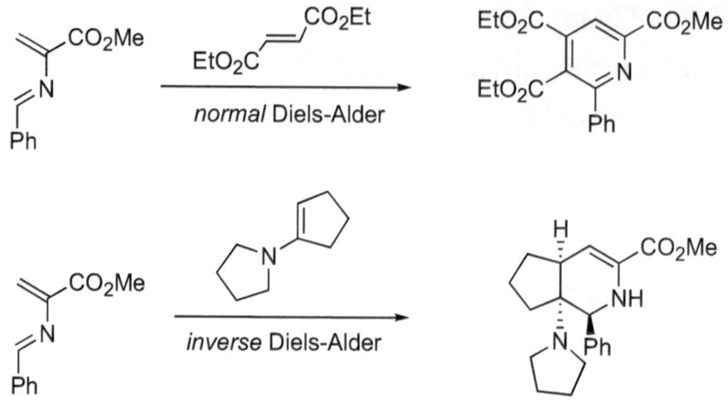

Scheme 18.1

A change of the substituent at the C1 position results in a radical modification of the reactivity of azadienes **1** towards dienophiles. Thus, azadiene **4** only participates in *normal* Diels-Alder processes, whereas azadienes **5** and **6** only take part in *inverse* cycloaddition reactions (Fig. 18.1).

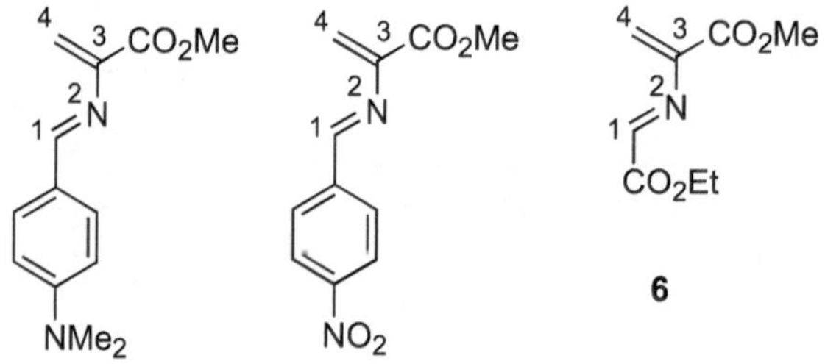

Figure 18.1

*Considering the Frontier Molecular Orbitals (FMO) involved justifies the unexpected normal Diels-Alder reactivity of 2-azadiene **1** toward diethylfumarate. Compare the reactivities of butadiene and 2-azadiene **1** and discuss the dual reactivity of this compound with respect to both, electron-deficient and electron-rich dienophiles. Justify the observed reactivity of dienes **4**, **5** and **6** in Diels-Alder reactions.*

Experimental Data

Frontier orbital energies (obtained from AM1 calculations)[1]

---

[1] The AM1 calculations described above were done in all cases considering the *s-cis* conformation of the diene.

Butadiene:	$E_{HOMO} = -9.35$ eV;	$E_{LUMO} = 0.46$ eV
2-Aza-1,3-butadiene:	$E_{HOMO} = -9.78$ eV;	$E_{LUMO} = 0.27$ eV
Azadiene **1**:	$E_{HOMO} = -9.44$ eV;	$E_{LUMO} = -0.61$ eV
Azadiene **4**:	$E_{HOMO} = -8.21$ eV;	$E_{LUMO} = -0.22$ eV
Azadiene **5**:	$E_{HOMO} = -10.13$ eV;	$E_{LUMO} = -1.60$ eV
Azadiene **6**:	$E_{HOMO} = -10.51$ eV;	$E_{LUMO} = -0.85$ eV
Diethyl fumarate:[2]	$E_{HOMO} = -12.70$ eV;	$E_{LUMO} = 1.49$ eV

Discussion

Reactivity of 2-azadiene 1 toward diethyl fumarate

The reactivity in Diels-Alder reactions is controlled by the HOMO-LUMO energy gap between the reagents. Generally, the small energy-separation is found between the HOMO of the diene and the LUMO of the dienophile and we talk about *normal* Diels-Alder reactivity. High HOMO dienes and low LUMO dienophiles are ideal substrates for these types of reactions. By contrast, electron-deficient dienes like **1** have low HOMOs and are more prone to participate in *inverse electron demand* Diels-Alder. In these cases the interaction HOMO dienophile-LUMO diene controls the process.

If we calculate the energy gap for the two possible HOMO-LUMO combinations between azadiene **1** and diethyl fumarate, the small difference in energy (10.93 eV) is found between the pair HOMO azadiene-LUMO fumarate (the difference between HOMO fumarate-LUMO azadiene is 12.09 eV). That is, despite that 2-azadiene **1** has a low energy HOMO it can react in a *normal* [4+2] cycloaddition process with a dienophile having a low energy LUMO.

Reactivity of 2-azadiene 1 compared with butadiene

We have mentioned that the HOMO-LUMO energy gap between the reagents determines the reactivity in Diels-Alder reactions. However, it is also well known that the energies of the Frontier Molecular Orbitals (FMO) are greatly affected by the substituents. Thus, electron-withdrawing groups lower the energy of both the HOMO and the LUMO, while the effect of electron-donating groups is to raise the energy of both FMO. On the other hand, substituents that add extra conjugation (like a phenyl group) raise the energy of the HOMO but lower the energy of the LUMO.

If we compare for example, the HOMO and LUMO energies of 1,3-butadiene and 2-aza-1,3-butadiene (Fig. 18.2) we will notice how the incorporation of a nitrogen atom in position 2 of the diene has lowered the energies of the frontier orbitals. In particular, the HOMO of the 2-aza-1,3-butadiene is now 0.43 eV lower in energy than the HOMO of 1,3-butadiene and for this reason, it must be less reactive than butadiene toward an electron-deficient dienophile.

---

[2] Grée R, Tonnard F, Carrié R (1975) *Bull. Soc. Chim.* 1325-1330.

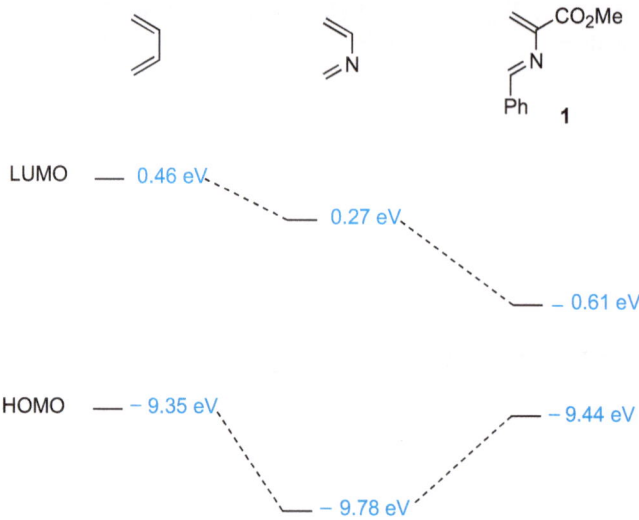

Figure 18.2

The addition of substituents to the 2-azadiene skeletons will certainly affect the FMO energies. This observation could give us the key to understanding the odd reactivity observed for azadiene **1** that reacts with electron-rich and electron-poor dienophiles. Indeed, the effect of the methoxycarbonyl group (electron-withdrawing) should be to lower both HOMO and LUMO energy levels, whereas the phenyl group (adds extra conjugation) should raise the HOMO and lower the LUMO energies. The average effect is shown in Figure 18.2. The HOMO-LUMO gap in **1** has been considerably reduced compared with 2-aza-1,3-butadiene. The HOMO in **1** is now very close in energy to that of butadiene (–9.44 and –9.35 eV, respectively). This explains the participation of azadiene **1** in *normal electron demand* Diels-Alder reactions that require dienes with relatively high energy HOMOs. On the other hand, the LUMO of **1** is low, considerably lower in energy than those of butadiene and 2-aza-1,3-butadiene. A low energy LUMO is required for the participation of a diene in *inverse* Diels-Alder reactions. This explains why azadiene **1** is also reactive towards electron-rich olefins (Fig. 18.3).

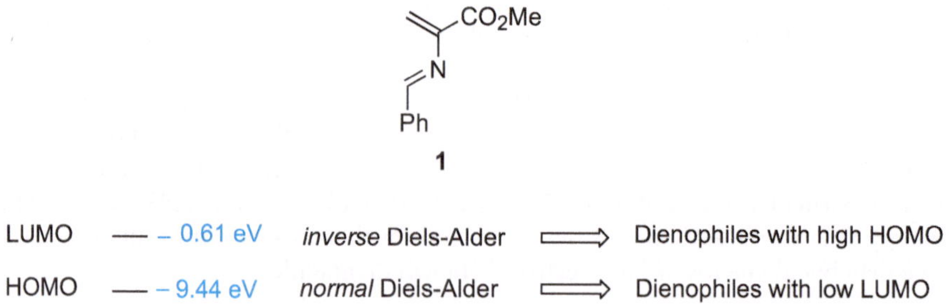

LUMO	— – 0.61 eV	*inverse* Diels-Alder ⟹	Dienophiles with high HOMO
HOMO	— – 9.44 eV	*normal* Diels-Alder ⟹	Dienophiles with low LUMO

Figure 18.3

Reactivity of substituted 2-azadienes **1, 4, 5** *and* **6**

The same type of arguments can be employed to interpret the behavior of substituted azadienes **4**, **5** and **6** in Diels-Alder reactions. The presence of an electron-donating substituent in position *para* of the aromatic ring should raise the energies of the FMO (Fig. 18.4). In fact, the calculated value for the HOMO of 4-*N,N*-dimethylaminophenyl azadiene **4** is –8.21 eV, considerably higher than that of azadiene **1** (–9.44 eV). Reasonably **4** should be more reactive than **1** toward low-LUMO electron-deficient dienophiles. This means that **4** should be a better diene than **1** in *normal* Diels-Alder reactions. By contrast, the significant increase of the LUMO energy in **4** (0.39 eV higher than **1**) makes this compound unable to react with electron-rich (high-HOMO) dienophiles, inhibiting its participation in *inverse* Diels-Alder processes.

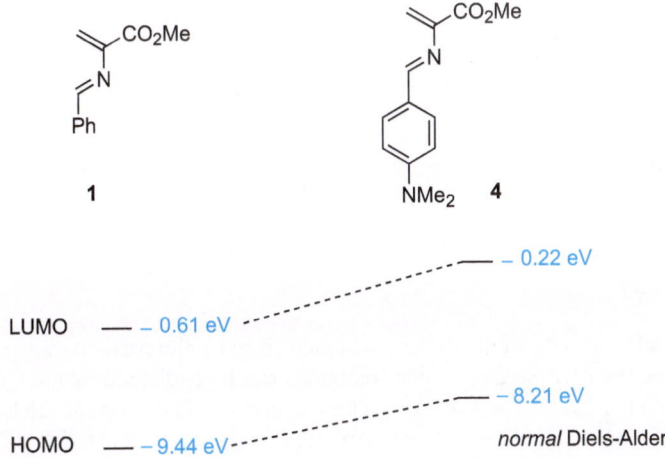

Figure 18.4

The effect of electron-withdrawing groups in the aromatic ring of azadiene **1** is exactly the opposite. As indicated in Fig. 18.5, the calculated HOMO energy of azadienes **5** and **6** (–10.13 and –10.51 eV respectively) are much lower than that of **1** (–9.44 eV). This explains the lack of reactivity of these compounds toward low-LUMO electron-deficient dienophiles. On the other hand, the LUMO energies of **5** and **6** (–1.60 and –0.85 eV, respectively) are considerably lower than that of **1** (–0.61 eV). That means that **5** and **6** can participate in *inverse* Diels-Alder reactions even better than azadiene **1**.

Considering all the AM1 calculated LUMO relative energies along the series of azadienes **1, 4, 5**, and **6** we can establish that their order of decreasing reactivity to participate in the *inverse electron demand Diels-Alder reactions* shall be **5 > 6 > 1 >> 4**. Only azadienes **1** and **4** have HOMOs that can participate in *normal Diels-Alder reactions*.

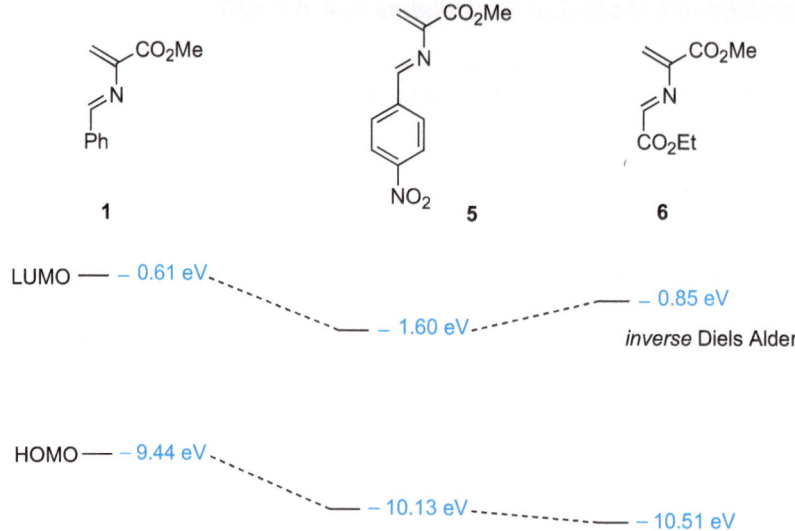

Figure 18.5

In Summary

The unusual reactivity of azadiene **1,** characterized by the participation in both the normal and the **Inverse** Diels-Alder reactions can be explained by the fact that the HOMO and the LUMO of **1** are very close in energy. The study of the FMO energies in other substituted 2-azadienes structurally related to **1** justifies their behavior either in HOMO-controlled (normal) or in LUMO-controlled (inverse) Diels-Alder cycloadditions.

Additional Comments

This problem is based on the work by Pinho e Melo TMVD, Fausto R, d'A Rocha Gonsalves AM, Gilchrist TL (1998) *J. Org. Chem.* 63:5350-5355.

Subjects of Revision

Cycloadditions: effect of the substituents on the energy of the FMO.

Level 2 – Case 19
Chelate-Controlled Carbonyl Addition Reactions

Key point: *Nucleophilic addition. Stereochemistry*

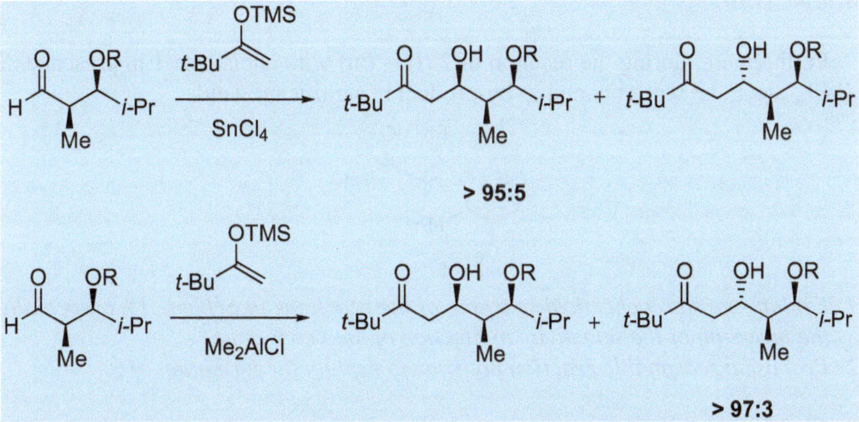

> 95:5

> 97:3

Chelate-Controlled Carbonyl Addition Reactions

The nucleophilic addition of pinacolone enolsilane **1** to *syn*-α-methyl-β-alkoxy aldehydes **2**, promoted by Lewis acids, is highly stereoselective. Thus, when SnCl$_4$ was employed, aldols **3** were obtained preferentially with selectivities higher than 95:5. Instead, the stereoisomers **4** were isolated with nearly total stereoselectivity when Me$_2$AlCl (2 eq) was used to promote the reaction (Scheme 19.1a and b).[1]

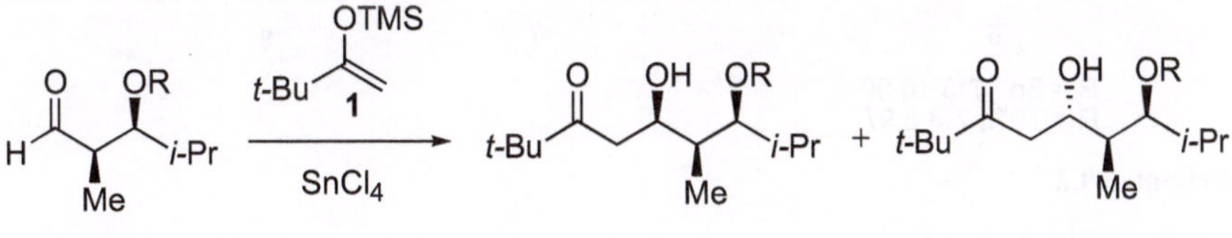

2

3

4

R = Bn, **3:4** 95:5
R = TBS, **3:4** 99:1

Scheme 19.1a

[1] Bn = Benzyl; TBS = *t*-BuMe$_2$Si; TMS = Me$_3$Si.

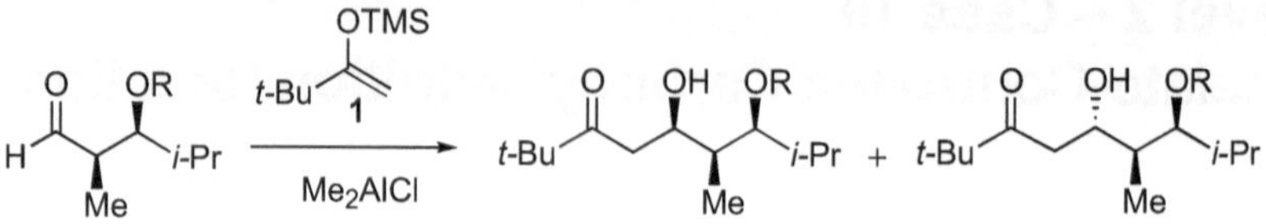

2

R = Bn, **3:4** 1:99
R = TBS, **3:4** 3:97

Scheme 19.1b

Furthermore, during the reaction of **2** (R = Bn) with enolsilane **1** in presence of SnCl$_4$, alcohol **5** was obtained as by-product in significant yield.

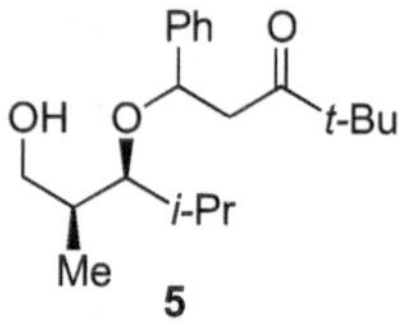

1. Explain the stereochemical outcome of the reactions in Scheme 19.1, specially the inversion of the selectivity in function of the Lewis acid.
2. Propose a reasonable reaction pathway to explain the formation of 5.

Experimental Data

Aldehyde **6** reacts with enolsilane **1** in the presence of Me$_2$AlCl to yield **8** as the major reaction products (the stereoselectivity **7:8** is higher than 1:9) (Scheme 19.2).

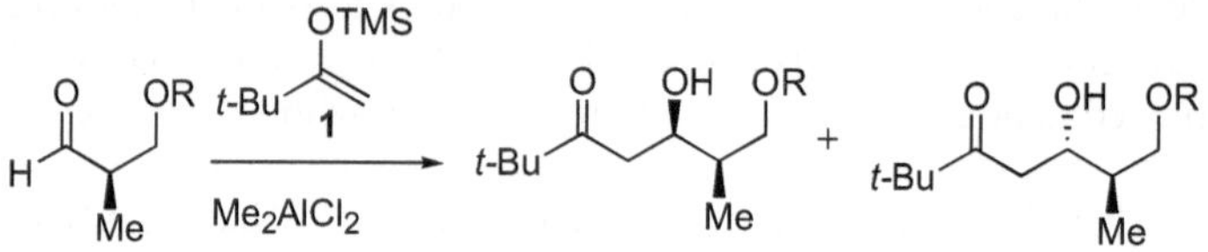

6

R = Bn, **7:8** 10:90
R = TBS, **7:8** 3:97

Scheme 19.2

Discussion

Stereochemistry

The experimental data in Scheme 19.1 point to the metal in the Lewis acid as directly responsible for the stereochemical outcome of the reaction. This metal-dependence suggests that other factors different from the facial selectivity of the aldehyde (the preferential attack of the nucleophile to one of the two faces of the aldehyde) are involved in the stereodiscrimination of the reaction. If the facial selectivity were solely responsible for the stereochemistry of the process, changes in the **3:4** diastereomeric ratio should be expected after changing the Lewis acid, but not the complete inversion that is experimentally observed.

Aldehyde **2** has a nicely positioned β-oxygen that may be able to coordinate the metal. Within this premise, the stereochemical outcome of the reaction would depend on the ability of the metal to coordinate both, the carbonyl oxygen and the β-oxygen, or just the carbonyl oxygen alone.

Before considering any additional chelation, it is useful to apply the standard Felkin-Anh's model to the reaction between aldehyde **2** and enolsilane **1**. According to this model, the reactive conformations of an aldehyde bearing an α-stereocenter, have the bonds to the L (large), M (medium) and S (small) substituents, staggered relative to the carbonyl group as in **9** (Fig. 19.1). In this model the L substituent is located at the least sterically hindered site. The nucleophile will attack the carbonyl group through the less-hindered face (between S and M groups) following a non-perpendicular trajectory called a Bürgi-Dunitz trajectory. This situation takes the nucleophile close to S in conformation **9** (Fig. 19.1).

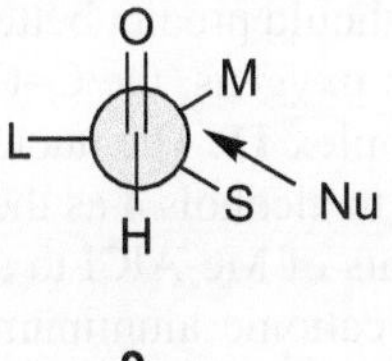

9

Figure 19.1

Application of this model to aldehyde **2** results in conformation **10** that would lead to aldols **3** (Fig. 19.2). Compounds **3** are called *Felkin products*. This is the situation for the tin-promoted reaction. Therefore it is not necessary to proceed for the reaction with SnCl₄, as the simplest model perfectly explains the observed diastereoselectivity.

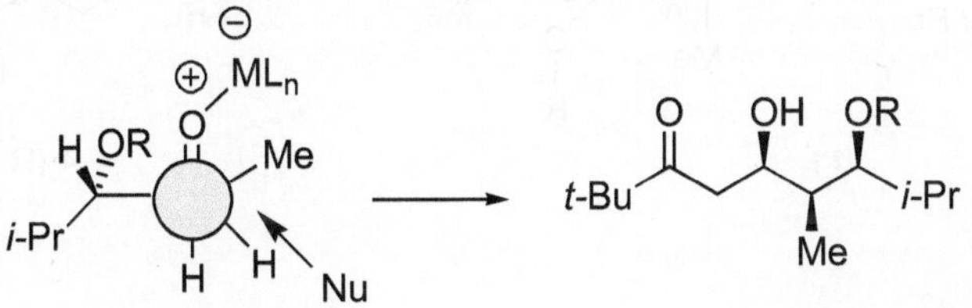

10   **3** (R = Bn, TBS)

Figure 19.2

The reaction with Me₂AlCl follows a different pathway. If we apply the Felkin-Anh's model in this case the situation has to be entirely analogous to that depicted in Fig. 19.2, as only the Lewis acid has changed. Therefore, the Felkin product **3** should have been obtained as the major reaction product, which is not the case. Furthermore, the configuration on the newly formed stereocenter in the sole reaction product **4** is exactly the opposite of that predicted by the Felkin-Anh's model. Compounds **4** are named *anti-Felkin products*. The conclusion is that, either the Felkin-Anh's model is not appropriate to predict the stereochemistry in this case or the reaction follows a completely different pathway.

Although hard to believe, it can be thought that the *anti-Felkin* selectivity is not due to the presence of an α-stereocenter (the one we have considered when discussing the Felkin-Anh's model) but to the β-stereocenter present in **2**. This possibility is easily discarded, simply by carrying out the analogous analysis in a substrate lacking the β-stereocenter as aldehyde **6**. This compound reacts with pinacolone enolsilane **1** in the presence of Me_2AlCl to yield the corresponding aldols **7** and **8** with stereoselectivities in the range of those obtained with aldehyde **2** under the same conditions. Therefore some kind of "dominant effect emanating from the β-stereocenter" should be ruled out (see Scheme 19.2).

These results leave us with a single option: to involve the β-oxygen in the reaction. In fact, Me_2AlCl is an exceptional chelating agent and it may form chelated transition states involving both, the carbonyl oxygen and the β-oxygen. In these cases the cyclic Cram's model should predict better the experimental results. Since the Al-reagent coordinates both oxygens, the C=O is fixed forming a six-membered ring cationic aluminum complex **11**. The nucleophilic attack on the less-hindered *si*-diastereoface would lead to alcohols **4** as the main reaction product (Scheme 19.3). The use of two equivalents of Me_2AlCl to achieve the desired stereocontrol is interpreted by formation of cationic aluminum intermediates such as **11**. The counterion of these complexes ($Me_2AlCl_2^-$) derives from the second molecule of the aluminum reagent already present in the reaction medium.

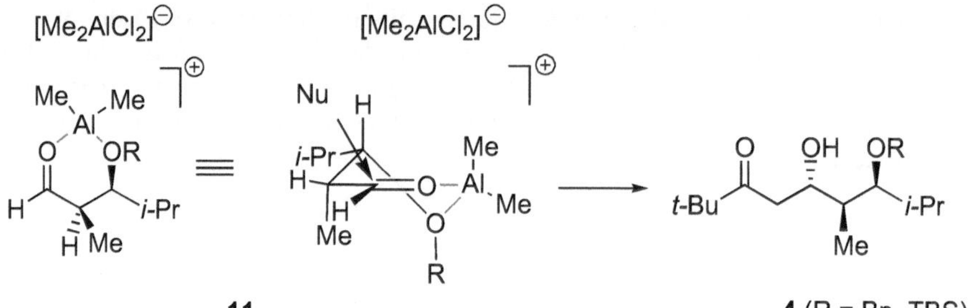

Scheme 19.3

Formation of alcohol 5

If we try to determine the changes in bonding patterns during the reaction between aldehyde **2** and enolsilane **1** in the presence of $SnCl_4$, to give alcohol **5** we would notice that one of the benzylic protons has moved to the aldehyde (1,5-H migration) (Scheme 19.4). Furthermore, the enolsilane fragment **1** (in the ketone form) is placed at the benzylic carbon in the reaction product. The stereochemistry of C2 and C3 in compound **2** has been maintained during the process, which probably means that they have not been involved in the reaction.

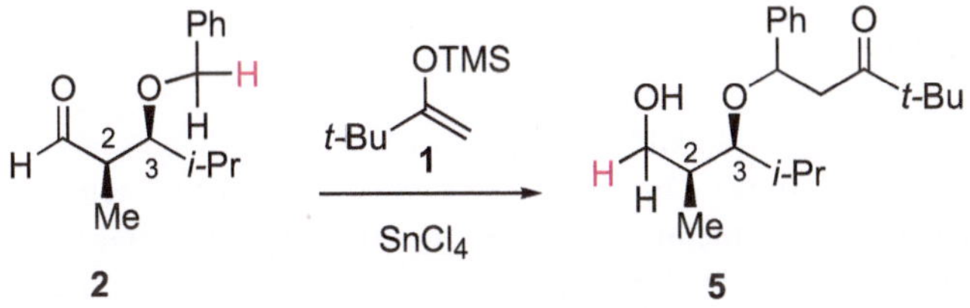

Scheme 19.4

It is now necessary to discuss this process step by step. We should point out that now *we are proposing a reaction mechanism to justify an experimental result, without any supportive experimental evidence.* Based on the previous discussion with SnCl$_4$ used as Lewis acid, it is reasonable to accept that coordination of the carbonyl group to the Lewis acid occurs during the first stage of the reaction. After this initial coordination, we could propose that a hydride transfer through a six-membered ring chair-like transition state **12** may occur. This hydride transfer would explain the 1,5-migration of the benzylic hydrogen, without compromising the integrity of the preexisting stereocenters. The new electrophilic species **13** thus formed, would react now with **1** to yield alcohol **5** (Scheme 19.5).

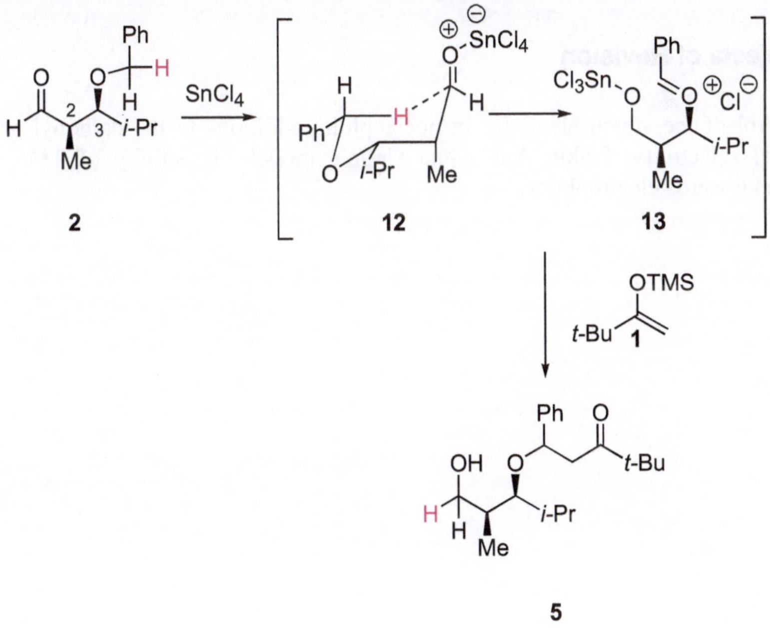

Scheme 19.5

This mechanism would explain the formation of the final product. However, this is only a mechanistic proposal and additional experimental evidence would need to be compiled to confirm its validity.

In Summary

The stereoselectivity of the addition of pinacolone enolsilane **1** to β-alkoxy aldehydes bearing two stereocenters depends on the ability of the metal to form intermediate chelates. Those metals that monocoordinate the carbonyl group form Felkin products and the stereochemistry of these aldols is predicted by the Felkin-Anh's model. For metals chelating both the carbonyl and alkoxy groups, anti-Felkin products are obtained. In these cases the cyclic-Cram's model has to be used to predict the stereochemical outcome of the reaction. Therefore, non-chelated (Felkin-Ahn) and chelated models (cyclic-Cram) have been successively applied to understand the stereochemistry of the final reaction products.

Additional Comments

This problem is based on the work by Evans DA, Allison BD, Yang MG (1999) *Tetrahedron Letters* 40:4457-4460 and by the work by Evans DA, Allison BD, Yang MG, Masse CE (2001) *J. Am. Chem. Soc.* 123:10840-10852.

Subjects of Revision

Control of the stereoselectivity in nucleophilic additions to the carbonyl group. Facial selectivity. Felkin Anh's and Cram's models. Reactivity of TMS enol ethers towards electrophiles.

Level 2 – Case 20
Esterification of Carboxylic Acids with Dimethyl Carbonate and DBU

Key point: *Nucleophile catalysis. Isotope labeling*

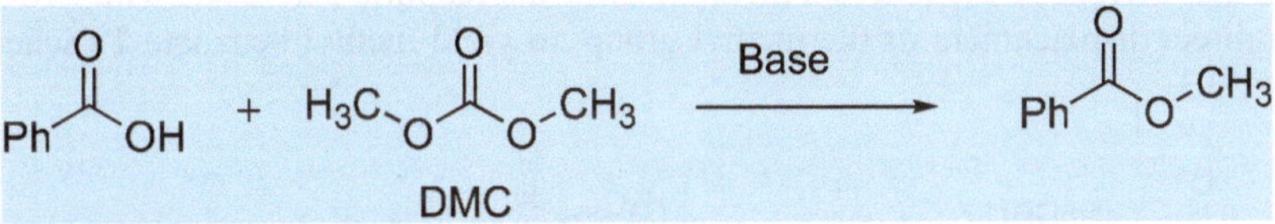

Esterification of Carboxylic Acids with Dimethyl Carbonate and DBU

Dimethyl carbonate (DMC) has attracted considerable attention in the last years as a nontoxic alternative to the traditional methylating agents methyl iodide or dimethyl sulfate. Although this reagent efficiently methylates carboxylic acids, phenols, anilines and activated methylenes, generally its use requires high temperatures and very frequently high pressures (Scheme 20.1).

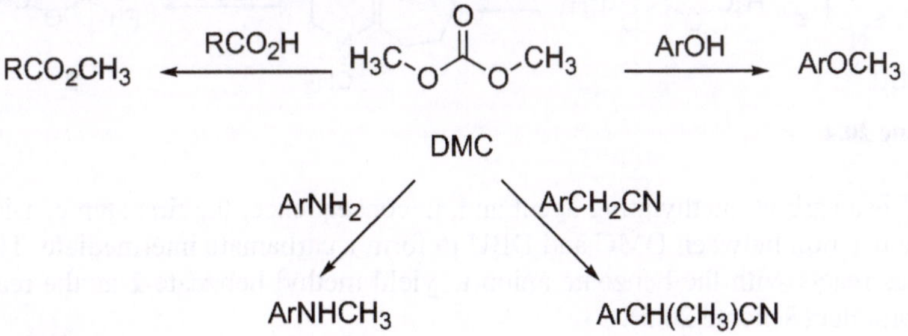

Scheme 20.1

Among the diverse mechanistic studies directed to prevent the use of high temperatures during the process, the esterification of benzoic acid **1** with DMC, at 90°C, in the presence of a variety of amine bases (1 equiv) was investigated (Scheme 20.2). Although different types of bases were studied, (e.g. *N*-methylmorpholine (NMM), tributylamine, 4-dimethylaminopyridine (DMAP), ammonium hydroxide) only 1,8-diazabicyclo[5.4.0]undec-7-ene (DBU) yielded methyl benzoate **2** in quantitative yield. With the rest of the amines tested, the ester **2** was not obtained even after prolonged reaction times.

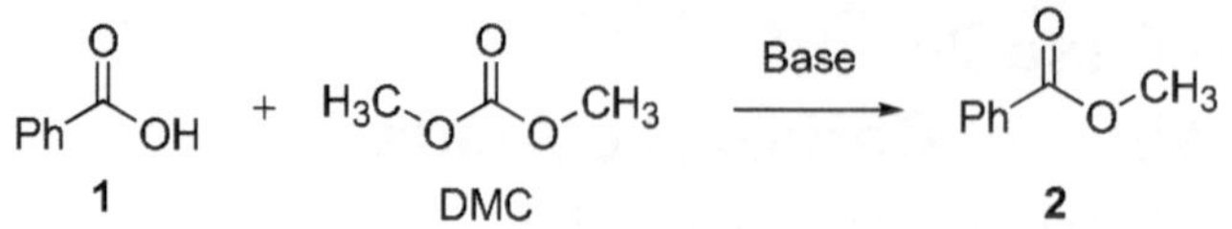

Scheme 20.2

Why does the reaction work with DBU and not with the rest of the amines studied?
This question was a subject of debate and led to different speculations about the role of DBU during the process. The following mechanistic proposals were raised:

1. DBU is a base that merely removes the proton from the acid. The benzoate anion thus formed in the reaction medium is the nucleophile that reacts with DMC by direct displacement of the methyl group, to yield methyl benzoate **2** (Scheme 20.3).

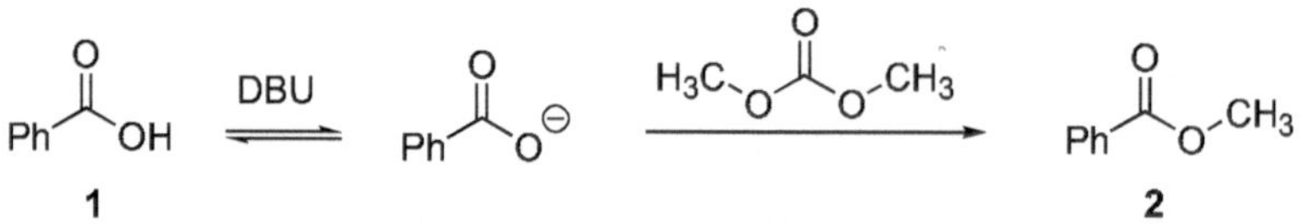

Scheme 20.3

2. DBU is a true reagent in the process. As DMC is a methylating agent, the first step of the reaction consists on the formation of an *N*-methylated-DBU salt that reacts with the benzoate anion to form the reaction product **2** (Scheme 20.4).

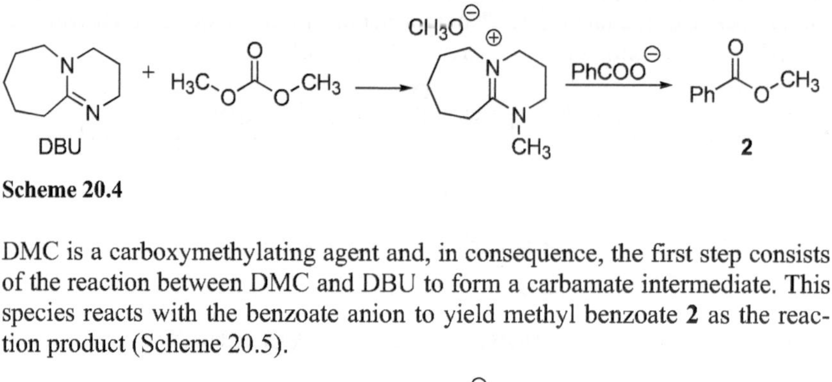

Scheme 20.4

3. DMC is a carboxymethylating agent and, in consequence, the first step consists of the reaction between DMC and DBU to form a carbamate intermediate. This species reacts with the benzoate anion to yield methyl benzoate **2** as the reaction product (Scheme 20.5).

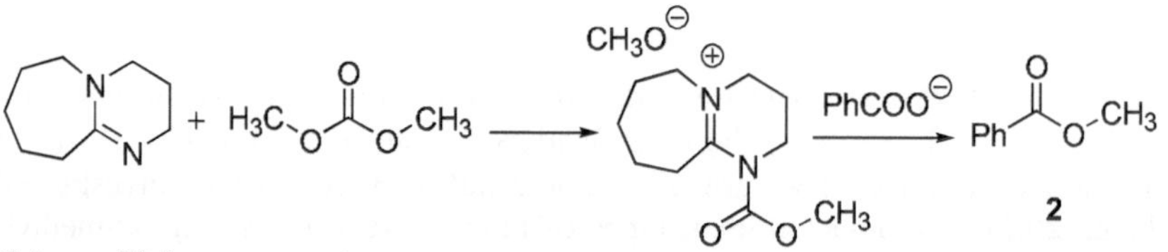

Scheme 20.5

4. Undoubtedly a carbamate intermediate is formed, but the reaction proceeds by attack of the benzoate anion to the carbamate carbonyl group to form an anhy-

dride intermediate, which in the presence of methanol gives methyl benzoate **2** (Scheme 20.6).

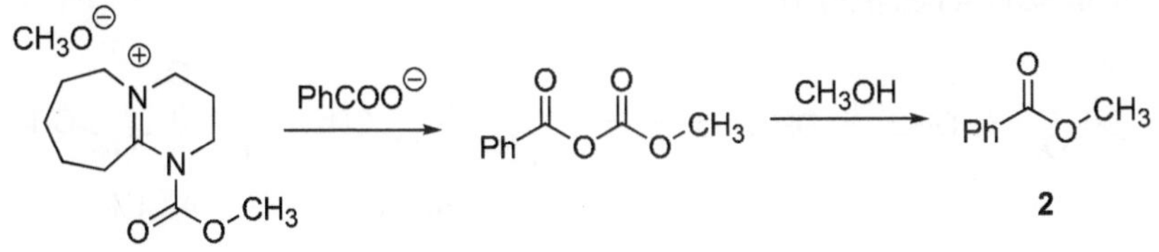

Scheme 20.6

Which of the previous statements is the right one? Comment all options and discuss the experimental data.

Experimental Data

1. The esterification of **1** with DMC at 90°C does not work in the absence of base.
2. The reaction rate strongly depends on the concentration of DBU.
3. DBU is recovered unaltered at the end of the reaction.
4. Labeled *N*-methylated salt **3** was obtained by reaction of DBU and $^{13}CH_3I$. A solution of benzoic acid, DBU and salt **3** in acetonitrile was heated to reflux overnight, but failed to produce any labeled methyl benzoate **4**. The total recovery of unaltered salt **3** was confirmed by NMR (Scheme 20.7).

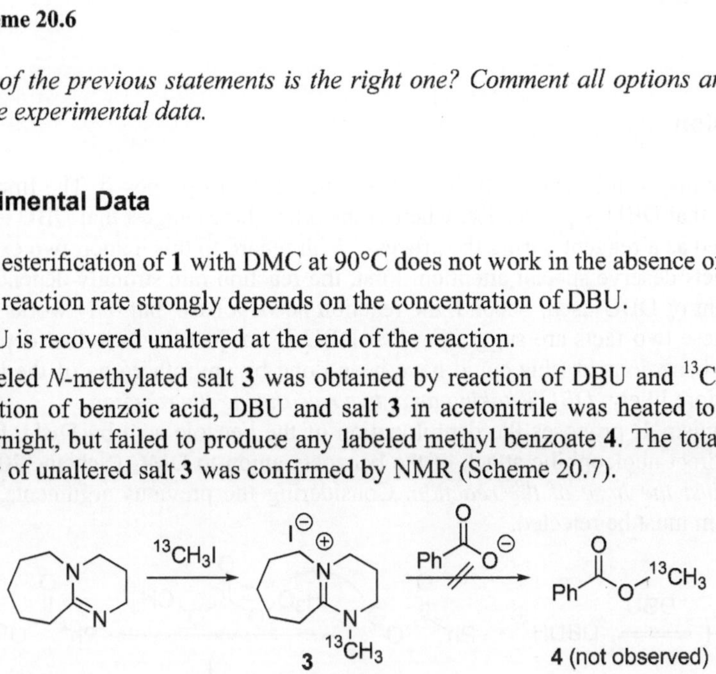

3   **4 (not observed)**

Scheme 20.7

5. Carbamate **5** was independently synthesized from methyl chloroformate and an excess of DBU. When a solution of benzoate anion was added to **5**, a voluminous evolution of gas started and methyl benzoate **2** was obtained (Scheme 20.8).

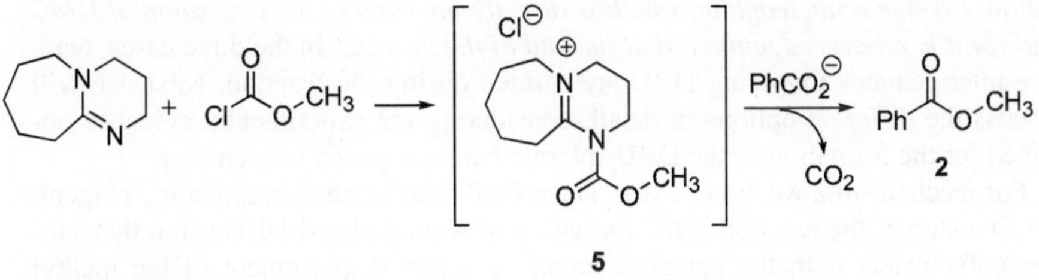

5

Scheme 20.8

6. Labeled benzoic acid was reacted with DBU and DMC at 90°C. The study indicated that an approximately 1:1 mixture of labeled methyl benzoates was obtained (Scheme 20.9).

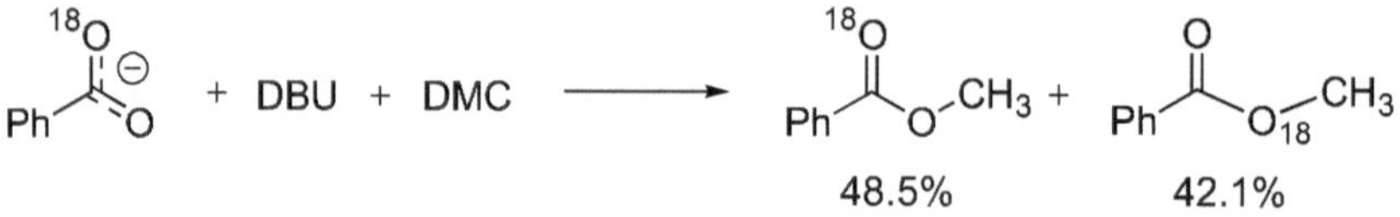

Scheme 20.9

Discussion

There is a major difference between the four mechanisms proposed. The first one considers that DBU is just a base, whereas the other three suggest that DBU is fully involved as a reagent during the process. With regard to this matter, two experimental facts deserve special attention. First, the reaction rate strongly depends on the amount of DBU used. Second, the reaction needs a base, but only works with DBU. These two facts are suggesting that DBU is not just the base that deprotonates the benzoic acid (this could have been done by any other one of the bases tested). More likely, *DBU is taking an active role during the reaction.*

Mechanism 1, proposes the deprotonation of the benzoic acid by DBU, followed by direct nucleophilic attack of the benzoate anion to DMC (Scheme 20.10). *DBU is just the base of the reaction.* Considering the previous arguments, this mechanism must be rejected.

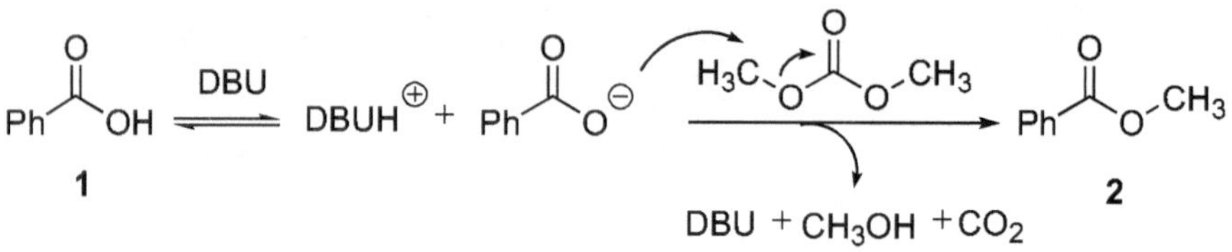

Scheme 20.10

The remaining alternatives have points in common and postulate three reasonable mechanisms for benzoate formation employing a combination of DMC and DBU. The three routes suggest that DBU is a *nucleophilic catalyst* for the esterification: *(a) it is a nucleophile, which is directly involved in the activation of DMC and (b) it is recovered unaltered at the end of the process.* In the three cases, reactive intermediates involving DBU are formed during the reaction. Next, we will discuss the different options in detail, considering the experimental evidence obtained for the formation of the DBU-intermediates proposed in each case.

For mechanism 2 we assume that, since DMC can be an *N*-methylating reagent, the first step of the reaction is the formation of *N*-methylated-DBU salt **6** that subsequently reacts with the benzoate anion by direct displacement of the methyl group (Scheme 20.11).

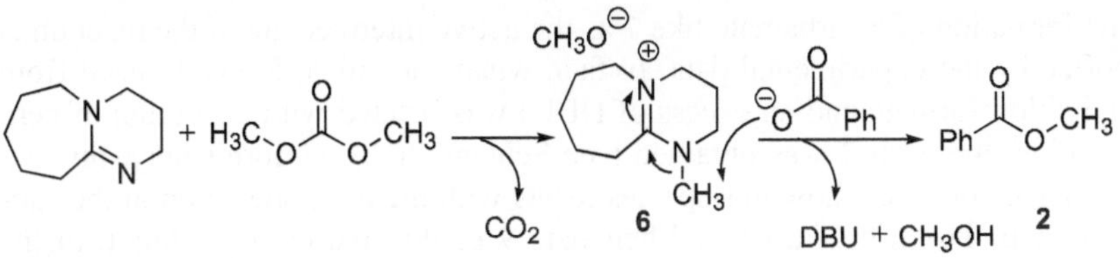

Scheme 20.11

However, if salt **6** were formed during the reaction, structurally related labeled *N*-methylated salt **3** should also react with the benzoate anion under the same esterification conditions. Experimentally **3** was recovered unaltered when treated with benzoic acid and DBU. This result indicates that salts **6** are not the active intermediate responsible for the formation of the methyl ester. Hence, mechanism 2 must be rejected (Scheme 20.12).

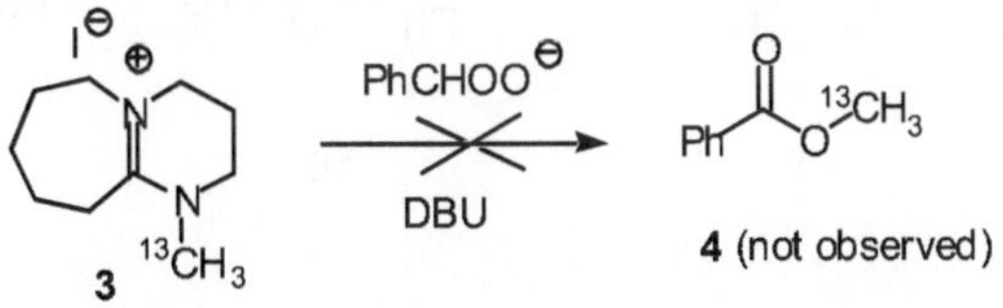

Scheme 20.12

This leaves us with only two options. Both consider DMC as a carboxymethylating reagent and postulate the formation of a carbamate intermediate **7** in the first step of the reaction. Mechanism 3 proposes the reaction of carbamate **7** with the benzoate anion by direct displacement of the methyl group (Scheme 20.13).

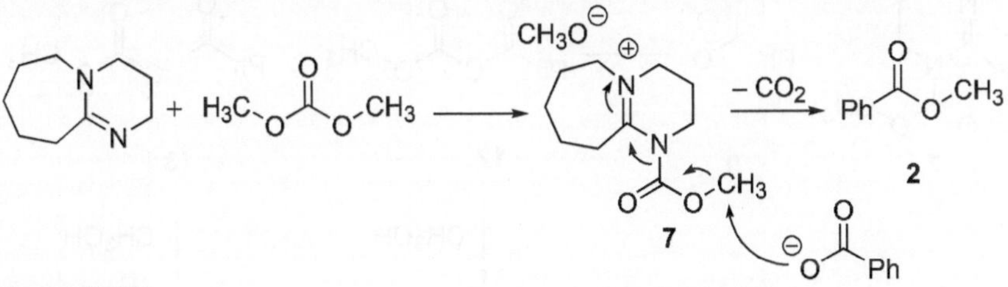

Scheme 20.13

On the other hand, mechanism 4 considers the nucleophilic attack of the benzoate anion to the carbonyl group in **7** to form an anhydride intermediate **8**. Reaction of **8** with methanol results in the formation of benzoate **2** (Scheme 20.14).

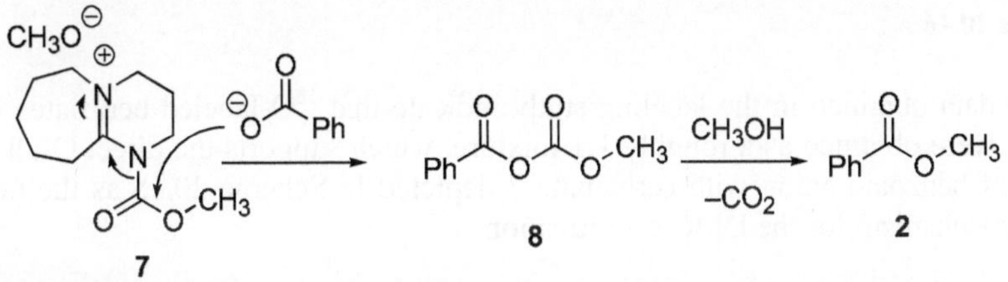

Scheme 20.14

The formation of a carbamate like **7** as the active intermediate of the reaction is supported by the experimental data. In fact, when carbamate **5** (synthesized from methyl chloroformate and an excess of DBU) was reacted with a solution of benzoate anion, the ester **2** was obtained (see Scheme 20.8). Unfortunately, this experiment cannot reveal how this species reacts with the benzoate anion in the subsequent step. The use of a labeled benzoate **9** could give us the solution to the problem.

If the direct methyl displacement proposed by mechanism 3 does occur, the reaction between carbamate **7** and labeled benzoate anion should give an equal mixture of ^{18}O-labeled benzoates **10** and **11** (Scheme 20.15).

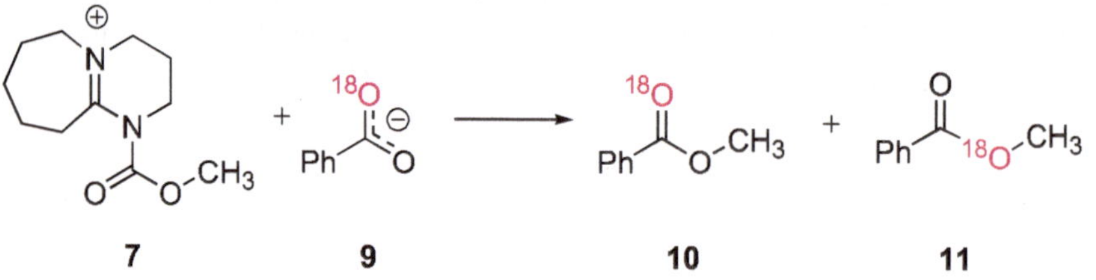

Scheme 20.15

However, if the benzoate anion adds to the C=O group in **7**, a 1:1 mixture of labeled anhydrides **12** and **13** should be formed instead. Subsequent nucleophilic attack of methanol on **12** would produce labeled benzoate **10**, whereas in the case of **13**, unlabeled benzoate **2** should be obtained (Scheme 20.16).

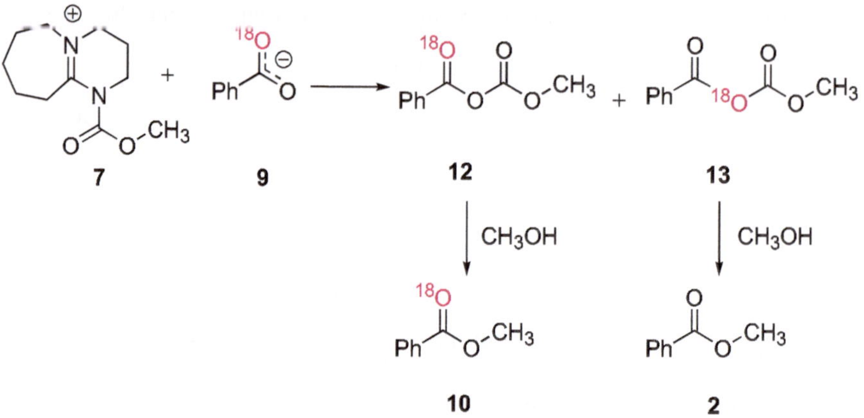

Scheme 20.16

The data obtained in the labeling study indicate that ^{18}O-labeled benzoates **10** and **11** were obtained as a roughly 1:1 mixture, which supports the direct O-alkylation of benzoate anion with carbamate **7** depicted in Scheme 20.15 as the preferred mechanism for the DMC esterification.

In Summary

DMC is an efficient methylating agent of carboxylic acids in the presence of DBU. The reaction pathway involves initial *N*-acylation of DBU to form a carbamate intermediate. This carbamate reacts with the carboxylate anion (*O*-alkylation) to afford the methyl ester. DBU not only is the base that forms the carboxylate anion but is also the nucleophile catalyst of the process.

Questions

The addition of methanol decreases the rate of the esterification of carboxylic acids with DMC-DBU. Without the added methanol methyl benzoate **2** is quantitatively obtained within 3 h, but with extra methanol added, methyl benzoate formation is still not complete after 24 h. Could you explain why?

Answer to the Question

Anion solvation by protic solvents like methanol takes place preferably through hydrogen bonding. The carboxylate anion will be highly solvated by methanol and hence its nucleophilicity will be reduced under these conditions, causing a decrease in the reaction rate.

Additional Comments

This problem is based on the work by Shieh W-C, Dell S, Repic O (2002) *J. Org. Chem.* 67:2188-2191.

Subjects of Revision

Detection of intermediates. Labeling experiments. Nucleophile catalysis.

Level 2 – Case 21
A Hammett Analysis in a Multistep Reaction: Rhodium(II)-Catalyzed Decomposition of α-Diazo Esters

Key point: *Hammett constants*

A Hammett Analysis in a Multistep Reaction: Rhodium(II)-Catalyzed Decomposition of α-Diazo Esters

Hammett analysis in multistep reactions is not always a simple task. The rate-determining step is not necessarily the first step of the reaction and in a reaction sequence, substituents that favor a step could favor or disfavor the next one. This situation sometimes complicates evaluation of the influence of electron-withdrawing or electron-donating groups on the overall reaction mechanism.

One of the parameters used to determine the influence of electronic effects in a reaction is the Hammett reaction constant ρ. However, the experimentally observed ρ values (ρ_{obs}) are the average of the contributions of the electronic effects on the rate-determining step and all the steps preceeding, which generally means that the magnitude of ρ_{obs} is not as large as expected. This fact adds an extra difficulty to the analysis of the influence of the electronic effects in the elucidation of multistep mechanisms, particularly when intermediates bearing charges of opposite signs are involved.

A good example for a detailed Hammett correlation study is the rhodium(II)-catalyzed decomposition of α-diazo esters. The most likely mechanism for the α-diazo ester decomposition process involves the initial complexation of the negatively polarized carbon of the diazocompound **1** to the axial site of the rhodium(II) catalyst **2** (which is coordinatively unsaturated) to form intermediate **3**. Subsequent extrusion of N_2 from **3** generates the Rh(II) carbene intermediate **4**, *the reactive species* that leads to the final products in a fast step. Among the reactions of wider application of such reagents in organic chemistry we should mention cyclopropanations, C-H insertions or generation of carbonyl ylides (Scheme 21.1).

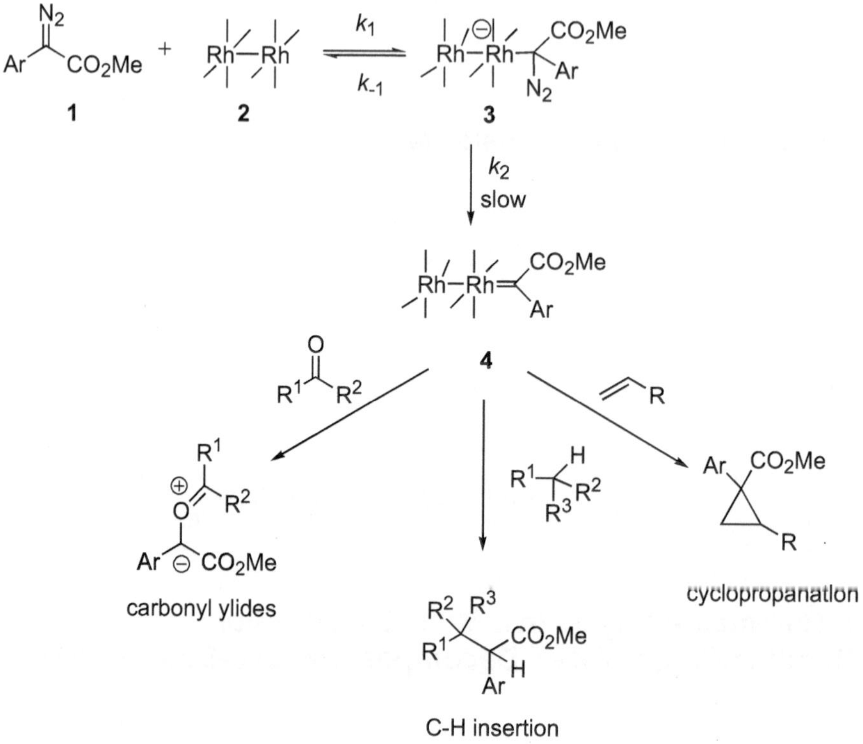

Scheme 21.1

Discuss the experimental data and determine the influence of the electronic effects on the two key steps of the reaction.

Experimental Data

1. Kinetic studies have demonstrated that the first step of the reaction is a fast preequilibrium ($K = k_1/k_{-1}$) and the second step (k_2) is rate limiting.
   The rate law will be then given by $v = k_2[3]$.
   As $K = [3]/[\text{diazoester}][\text{catalyst}]$, we can write:
   $[3] = K[\text{diazoester}][\text{catalyst}]$.
   Hence, $v = K k_2[\text{diazoester}][\text{catalyst}]$.

Then, the overall kinetic law is pseudo-first order and could be written as:

$$v = k_{obs}[\text{diazoester}] \qquad\qquad \text{where } k_{obs} = K\, k_2[\text{catalyst}] \qquad (21.1)$$

2. A Hammett correlation analysis was made on the Si–H insertion of aryl diazo-acetates **1** with triethylsilane in the presence of Rh(II) catalysts, as shown in Scheme 21.2.

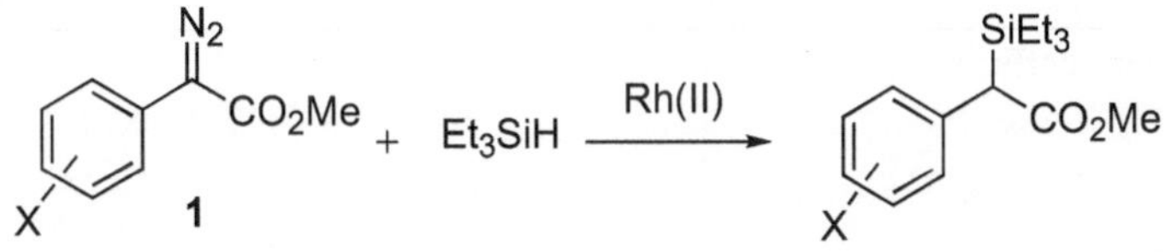

Scheme 21.2

The Hammett correlation study was carried out with a series of substituted diazo compounds **1** (X= p-NO$_2$, p-Br, m-MeO, p-Ph, m-Me, p-MeO and H) with different rhodium(II) catalysts (Rh$_2$(OAc)$_4$, Rh$_2$(Ooct)$_4$, Rh$_2$(tfa)$_4$). The relative rate (k_{rel}) constants were calculated for each substituted diazo phenylacetate ($k_{rel(X)} = k_{obs(X)} / k_{obs(H)}$). The $k_{obs(X)}$ values were obtained for each substituent according to pseudo-first order Eq. 21.1. The data are collected in Table 21.1.

The $k_{rel(X)}$ values show that diazo substrates **1** with electron-donating substituents in the aromatic ring decompose *faster* than those with electron-withdrawing substituents. The Hammett plots indicate in all cases a better correlation with the substituent constant σ^+ than with σ or σ^- (it can be deduced from the highest values of the correlation coefficients r when σ^+ values are employed).

The values of the reaction constants ρ_{obs} for all the catalysts tested are summarized in Table 21.2.

Table 21. 1

X	$k_{obs(x)}$		
	Rh$_2$(Oac)$_4$	Rh$_2$(Ooct)$_4$	Rh$_2$(tfa)$_4$
p-MeO	10.5	12.3	16.4
p-Ph	1.35	1.73	1.32
m-Me	1.20	1.19	1.54
H	1.00	1.00	1.00
m-MeO	1.00	0.62	0.98
p-Br	0.84	0.90	0.65
p-NO$_2$	0.084	0.102	0.072

OAc = acetate; Ooct = octanoate; tfa = trifluoroacetate

Table 21. 2

Substituent constants		$\rho_{obs}(r)$	
	$Rh_2(Oac)_4$	$Rh_2(Ooct)_4$	$Rh_2(tfa)_4$
σ^+	$-1.29(-0.99)$	$-1.31(-0.99)$	$-1.46(-0.99)$
σ	$-1.72(-0.94)$	$-1.69(-0.92)$	$-1.94(-0.94)$
σ^-	$-1.11(-0.91)$	$-1.07(-0.87)$	$-1.23(-0.89)$

r = correlation coefficient.

Discussion

The experimental fact is that the decomposition of diazo compounds **1** is accelerated by electron-donating substituents in the aromatic ring, independently of the catalyst employed. This is in full agreement with the Hammett correlation study that gives negative values of the reaction constants ρ_{obs} in all cases studied (Table 21.2).

Can we then conclude that *each of the two key reaction steps is favored by electron-donating groups?* Not necessarily. To make an accurate analysis of the effect of the substituents upon a reaction mechanism it is essential to consider each step of the reaction independently.

Step 1 involves the complexation of **1** with the Rh(II) catalyst and step 2 (the rate-determining step) is the generation of the rhodium carbene intermediate **4** by decomposition of complex **3**. We should point out that step 3, the reaction of the carbene **4** with triethylsilane, is a *fast* step that has no influence on the current study.

Starting with the Hammett equation for the rhodium(II)-catalyzed decomposition of α-diazo esters **1**, the general form of the Hammett equation in this case is:

$$\log (k_{obs(X)} / k_{obs(H)}) = \rho_{obs}\, \sigma \tag{21.2}$$

Considering that $k_{obs} = K\, k_2$ [catalyst] (Eq. 21.1), the Hammett equation could be expressed in the following way:

$$\log (k_{obs(X)} / k_{obs(H)}) = \log [(K_{(X)}\, k_{2(X)} / K_{(H)}\, k_{2(H)}] = \rho_{obs}\, \sigma \tag{21.3}$$

Equation 21.3 can also be written as:

$$\log (k_{obs(X)} / k_{obs(H)}) = \log [(K_{(X)} / K_{(H)}] + \log [k_{2(H)} / k_{2(X)}] = \rho_{obs}\, \sigma \tag{21.4}$$

or as:

$$\log (k_{obs(X)} / k_{obs(H)}) = \rho_1\, \sigma + \rho_2\, \sigma \text{ and } \rho = \rho_1 + \rho_2 \tag{21.5}$$

Therefore, the measured ρ values (ρ_{obs}) are the sum for the two individual steps ρ_1 and ρ_2.

Next, we will discuss if the two steps have the same electronic requirements.

Step 1 is a pre-equilibrium. In diazocompounds **1** the carbon bearing the diazo group is negatively polarized, as represented in the canonical forms depicted in Scheme 21.3. In this step, an electron-withdrawing group like NO_2 will stabilize **1** but also will make it less reactive. An electron-donating group will operate in the opposite manner. Thus, compounds **1** bearing electron-withdrawing substituents, will decompose more slowly than those bearing electron-donating substituents.

Consequently, in step 1, electron-donating groups push the equilibrium to the right favoring the formation of **3** and hence ρ_1 should be negative.

step 1

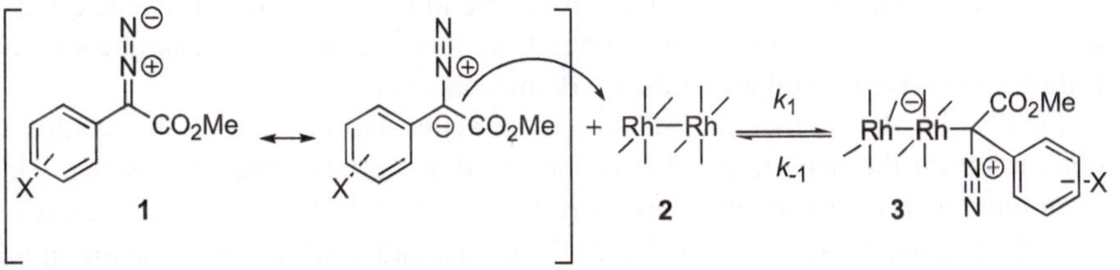

Scheme 21.3

During the second step, carbene intermediate **4** is formed by irreversible extrusion of N_2 from intermediate **3**. Rh(II) carbene complexes **4** have the Ph=C bond highly polarized, with the negative charge in the metal and the carbene carbon being positively charged (**4B** in Scheme 21.4). Hence, electron-donating groups should stabilize this intermediate, speeding up the nitrogen extrusion (Scheme 21.4).

step 2

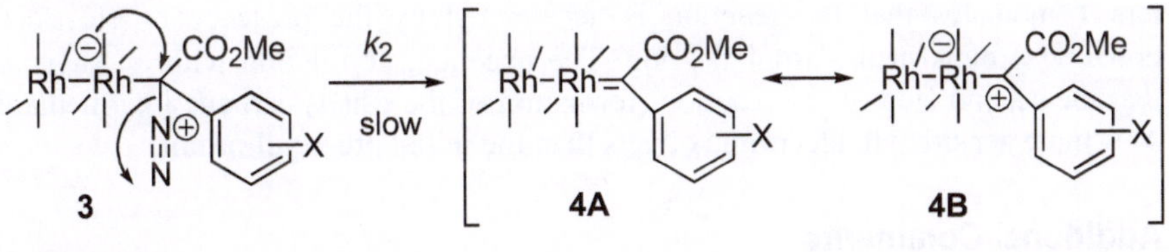

Scheme 21.4

Therefore, in step 2 electron-donating groups accelerate the decomposition of **3** to **4** and hence ρ_2 should be negative.

The measured ρ values are the sum of the two individual steps. If ρ_1 and ρ_2 are expected to be negative, ρ_{obs} must be negative as well, and in fact it is. Then, in this case, it will be correct to say that all, the overall reaction and each individual step are accelerated by electron-donating groups.

The case studied is a multistep reaction in which all the key steps of the reaction are accelerated by the same type of substituents. In other reactions however, the sign of ρ for each individual step is opposite.

We can use a similar reasoning to discuss the good correlation of the reaction Hammett plots with the substituent constant σ^+. We should remember that a good correlation with σ^+ suggests that a positive charge (or partial positive charge) is being developed in the transition state of the reaction. In this case, in step 1 the partial *negative* charge that was placed in the carbon next to the aromatic ring in **1** disappears when the substrate proceeds to complex with the catalyst to form **3**. In this situation we could speculate that a better correlation with the substituent constant σ^- rather than with σ or σ^+ should be observed. On the other hand, a better correlation with σ^+ than with σ should be obtained during step 2, as a partial positive charge develops in the transition state. Considering that the observed effect is an excellent correlation with σ^+ it is reasonable to think that the electronic effects in the decomposition of diazocompounds **1** are more significant during the second (rate-limiting) step. In other words, ρ_2 dominates over ρ_1.

Finally, we should make a little comment about the information that could be obtained from the magnitude of ρ in the reaction we have been discussing. The magnitude of ρ expresses the sensitivity of a reaction to the electronic effects of the substituents. As shown in Table 21.2 the magnitude of ρ_{obs} is moderate in all cases, (ranging from 1.29 to 1.46) which indicates that the rhodium(II)-catalyzed decomposition of compounds **1** *is not very sensitive* to the introduction of electron-donating substituents. This fact could be interpreted considering that either the positive charge buildup in the carbene **4** is small, or that the N_2 extrusion step has a relatively earlier (reactant-like) transition state. In other words, the transition state of the slow step of the reaction resembles more diazocarbene **3** than the carbene complex **4**.

In Summary

The Hammett correlation study on the Rh(II)-catalyzed decomposition of diazo esters **1** indicates that the reaction is accelerated by the presence of electron-donating groups in the aromatic ring. The excellent correlation with σ^+ suggest that the second step of the reaction (formation of the Rh(II) carbene intermediate **4**) is more sensitive to electronic effects than the initial pre-equilibrium.

Additional Comments

An example of a stepwise mechanism in which the sign of ρ is opposite for each individual step is the acid-catalyzed hydrolysis of benzoates ($\rho_{obs} = +0.14$) which shows a negative ρ_1 in the initial protonation step and a positive ρ_2 during the nucleophilic attack of water.

This problem is based on the work by Qu Z, Shi W, Wang J (2001) *J. Org. Chem.* 66:8139-8144.

Subjects of Revision

Hammett equation. Hammett parameters.

Level 2 – Case 22
Tandem Cycloadditions with Nitronates

Key point: *Cycloadditions*

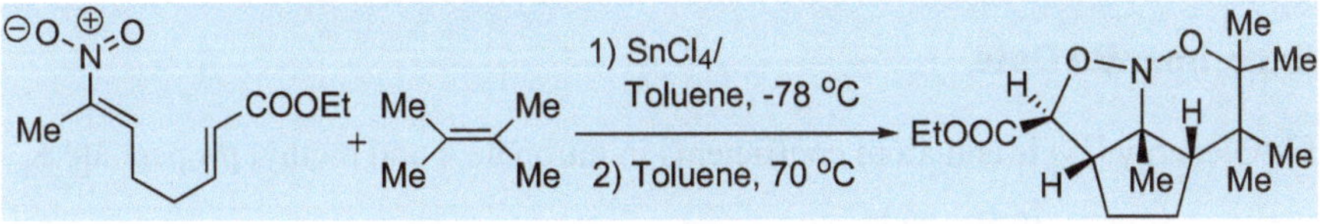

Tandem Cycloadditions with Nitronates

Nitroalkenes **1** behave as heterodienes in [4+2] *inverse electron demand* cycloadditions with simple unactivated alkenes, enamines, or enol ethers (**2**) as dienophiles. These reactions require the presence of a Lewis acid to enhance the reactivity of the nitroalkene and accelerate the process. The products obtained in such reactions are six-membered cyclic compounds called **nitronates** (**3**) (Scheme 22.1). These compounds can be used in turn, as 1,3-dipoles in [3+2] cycloaddition reactions.

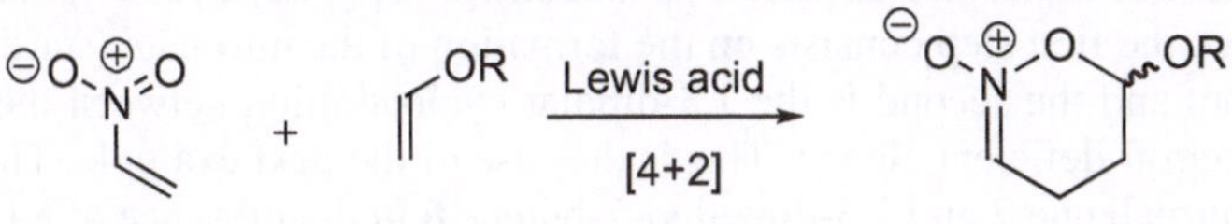

Scheme 22.1

Part 1

It has been established that [3+2] cycloadditions with nitronates are HOMO nitronate (dipole)-LUMO dipolarophile controlled and show a clear preference for the *exo* approach of the dipolarophile (versus the *endo* mode).

Based on these affirmations and using the Frontier Molecular Orbital (FMO) theory, justify the regio- and stereochemistry of compound **6**, obtained in the reaction between nitronate **4** and methyl propenoate **5** (Scheme 22.2).

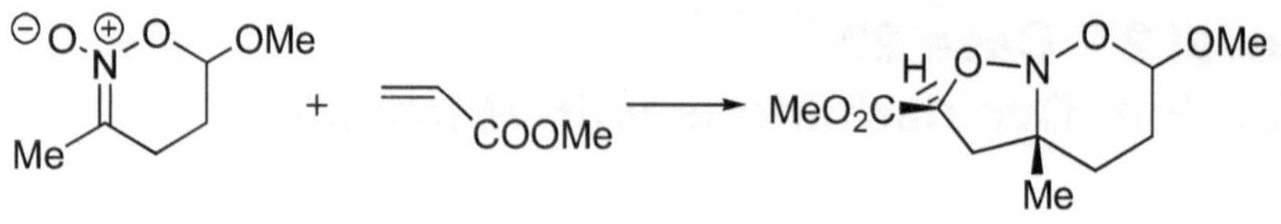

Scheme 22.2

Experimental Data

FMO energy levels and atom coefficients in nitronate **4** and methyl propenoate **5**.

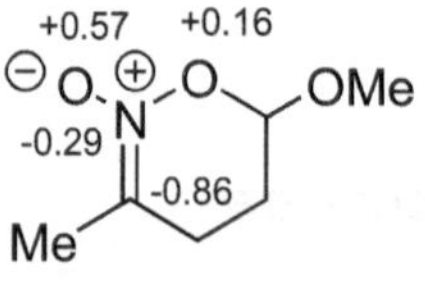

4 HOMO (-8.98 eV) 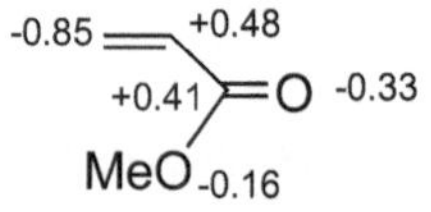   **5** LUMO (-0.01 eV)

Part 2

Sometimes, nitronates are involved in tandem [4+2]/[3+2] cycloaddition processes. That is, the first step consists on the formation of the nitronate (starting from a
nitroalkene) and the second is the 1,3-dipolar cycloaddition between the nitronate
and an electron-deficient alkene. This is the case of the next example. The reaction
between nitroalkene **7** and 2,3-dimethyl-2-butene **8** in the presence of a Lewis acid
and under the conditions indicated in Scheme 22.3 yields tricyclic adduct **9**.

   Draw a mechanism to explain how compound **9** *is formed and justify the stereochemistry of the product.*

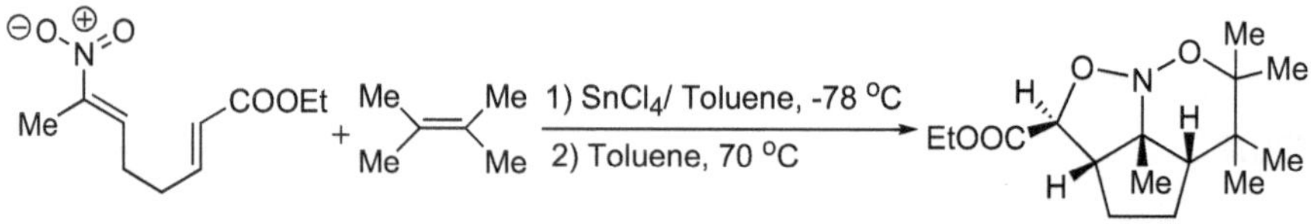

Scheme 22.3

Part 3

Treatment of vinylsilane **10** with 2 equiv of trimethylsilyl chloride (TMSCl) and Et$_3$N, afforded bicyclic isoxazolines **11** and **12** as a 4.5:1 mixture of stereoisomers (Scheme 22.4).

Propose a reasonable mechanism to account for the formation of compounds **11** *and* **12**, *justifying the diastereoselectivity of the reaction.*

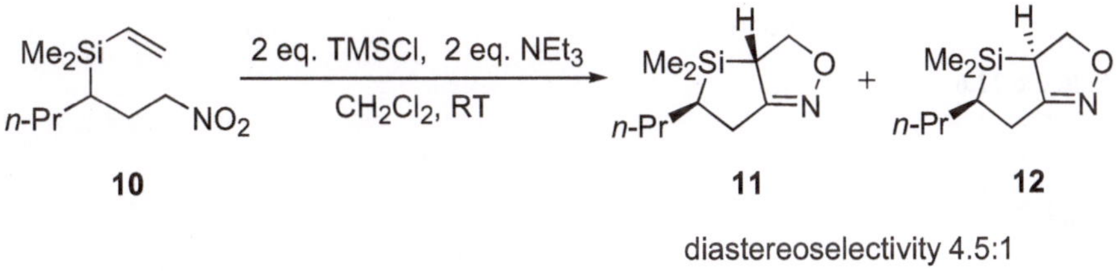

diastereoselectivity 4.5:1

Scheme 22.4

Discussion

Part 1

The aim of the first question is to discuss the regio- and stereoselectivity of the [3+2] cycloadditions with nitronates as 1,3-dipoles. The regiochemistry of the reaction between nitronate **4** and methyl propenoate **5** can be understood by examining the coefficients of the individual atoms in the FMO involved, the HOMO of **4** and the LUMO of **5**. We should remind ourselves that overlapping between two atoms requires coefficients of the same sign and that the overlapping will be more effective if they have similar sizes. In this case, the optimal overlapping between the dipole and the dipolarophile occurs when the α and β carbons of acrylate **5** become attached respectively to the oxygen and the C=N carbon in the nitronate **4**, leading to the reaction product **6** with the regiochemistry indicated in Scheme 22.5.

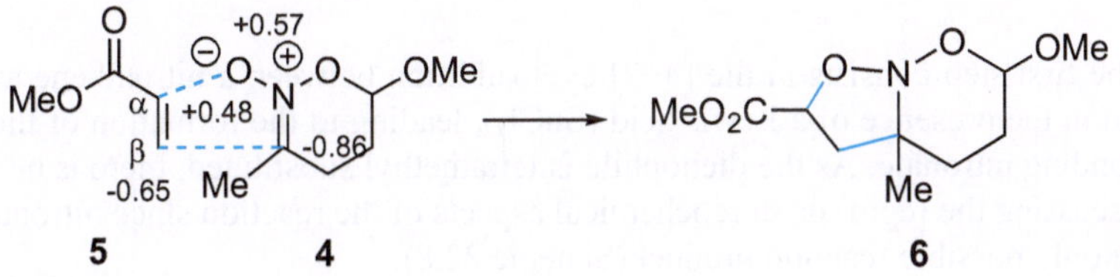

Scheme 22.5

Finally, the stereochemistry of **6** will be determined by the preference for the *exo* approach of the diene, as shown in Scheme 22.6.

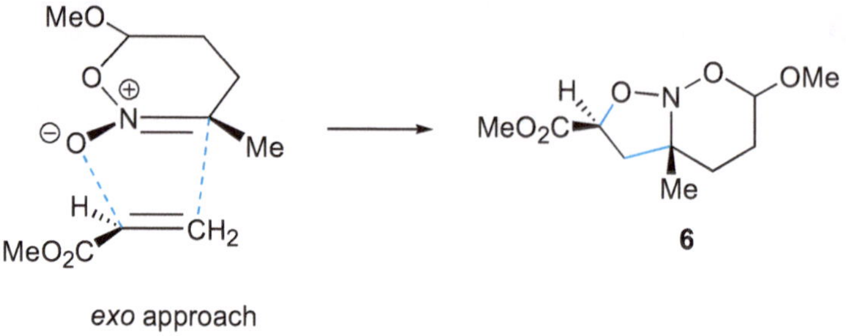

exo approach

Scheme 22.6

Part 2

Once we understand the regio- and stereochemistry of the [3+2] cycloaddition reactions involving nitronates, the formulation of the tandem process proposed in
Part 2 is not difficult. The starting compounds are a nitroalkene **7** (tethered to an
α,β-unsaturated ester) and 2,3-dimethyl-2-butene **8**, a simple unactivated alkene.
The fragments of the starting materials can be easily recognized in the structure of
the reaction product **9**, as indicated in Scheme 22.7. The nitroalkene skeleton has
been drawn in red and the 2,3-dimethyl-2-butene in blue. Disconnection of bonds
a and **b** in **9** leads back to nitronate intermediate **13**. Disconnection of **c** and **d**
bonds in **13** leads back to nitroalkene **7** and 2,3-dimethyl-2-butene **8** (Scheme
22.7). The overall reaction could be interpreted as a tandem intermolecular [4+2]/
intramolecular [3+2] cycloaddition. Next we will formulate the process step by
step.

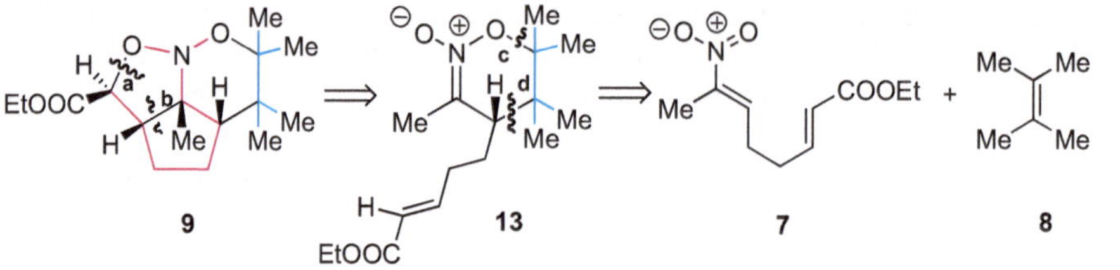

Scheme 22.7

The first step consists in the [4+2] cycloaddition between a nitroalkene and an
olefin in the presence of a Lewis acid (SnCl₄), leading to the formation of the corresponding nitronate. As the dienophile is tetramethyl substituted, there is no point
in discussing the regio- or stereochemical aspects of the reaction since nitronate **13**
is the only possible reaction product (Scheme 22.8).

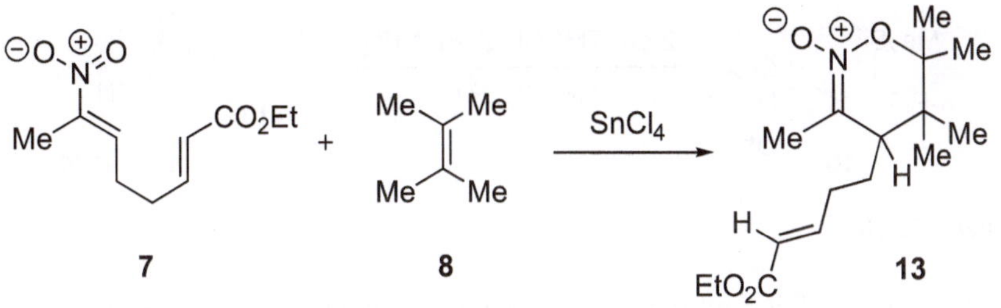

Scheme 22.8

The structure of **13** is particularly suitable for an intramolecular cycloaddition since it contains a 1,3-dipole (the nitronate fragment) and a dipolarophile (the acrylate fragment) linked by a two-methylene tether. The second step of the tandem reaction is then clear. Once nitronate **13** is formed, an intramolecular [3+2] cycloaddition takes place to yield tricyclic nitroso acetal **9** (Scheme 22.9).

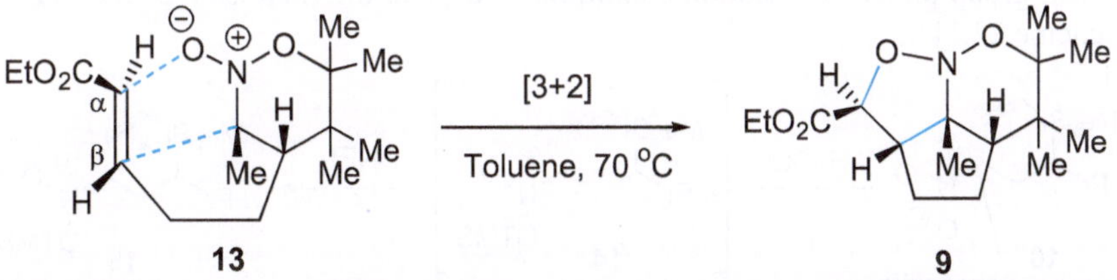

Scheme 22.9

The regio- and stereochemistry of the reaction will follow the patterns previously discussed. The α and β carbons of the acrylate fragment will become attached respectively to the oxygen and the C=N carbon in the nitronate moiety, and the preference for the *exo* approach of the alkene will lead to the reaction product **9** with the carboxylate group, the β-H and the bridged Me in a *cis*-arrangement.

Part 3

If we have a look to the structures of compounds **10**, **11** and **12** we will find some similarities with the tandem process we have just discussed. First, bicyclic isoxazolines **11–12** have been obtained through an intramolecular cyclization process. Second, the skeleton of the starting material is easily recognizable in the final products, although some bonds have changed. The nitro group has been modified and the double bond that was placed at the end of the chain has disappeared (bonds in red) but new C=N and C-C bonds have been incorporated (bonds in blue) (Scheme 22.10). However, all the reactions we have discussed previously have in common a nitroalkene as starting material. **Where is the nitroalkene in this case?**

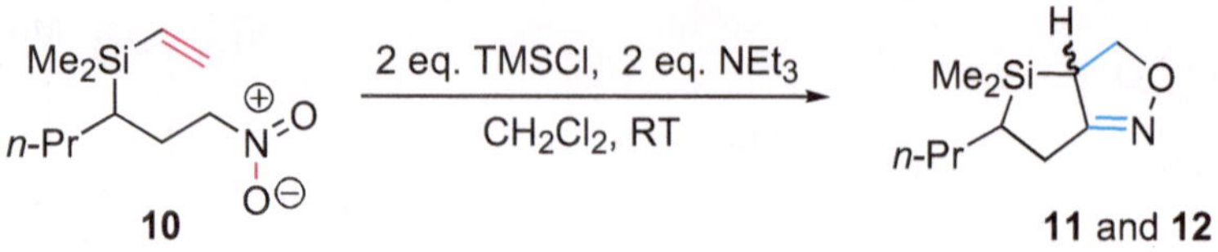

Scheme 22.10

Vinyl silane **10** is a nitro compound and it is well known that the α protons of a nitro group are very acidic ($pK_a \approx 10$). Although the Et₃N is a weak base ($pK_a \approx$ 10–11) the deprotonation could occur to some extent. As the reaction is carried out with TMSCl, it seems reasonable to think that prior to the removal of the acidic proton, one of the oxygens of the NO₂ group has been silylated to yield compound **14**. Formation of intermediate **15** in the presence of NEt₃ is immediate. This species is a silylnitronate that could behave as a 1,3-dipole reacting with the vinylic double bond in a [3+2] intramolecular fashion to give **16**. Elimination of the OTMS group under the reaction conditions will yield the final isoxazolines **11–12** (Scheme 22.11).

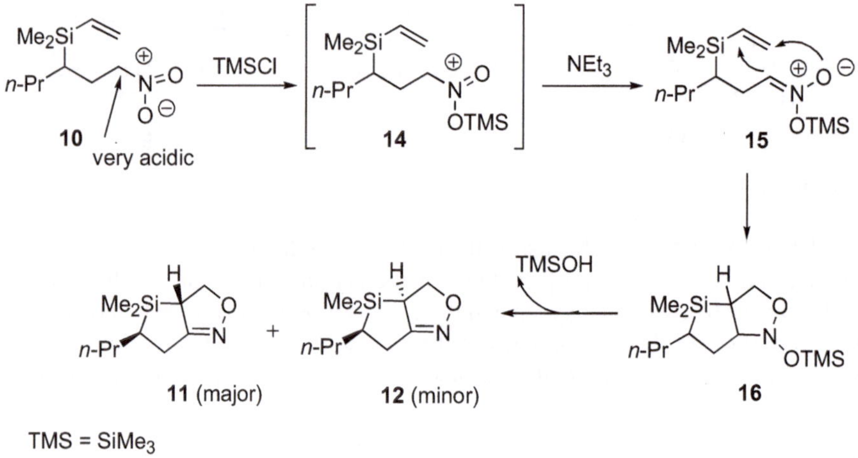

TMS = SiMe₃

Scheme 22.11

Can we explain the diastereoselectivity of the reaction?

The 4.5:1 selectivity observed for compounds **10** and **11** must arise from two transition states of different stability in the cyclization step (Scheme 22.12). Transition state **I**, precursor of the major reaction product **11** seems to be more stable than **II**, precursor of the minor isomer **12**. Very likely, **II** is less stable than **I** by unfavorable interactions due to allylic 1,3-strain involving the N-O bonds. These interactions are minimized in **I** and consequently, compound **11** is formed in preference.

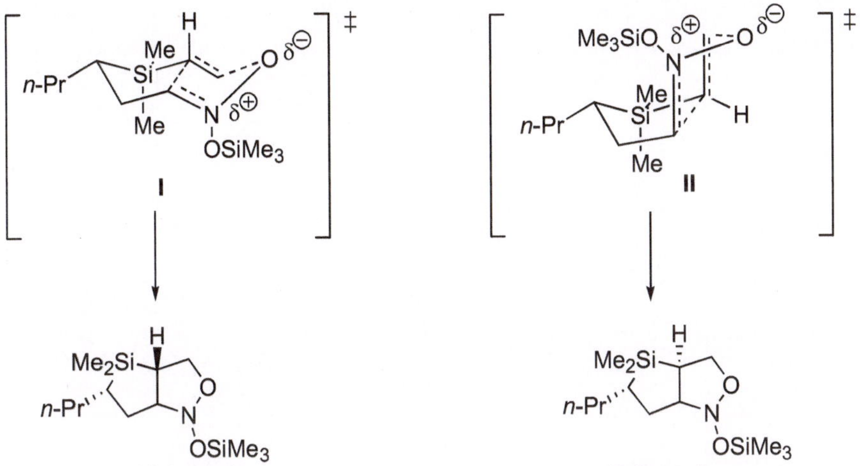

11 (major) **12** (minor)

Scheme 22.12

In Summary

Nitronates are valuable 1,3-dipoles that can be easily obtained by reaction of a nitroalkene and an olefin in the presence of a Lewis acid. Frequently, nitronates are involved in tandem [4+2]/[3+2] cycloaddion processes in which they are not necessarily isolated, but reacted with a dipolarophile either intra- or intramolecularly to yield polycyclic compounds.

Additional Comments

This problem is based on the work by Denmark SE, Moon Y-C, Senanayake CBW (1990) *J. Am. Chem. Soc.* 112:311-315 and by Young DGJ, Gómez-Bengoa E, Hoveyda AH (1999) *J. Org. Chem.* 64:692-693.

Subjects of Revision

[4+2] and [3+2] Cycloadditions. Tandem reactions. FMO theory.

Level 2 – Case 23
Hydrolysis of 2-Aminobenzoate Esters

Key point: *Intramolecular general base catalysis*

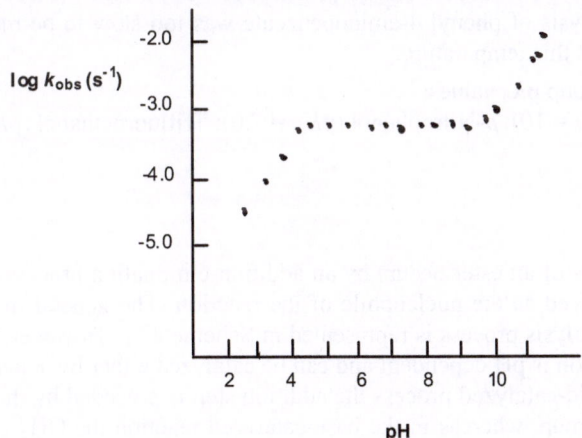

$R = C_6H_5,\ p\text{-}NO_2C_6H_4,\ CH_2CF_3$

Hydrolysis of 2-Aminobenzoate Esters

The hydrolysis of phenyl 2-aminobenzoate is pH dependent. The reaction follows a pseudo-first order rate law and the plot of log k_{obs} versus pH shows OH$^-$ catalysis at pH > 9 and a large pH independent region between pH 4 and pH 8.5. The plot bends downward near pH 4 to give a slope of 1.0. The downward bend corresponds with the pK_a of the amino group (Fig. 23.1).

Figure 23.1

To determine the mechanism of the reaction in the pH-independent region, the hydrolysis of phenyl, *p*-nitrophenyl and trifluoroethyl 2-aminobenzoates **1–3** has been studied (Scheme 23.1). The three esters hydrolyze with similar rate constants and these reactions are approximately two-fold slower in D_2O than in H_2O ($k_{H_2O}/k_{D_2O} = 2$).

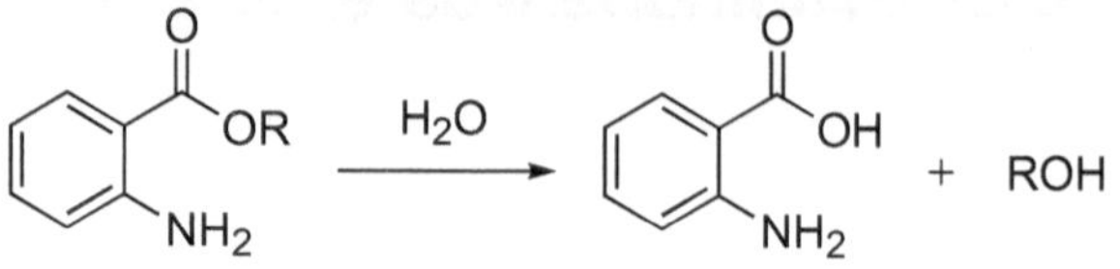

1 R = C_6H_5,
2 R = p-$NO_2C_6H_4$
3 R = CH_2CF_3

Scheme 23.1

*Considering the following experimental data propose a mechanism for the hydrolysis of 2-aminobenzoates **1–3** in the range between pH 4 and pH 8.5 (pH-independent region).*

Experimental Data

1. Rate constants for the hydrolysis of esters at 50°C in the pH-independent region (from pH 4 to 8.5):
   Phenyl benzoate ($k = 3 \times 10^{-5}$ s^{-1})
   Phenyl 2 aminobenzoate **1** ($k = 3.2 \times 10^{-4}$ s^{-1})
   p-Nitrophenyl 2-aminobenzoate **2** ($k = 2 \times 10^{-3}$ s^{-1})
   Trifluoroethyl 2-aminobenzoate **3** ($k = 4.7 \times 10^{-4}$ s^{-1})
   The hydrolysis of phenyl 4-aminobenzoate was too slow to be measured conveniently at this temperature.

2. Leaving group pK_a values:
   Phenol ($pK_a = 10$); p-Nitrophenol ($pK_a = 7.0$); Trifluoroethanol ($pK_a = 12.4$)

Discussion

The hydrolysis of an ester occurs by an addition-elimination process in which the water is involved as the nucleophile of the reaction. The general mechanism for the ester hydrolysis process is represented in Scheme 23.2. However, the ester hydrolysis reaction is pH dependent and can be catalyzed either by acids or by bases. During the acid-catalyzed process the addition step is preceded by the protonation of the C=O group, whereas in the base-catalyzed reaction the OH⁻ is the nucleophile of the reaction.

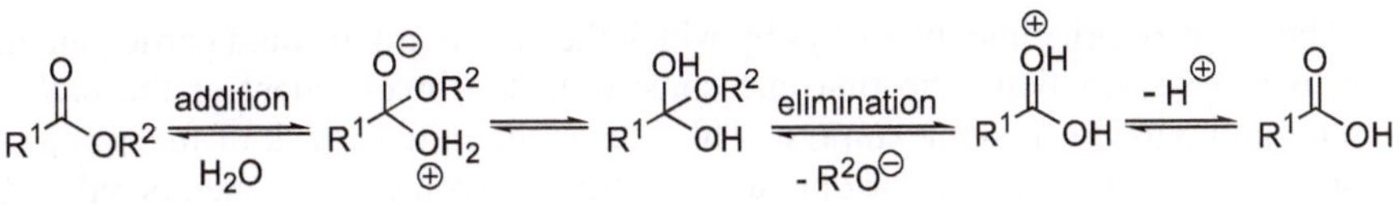

Scheme 23.2

Perhaps the most striking feature of the plot of log k_{obs} versus pH for the hydrolysis of phenyl 2-aminobenzoate **1** (Fig. 23.1) is the large pH-independent region extending from pH 4 to 8.5. At higher pH values the reaction rate increases due to the base (OH⁻) catalysis. However, near pH 4 (the pK_a of the amino group), the rate of the hydrolysis strongly decreases. This is significant since the protonation of the carbonyl group of an ester in acidic media enhances its electrophilicity, accelerating the hydrolysis rate (acid catalysis). The reduction in the reaction rate below pH 4 must be then related in some way to the protonation of the amino group.

By protonation, the amine is turned into an ammonium, an electron-withdrawing group that should enhance the ease of attack of a water molecule to the ester function (the ester C=O will become more electrophile). Then, the protonation of the amine should increase the hydrolysis rate. Nevertheless, the experimental fact is a *rate retardation* of the reaction.

The second effect of the protonation of the amino group is that the electron pair of the nitrogen is no longer free. In other words, any mechanism involving the nitrogen electron lone pair is unlikely to occur below pH 4. From these observations we can conclude that the mechanism of hydrolysis of esters **1** is not a simple nucleophilic attack of water to the carbonyl group of the ester. Apparently, it also requires a *free* amino group to occur.

The neutral amine must be playing an important role in the process.

The importance of the *free* amino group in the process also comes across when comparing the rate constants of hydrolysis of phenyl benzoate and phenyl 2-aminobenzoate **1** in the same pH 4-8.5 range. As depicted in Fig. 23.2, the hydrolysis of phenyl benzoate is considerably slower than that of **1**. As both esters only differ by the presence of the amino group in the aromatic ring, the differences in reaction rates again point to the involvement of the amino group in the reaction.

Interestingly, the hydrolysis of phenyl 2-aminobenzoate **1** is much faster than the hydrolysis of phenyl 4-amino benzoate. This suggests that to be involved in the process, the amino group must be placed at the *ortho* position of the aromatic ring.

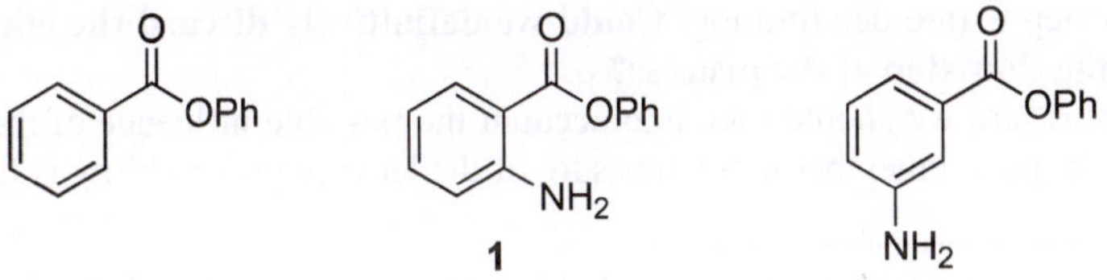

$$k = 3 \times 10^{-5} \text{ s}^{-1}$$

1

$$k = 3.2 \times 10^{-4} \text{ s}^{-1}$$

very slow

Figure 23.2

There are, in principle, two ways by which the amino group could participate in the hydrolysis reaction. The first one consists in the involvement of the *ortho*-amino group as *nucleophile catalyst*. This means the initial attack of the nitrogen atom to the carbonyl group, leading to sterically unfavorable four-membered ring intermediate **4**. Subsequent ring opening in **4** by nucleophilic attack of water yields the hydrolysis products (Scheme 23.3).

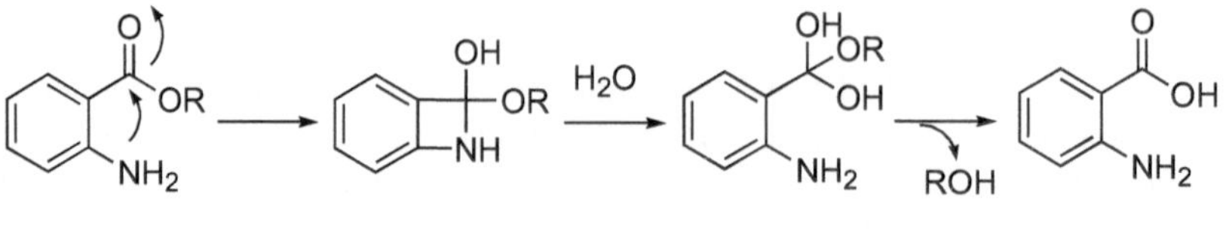

4

Scheme 23.3

Arguments against this alternative are the high strain of **4**[1] and the solvent isotope effect ($k_{H_2O}/k_{D_2O} = 2$) that clearly indicates a proton transfer involving the water molecule in the transition state of the slow step of the reaction. In the mechanism depicted in Scheme 23.3 the nucleophilic attack of H_2O (D_2O) to intermediate **4** does not require any previous or simultaneous proton transfer.

The second alternative involves the amino group as a *base catalyst* in the reaction. The nitrogen atom has an electron lone pair and there could be a hydrogen bonding interaction between the amine and the attacking water molecule. The nitrogen atom would remove a proton from the water as it attacks the carbonyl group, as indicated by transition state **5** (Scheme 23.4). This should be the slow step of the reaction, which is fully consistent with the observed solvent isotope effect ($k_{H_2O}/k_{D_2O} = 2$). The reaction could be then regarded as a case of intramolecular base catalysis by the neighboring amino group.

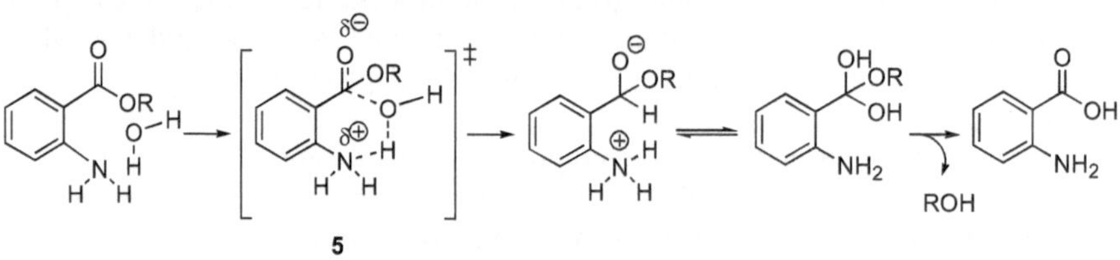

5

Scheme 23.4

All previous considerations have been made presuming that the (first) water addition step is rate-determining. **Could we definitively discard the elimination step as the slow step of the process?**

In this regard we should take into account the possible influence of the leaving group. We must remember that a measure of the leaving group ability is the pK_a of

---

[1] The energy of the hypothetical tetrahedral intermediate **4** from the density functional or ab initio calculations is much higher than that of the reactant **1** ($\Delta = 50.4$ kcal/mol). The C-N-C bond angle in **4** is 88.82°.

the conjugated acid of the group considered. The higher the pK_a value the better its leaving group ability. As the hydrolysis of an ester is an addition-elimination process we should observe some dependence of the reaction rate with the OR leaving group ability if the elimination step were rate-determining.

Esters **1-3** hydrolyze at comparable rates even when the leaving group abilities of nitrophenol ($pK_a = 7.0$), phenol ($pK_a = 10$) and trifluoroethanol ($pK_a = 12.4$) are very different. Hence, the nature of the OR group has little effect on these hydrolysis reactions, as large differences on the rate constants should be expected when passing from the better (nitrophenol) to the poorer leaving group (trifluoroethanol) (they differ by 5.4 pK_a units!). This experimental evidence is consistent with a fast elimination step.

In Summary

The hydrolysis of 2-aminobenzoate esters in the range of pH 4 to 8.5 occurs by intramolecular general base catalysis of the 2-amino group. The amine enhances the reactivity of the water by partially removing a proton as the water attacks the carbonyl group.

Additional Comments

This problem is based on the work by Fife TH, Singh R, Bembi R (2002) *J. Org. Chem.* 67:3179-3183.

Subjects of Revision

General base and nucleophile catalysis. Solvent isotope effects. Leaving group ability.

Level 2 – Case 24
Rearrangements of Cyclobutenones

Key point: *Electrocyclic ring closures and ring openings*

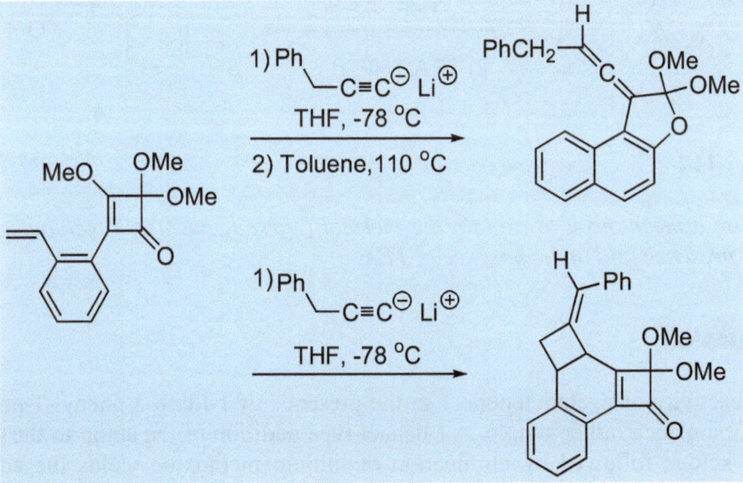

Rearrangements of Cyclobutenones

Treatment of **1** with 1-lithio-3-phenyl-1-propyne in THF at –78°C, yields 3-alky-nyl cyclobutenone **2** that after being refluxed in toluene gives naphthofuran deri-vative **3** in 45% yield (Scheme 24.1).

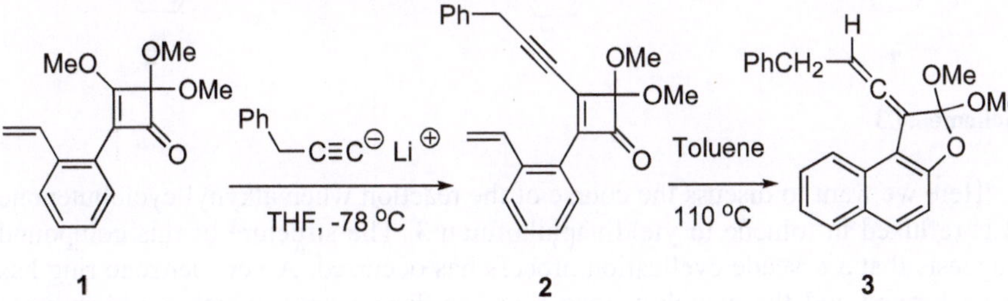

Scheme 24.1

However, treatment of cyclobutenone **1** with 1-lithio-3-phenyl-1-propyne in THF at –78°C followed by addition of triethylamine (TEA) at room temperature, produces bicyclo[4.2.0]octadienyl-fused cyclobutenone **4** in 44% yield (Scheme 24.2).

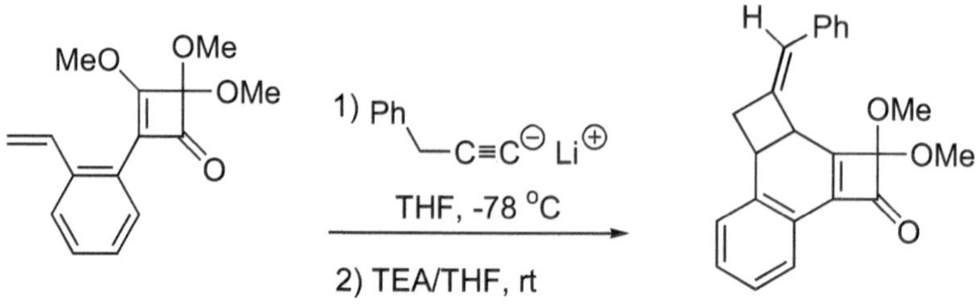

Scheme 24.2

*Propose a mechanism to explain the thermal rearrangements of cyclobutenone **1** in the presence and in the absence of TEA.*

Discussion

The reactivity of cyclobutenone **1** in the presence of 1-lithio-3-phenyl-1-propyne is at first glance rather simple. A Michael-type addition of the anion to the conjugated ketone followed by elimination of lithium methoxide yields the addition compound **2** (Scheme 24.3).

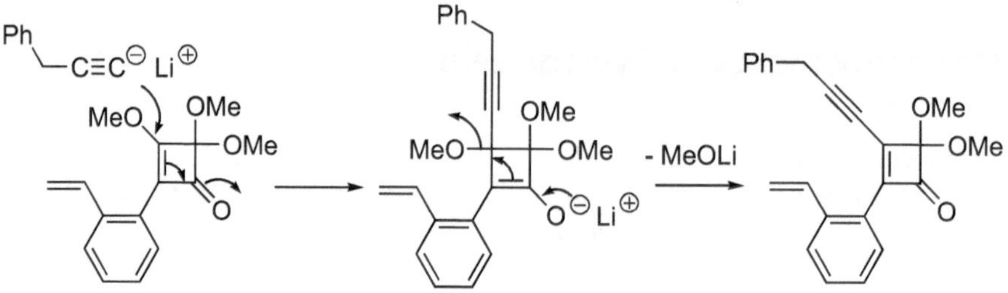

Scheme 24.3

Here we want to discuss the course of the reaction when alkynyl cyclobutenone **2** is refluxed in toluene to yield naphthofuran **3**. The structure of this compound suggests that a cascade cyclization process has occurred. A new benzene ring has been formed and the cyclobutenone ring has disappeared, which could indicate that this is the key reacting group.

The thermal ring opening of cyclobutenes to 1,3-butadienes is a well-known process. In this case, the electrocyclic 4π conrotatory ring opening of the cyclobutenone in **2** would yield vinylketene intermediate **5**, which in turn, could cyclize to naphthoquinone **6** by means of a 6π disrotatory electrocyclic ring closure. Quinone **6** tautomerizes to naphthol **7**, which by nucleophilic addition of the oxygen atom in the naphthol to the nearby conjugated enyne yields the observed reaction product **3** (Scheme 24.4)

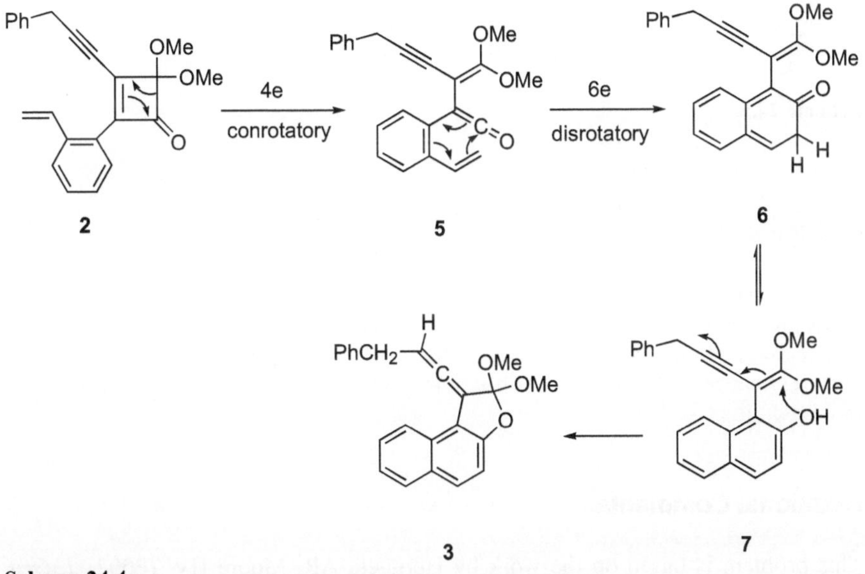

Scheme 24.4

Why is the outcome of the reaction different in the presence of TEA?

Although in this case alkynylcyclobutenone **2** is not isolated, it is reasonable to assume that this compound is already formed in the first step of the reaction. However, just having a look at the structures of products **2** and **4** makes us suspect that again a cascade cyclization process must have occurred after **2** was formed. TEA is a base that is able to remove the benzylic proton of the initially formed cyclobutenone **2**, promoting the isomerization of the alkynyl group in **2** to the corresponding allenyl derivative **8** (Scheme 24.5). The overall process will be an alkyne-allene base-catalyzed isomerization. Compound **8** has a conjugated π system nicely arranged to undergo an electrocyclic reaction. By means of an 8π conrotatory ring closure, cyclooctatetraene intermediate **9** should be formed. A new 6π disrotatory ring closure in **9** would finally provide the isolated product **4** with the expected *cis*-stereochemistry in the 6-4 ring fusion.

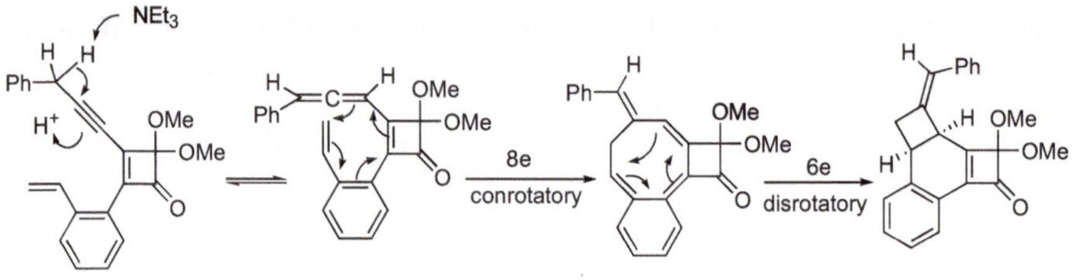

Scheme 24.5

In Summary

Thermal rearrangements of cyclobutenone **1** yield naphthofuran **3** or bicyclo-[4.2.0]octadienyl-fused cyclobutenone **4** depending on the reaction conditions. The rearrangements involve a series of cascade electrocyclic ring opening and ring closures that require the initial formation of alkynyl cyclobutenone **2** in the reaction medium.

Additional Comments

This problem is based on the work by Hergueta AR, Moore HW (2002) *J. Org. Chem.* 67:1388-1391.

Subjects of Revision

Electrocyclic ring closure and ring opening processes.

Level 2 – Case 25
Epoxi Ester-Orthoester Rearrangement

Key point: *Isotope labeling.*
Neighboring group participation

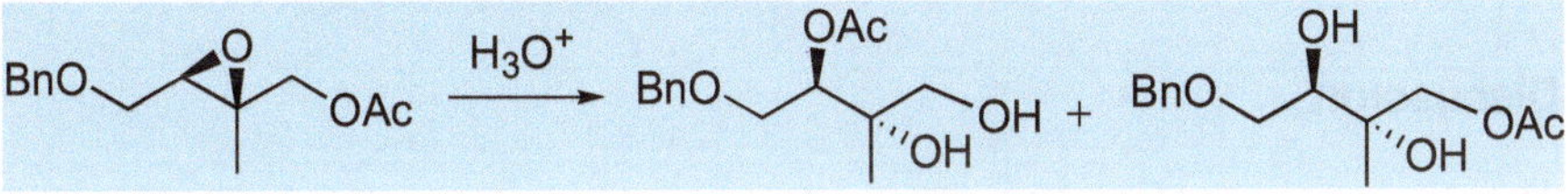

Epoxi Ester-Orthoester Rearrangement

Epoxi ester **1** effectively rearranges to a 1:1 mixture of acetates **2** and **3** under acidic conditions (0.5 M H_2SO_4 in THF/H_2O 9/1). It is believed that the reaction occurs through the formation of an orthoester intermediate **4** that is subsequently hydrolyzed in the acidic medium (Scheme 25.1). This route provides a convenient method for the synthesis of polyhydroxylated compounds with control of the stereochemistry during the epoxide ring-opening step.

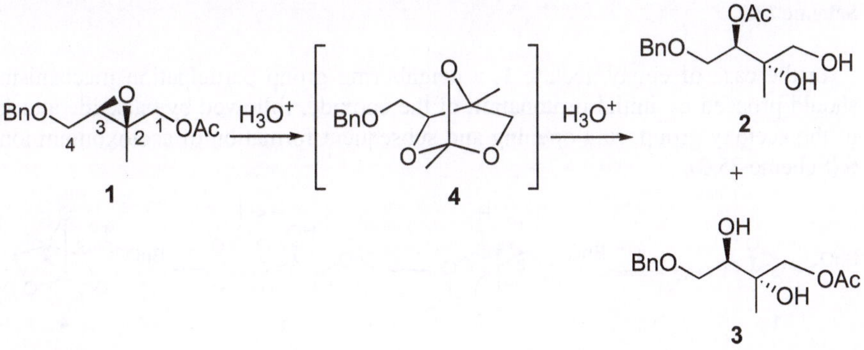

Scheme 25.1

*Considering the ability of the acetoxy substituent as neighboring group, propose a mechanism for the formation of orthoester **4**. Why do orthoesters hydrolyze so easily in acidic media? Explain the mechanism of hydrolysis.*

Experimental Data

Orthoester **4** has been characterized by NMR upon treatment of **1** with 0.5% trifluoroacetic acid (TFA)/CDCl$_3$ for 1 h. The study of the reaction mechanism was carried out in TFA/CDCl$_3$ starting from epoxy ester **1** labeled with ^{18}O in the carbonyl oxygen. The location of the label in the orthoester intermediate was determined by ^{13}C-NMR spectrometry, through the characteristic *upfield shifts* of the carbon signals when bonded to the heavy isotope. The ^{13}C NMR of labeled orthoester **4** showed upfield shifts of the signals corresponding to the orthoester carbon (120.5 ppm, $\Delta\delta$ = 25 ppb) and the C-2 position (85.5 ppm, $\Delta\delta$ = 34 ppb).

Discussion

The acetoxy group is probably one of the better-known *neighboring groups* in organic chemistry. To take part in a substitution process it must be placed next to the leaving group, assisting its departure by forming an acetoxonium ion intermediate, which generally is opened by nucleophilic attack, leading to the final substitution products (Scheme 25.2).

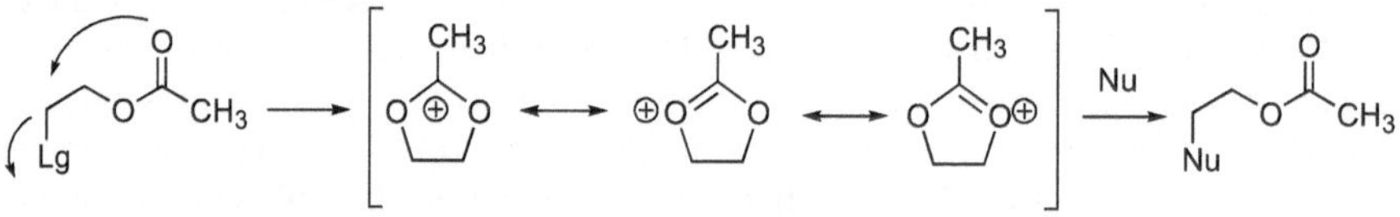

CH$_3$ CH$_3$ CH$_3$ Nu

Lg = leaving group   acetoxonium ion

Scheme 25.2

In the case of epoxy acetate **1**, a neighboring group participation mechanism should proceed by initial protonation of the epoxide, followed by backside attack of the acetoxy group, ring opening and subsequent formation of acetoxonium ion **5** (Scheme 25.3).

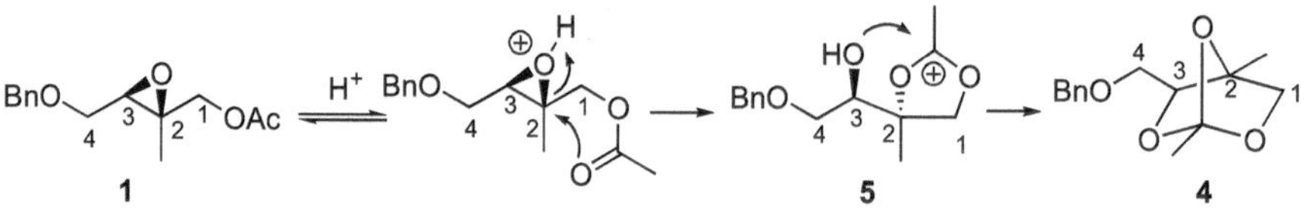

Scheme 25.3

This time the intermediate is not opened by nucleophilic attack. Instead, intramolecular quenching of cation **5** by the newly generated hydroxyl group at C-3 yields the bicyclic orthoester **4**.

The mechanism in Scheme 25.3 is supported by the results obtained in experiments carried out with ^{18}O-labeled epoxy acetate **1**. In Scheme 25.4 we have depicted the proposed pathway for the formation of the orthoester **4**-^{18}**O** starting from labeled **1** (heavy atom colored red). The black dots at the orthoester carbon and at C-2 indicate that these are the positions placed next to the heavy atom and hence, their NMR signals should be more shielded than in the unlabeled compound. This is in agreement with the experimental observations.

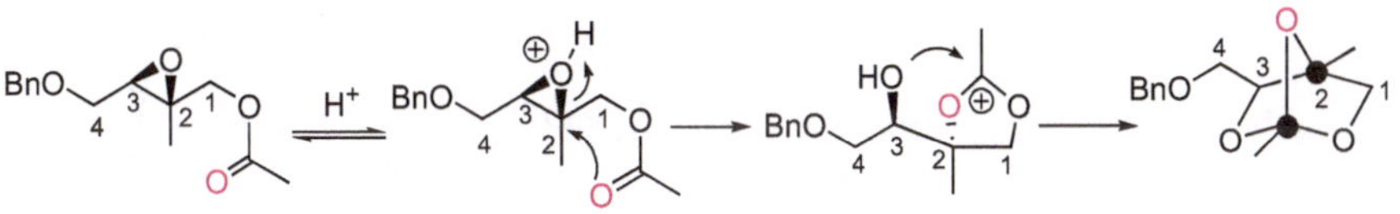

1-^{18}O 5-*exo* cyclization 5 4-^{18}O

O = ^{18}O

Scheme 25.4

Intermediate **5** in Scheme 25.4 is formed through a 5-*exo* cyclization process, *favored* following the Baldwin rules. However, we could have also considered the possibility of an alternative 6-*endo* cyclization, a competing pathway that is also allowed. Through the 6-*endo* cyclization mode, neighboring group-assisted opening of the epoxide ring in **1** should lead to dioxycarbenium ion **6**. Intramolecular quenching of the cation by the hydroxyl group at C-2 would form orthoester **7**, the enantiomer of compound **4** (Scheme 25.5).

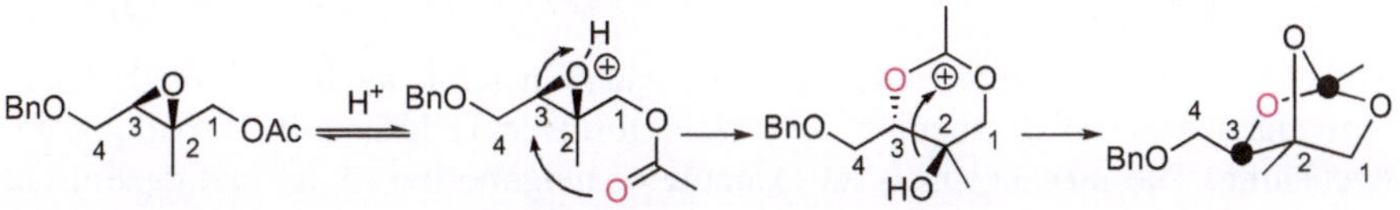

1 6-*endo* cyclization 6 7

O = ^{18}O

Scheme 25.5

Labeled **4** and **7** can be distinguished by ^{13}C NMR. Orthoester **7**, obtained through the alternative 6-*endo* cyclization mode places carbon C-3 next to the heavy atom and, in consequence, the NMR chemical shift of this carbon should be upfield (shielded) with respect to that of the unlabeled product. We should remember that the experimental data indicate that only the orthoester and the C-2 carbons modify their ^{13}C chemical shifts during the labeling experiments. As position C-3 is not affected, we can conclude that the 6-*endo* cyclization does not occur in this case.

Once we have discussed how the orthoester **4** is formed, we will go a step further considering why this compound cannot be isolated when the acid-catalyzed rearrangement of **1** is carried out in an aqueous medium.

The structure of an orthoester reminds us of that of acetals and ketals, compounds that are easily cleaved by dilute acids. The mechanism of hydrolysis for orthoester **4** should be similar to that of the acetals and ketals, starting with the protonation of one of the oxygen atoms and followed by the formation of a carbocation greatly stabilized by resonance. The overall process is a S_N1 mechanism that ends up by addition of water. Two different species **8** and **10** can be formed by protonation of orthoester **4**. Ring opening leads to cations **9** and **11** precursors of the acetates **2** and **3**, respectively, by nucleophilic attack of water (Scheme 25.6).

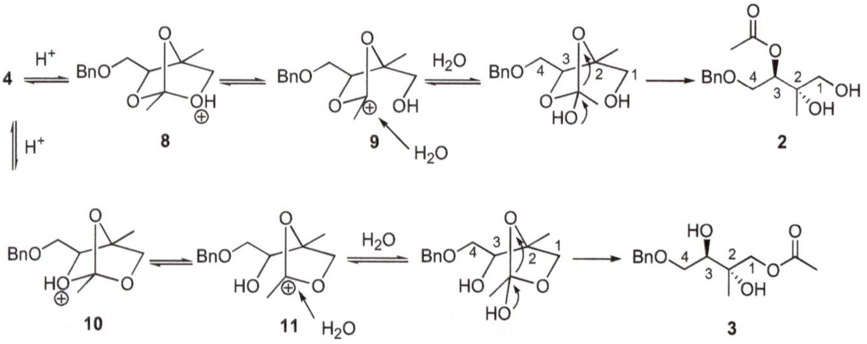

Scheme 25.6

In Summary

The epoxy ester-orthoester rearrangement occurs in acidic medium through the intermediacy of a dioxycarbenium ion, which forms an orthoester by intramolecular quenching. The mechanism is an example of neighboring group participation of the ester group.

Questions

The acid rearrangement of the epoxy ester moiety of **12** (Fig. 25.1) is the key step in the synthesis of orthoesterol B, a marine natural product with antiviral activity. Based on the mechanism proposed for the formation of **4** formulate the structure of the orthoester that should be obtained in the rearrangement of **12**. Interestingly, when diastereomer **13** was treated under the same conditions, it remained unaltered for months. Could you explain why?

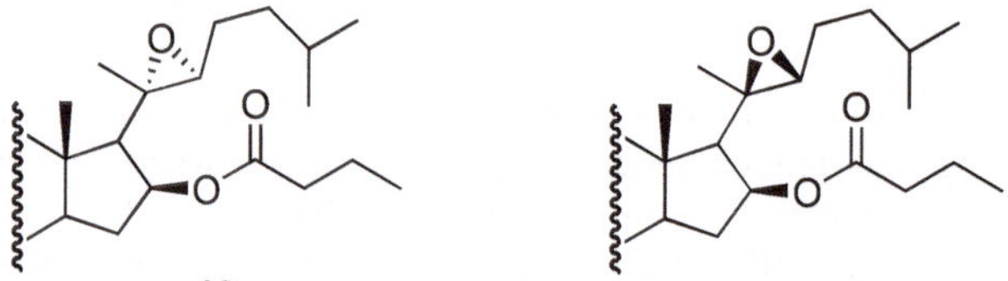

Figure 25.1

Answer to the Question

As we have discussed previously, the protonation of the oxygen in the epoxide ring in **12** would lead to **14** which cyclizes to dioxycarbenium ion **15** with inversion at the proximal center of the epoxide ring. This time the cyclization is a 6-*exo* process. Intramolecular quenching of the cation by the hydroxyl group yields the bicyclic orthoester **16** (Scheme 25.7).

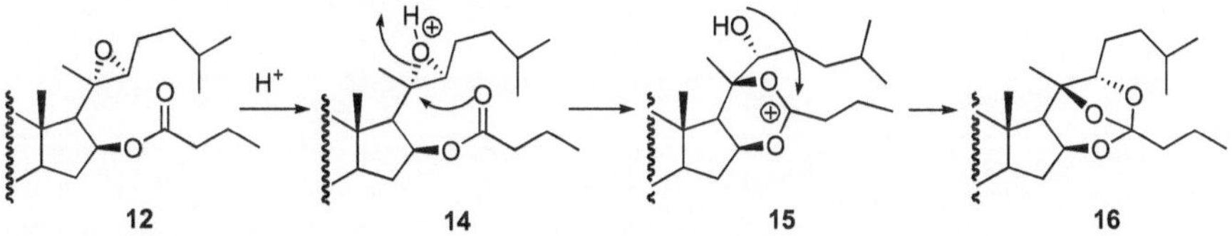

Scheme 25.7

The formation of the orthoester requires the backside attack of the C=O group to the protonated epoxide ring. Clearly this is not possible in isomeric compound **13** which has the epoxide ring directed towards the acetoxy group. Therefore, compound **13** remains unaltered under acidic conditions.

Additional Comments

This problem is based on the work by Giner JL, Ferris Jr WV, Mullins JJ (2002) *J. Org. Chem.* 67:4856-4859 and on the work by Giner JL, Faraldos JA (2002) *J. Org. Chem.* 67:2717-2720.

Subjects of Revision

Neigboring group participation. Hydrolysis of acetals. ^{13}C NMR in labeling experiments. Baldwin rules.

Level 2 – Case 26
2-Chloro-1,3,5-triazines as Activating Groups of Carboxylic Acids in the Formation of Peptide Bonds

Key point: *Kinetic isotope effects.*
Concerted versus stepwise mechanism

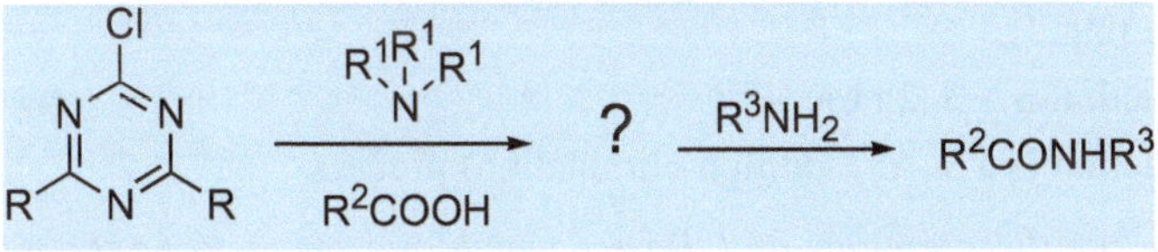

2-Chloro-1,3,5-triazines as Activating Groups of Carboxylic Acids in the Formation of Peptide Bonds

2-Chloro-4,6-disubstituted-1,3,5-triazines **1** have been widely used in the activation of carboxylic acids in the formation of peptide bonds. The reaction requires the presence of a tertiary amine in the reaction medium and the participation of quaternary triazinyl ammonium salts **2** as intermediates of the reaction has been demonstrated. The role of these species in the activation of the carboxylic acid involves substitution of the amine-leaving group in **2** by the carboxylate ion, to afford triazine *super active esters* **3**, the reagents employed in the peptide bond formation step (Scheme 26.1).

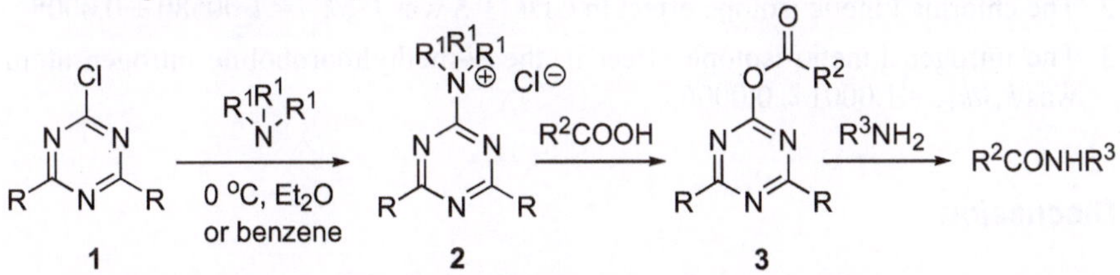

Scheme 26.1

The common trend of intermediates **2** is to decompose rapidly at room temperature. Nevertheless, triazinyl ammonium derivative **4**, obtained from 2-chloro-4,6-dimethoxy-1,3,5-triazine **5** (CDMT) and *N*-methylmorpholine **6**, resulted to be sufficiently stable to be studied (Scheme 26.2).

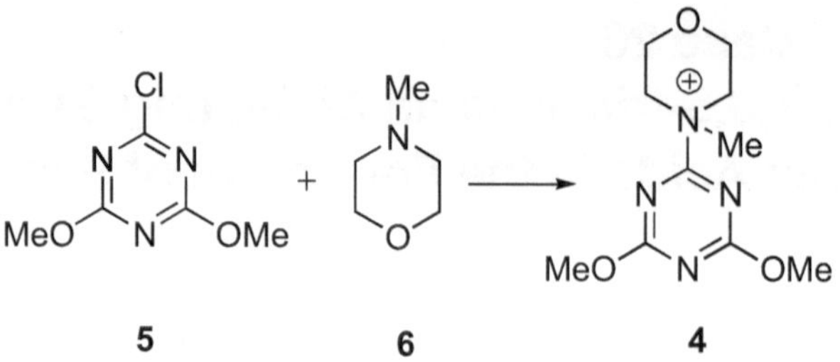

Scheme 26.2

Three alternative pathways have been suggested to explain the mechanism of formation of the intermediate morpholinium salt **4**:

1. The mechanism proceeds through the initial formation of a carbocation in the triazine ring.
2. The reaction is a S_N2 process.
3. The reaction is a S_NAr (addition-elimination) process.

Formulate the three options and discuss which one seems to be the more likely, based on the following experimental data.

Experimental Data

1. A study of the activation of several acids by means of CDMT **5** in the presence of tertiary amines, has revealed that the ability of tertiary amines to promote the activation does not correlate with their basicity. On the contrary, the reaction rate, decreases dramatically with the increase of steric hindrance of amine substituents. For example, triethyl amine (pK_a = 10.87), tributylamine (pK_a = 10.63) and N,N-dimethyl aniline (pK_a = 5.06) are inactive amines, whereas trimethyl amine (pK_a = 9.81), 4-(N,N-dimethylamino)pyridine (DMAP) (pK_a = 9.61) or N-methylmorpholine (pK_a = 7.42) are reactive amines.

2. The chlorine kinetic isotope effect in CDMT **5** was k_{35}/k_{37} = 1.00580 ± 0.0005.

3. The nitrogen kinetic isotope effect in the N-methylmorpholine nitrogen atom was k_{14}/k_{15} = 1.0001 ± 0.0006.

Discussion

First we will write the three possible alternative mechanisms proposed for the reaction.

Mechanism 1 involves the formation of a triazine carbocation **7** by departure of the chlorine atom in the first step. The ammonium salt **4** is obtained by nucleophilic attack of the morpholine nitrogen atom to the cation intermediate **7** (Scheme 26.3).

MECHANISM 1: Carbocation mechanism

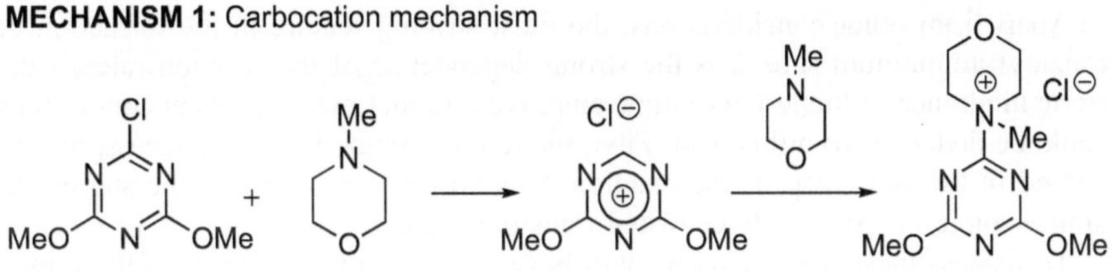

Scheme 26.3

Mechanism 2 in turn, proposes that the triazinyl ammonium salt **4** is formed in one step (as in a S_N2 process), by simultaneous rupture of the triazine C-Cl bond and formation of the triazine-morpholine bond, as represented in the transition state **8** (Scheme 26.4).

MECHANISM 2: S_N2 type

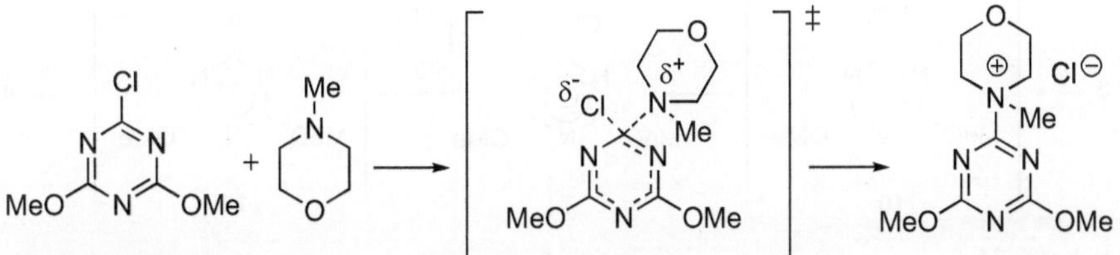

Scheme 26.4

Finally, Mechanism 3 proposes the classic stepwise S_NAr process, involving the reversible **addition** of the nucleophile (*N*-methyl morpholine) to the C2 position in the triazine ring to yield the intermediate **9**, and the subsequent departure of the chlorine atom in **9** during the **elimination** step (Scheme 26.5).

MECHANISM 3: Addition-Elimination mechanism (S_NAr)

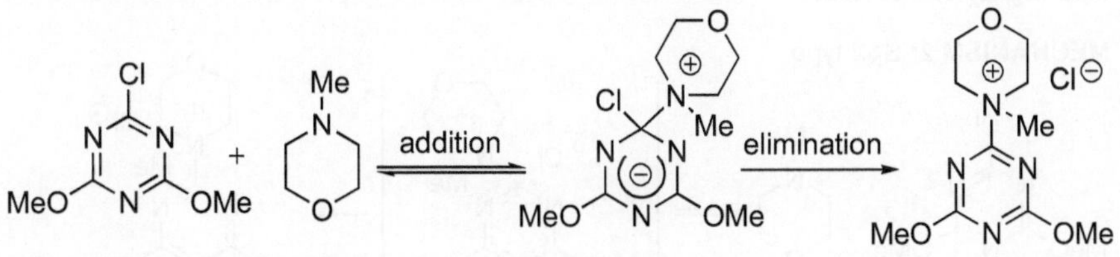

Scheme 26.5

Apart from other considerations, the most striking feature in the formation of triazinyl ammonium salts **2** is the strong dependence of the reaction rate on the steric hindrance of the tertiary amine employed. In fact, two important conclusions could be deduced from this data: First, the tertiary amine has to be necessarily involved in the slow step of the reaction. Second, the sensitivity to the steric hindrance suggests a **sterically crowded** transition state.

To discuss these arguments we will have a look at the structures of the transition states for the three mechanisms previously proposed. Mechanism 1 involves two steps and in consequence two transition states (**10** and **11**) should be considered, but none of them is particularly crowded. The first **10**, involves the C2-Cl bond breaking of triazine **5** (dissociative process) leading to triazine cation **7**. The amine is involved in the second step, consisting on the formation of a new bond between an unsubstituted position of the triazine ring (C2) and the morpholine nitrogen atom (Scheme 26.6).

MECHANISM 1: Carbocation mechanism

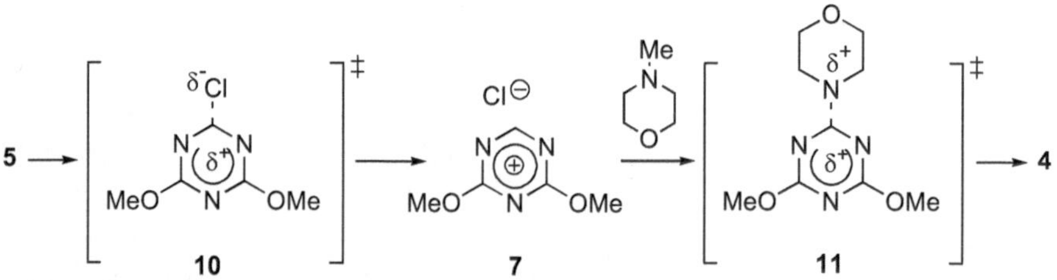

Scheme 26.6

The situation is very different for Mechanisms 2 and 3. The S_N2-like process proposed in Mechanism 2 requires the simultaneous bond breaking-bond formation represented in the crowded transition state **8**. The result is an increase of the steric repulsion due to the quaternization of the nitrogen atom of the morpholine as well as an increase of the steric hindrance caused by the hybridization change from sp^2 to sp^3 on the C2 position of the triazine ring (Scheme 26.7a, b). Similar arguments could be applied to the stepwise Mechanism 3 that involves transition states **12** and **13**. Both are structurally related to **8** and in consequence, they are also highly congested.

MECHANISM 2: S_N2 type

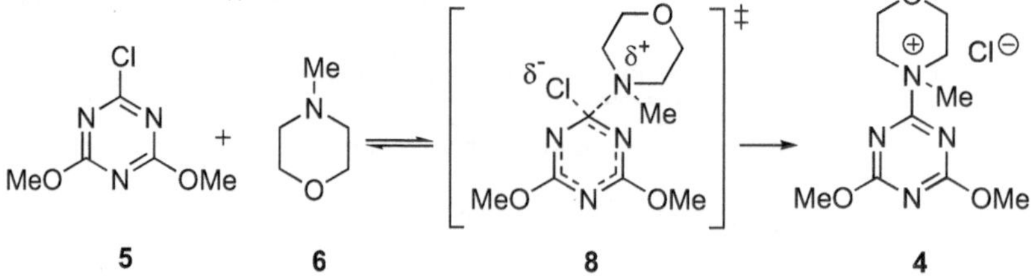

Scheme 26.7a

MECHANISM 3: Addition-Elimination mechanism (S_NAr)

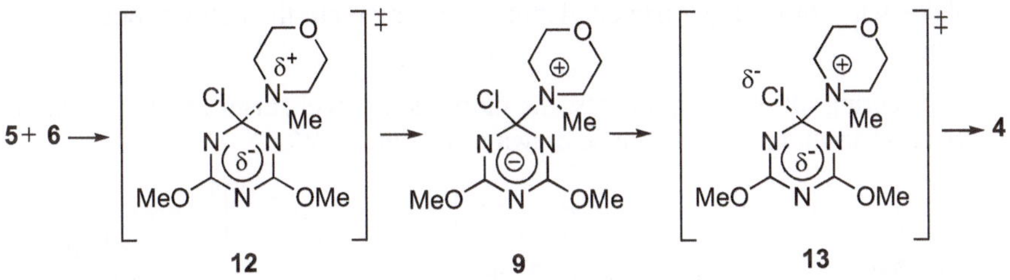

Scheme 26.7b

If we consider the three possible reaction mechanisms, only Mechanisms 2 and 3 propose sterically crowded transition states and, in consequence, Mechanism 1 should be discarded.

The study of the heavy atom kinetic isotope effects in the reaction between labeled CDMT and N-methylmorpholine could help us to decide between the remaining options. A significant chlorine KIE (k_{35}/k_{37} = 1.00580 ± 0.0005) has been observed when Cl-labeled CDMT was employed. However, during the reaction with N-labeled N-methylmorpholine, no KIE (k_{14}/k_{15} = 1.0001 ± 0.0006) was detected. We should remember that heavy atom effects are observed when bonds involving the heavy atom are broken or made **in the transition state of the rate-determining step of the reaction**. In this case, these results could be explained on the basis of a mechanism in which the C-Cl bond breakage occurs during the transition state of the slow step, but neither bond-breaking nor bond-forming processes involving a nitrogen atom are taking place.

The alternative proposed by Mechanism 2 (S_N2 type) suggests a transition state **8** in which the C-Cl bond breakage and the C-N morpholine bond formation occur simultaneously. Under these circumstances the nitrogen heavy atom isotope effect should have been also observed, which is not the case.

In consequence, based on the absence of nitrogen heavy atom KIE, Mechanism 2 must be discarded.

There is a single remaining alternative: the stepwise mechanism proposed as Mechanism 3. Let us check if it is in agreement with all the experimental data.

1. *The C-Cl bond must be broken during the slow step of the reaction.* This experimental observation implies that the elimination step is rate determining (see transition state **13** in Scheme 26.7).

2. *The C-N bond is not formed in the slow step of the reaction*, as it has been stated by the absence of nitrogen heavy atom KIE.

3. *The reaction rate is not dependent on the basicity of the tertiary amine.* In fact, since the C-N bond is formed during the first (fast) step of the reaction, the basicity (nucleophilicity) of the amine is irrelevant.

4. *The transition state of the rate-determining step is sterically crowded* and hence very sensitive to the steric hindrance of the substituents in the amine. This explains why for example trimethylamine is reactive but the bulky tributyl amine is not.

Considering all these arguments, the more likely mechanism for the formation of triazinyl ammonium salts can be drawn as in Scheme 26.8.[1]

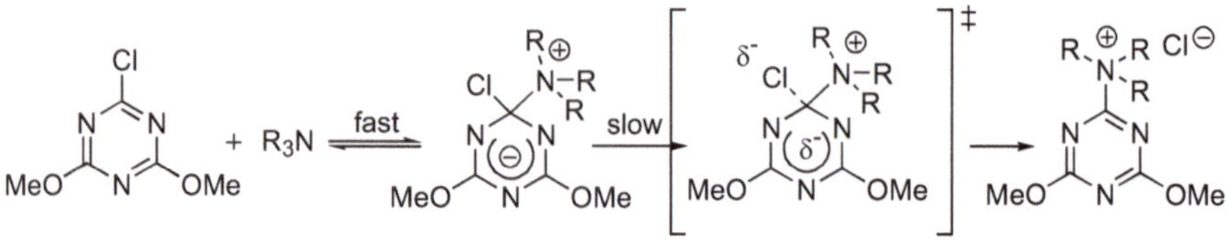

5

Scheme 26.8

In Summary

The activation of carboxylic acids by means of 2-chloro-1,3,5-triazines **1** proceeds via triazinyl ammonium salts **2** that are formed in situ in the presence of the appropriate tertiary amine. The formation of salts **2** occurs by an addition-elimination S_NAr mechanism and the reaction rates strongly depend on the steric hindrance of the substituents in the amine employed.

Questions

Examination of the mixture of compounds yielded by the reaction of CDMT **5** and *N*-methylmorpholine **6** in a chloroform solution at low temperature, revealed the presence of a new species. The plot of concentrations versus time is represented in Fig. 26.1.

Comment the pattern of the plot represented in Fig. 26.1. Plot **a** (red), *N*-methyl morpholine **6**; Plot **b** (black) reaction intermediate **9**; Plot **c** (blue) ammonium salt **4**.

---

[1] All attempts to isolate intermediate **9** were unsuccessful because of its rapid decomposition to CDMT at temperatures exceeding 5°C. However, it was possible to cumulate intermediate **9** to reach 50% of the concentration of the starting material. The structure of **9** was determined by a COSY experiment at –30 to –50°C.

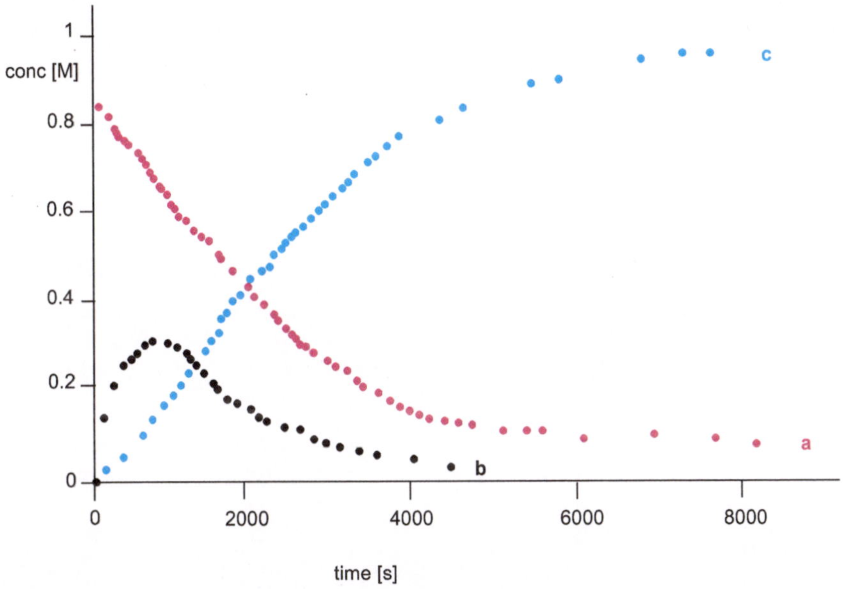

Figure 26.1

Answer to the Question

The concentration-time relationship illustrated in Fig. 26.1 is characteristic for a consecutive reaction that involves the participation of an intermediate and further confirms the mechanism in two subsequent reaction steps proposed in Scheme 26.8. As the reaction proceeds, concentration of N-methylmorpholine decreases (plot a) as concentration of intermediate **9** increases (plot b), reaches a maximum and finally decreases again. The shape of the curve for the ammonium salt **4** (plot c) is characteristic of this type of complex reaction. During an initial induction period the rate of formation of **4** is low, then increases and there is a period in which the rate of formation of **4** is approximately linear and finally, it slows down again and tends to its asymptotic value (concentration at the complexion of the reaction, $t = \infty$).

Additional Comments

This problem is based on the work by Kaminski ZJ, Paneth P, Rudzinski J (1998) *J. Org. Chem.* 63:4248-4255.

Subjects of Revision

Heavy atom kinetic isotopic effects. Kinetics of consecutive reactions. Nucleophilic substitution. S_NAr mechanism.

Level 2 – Case 27
Acid-Catalyzed Isomerization of Imines

Key point: *Nucleophile catalysis versus bond rotation*

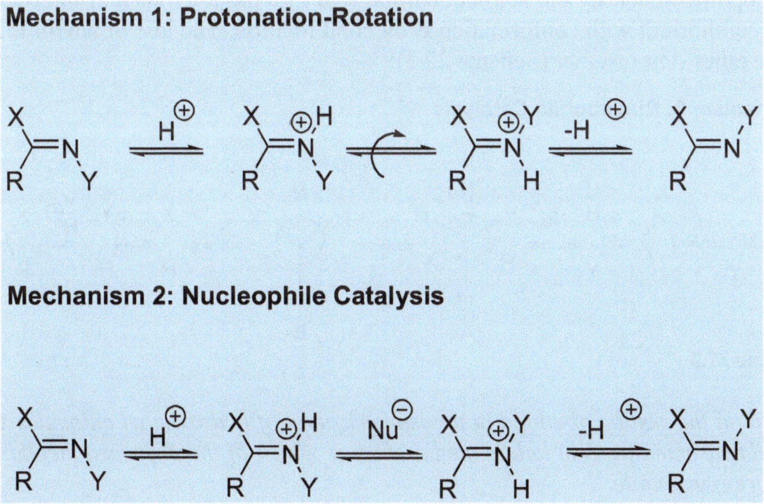

Mechanism 1: Protonation-Rotation

Mechanism 2: Nucleophile Catalysis

Acid-Catalyzed Isomerization of Imines

Despite the fact that the acid-catalyzed *Z/E* isomerization of compounds containing the C=N bond is a well-known process, the pathway for most of the isomerizations that appear in the literature is not clear (Scheme 27.1).

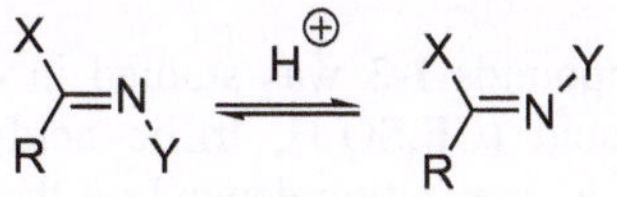

Scheme 27.1

In simple imines there are two reasonable mechanisms to explaining the *Z/E* isomerization. **Mechanism 1** (protonation-rotation) consists on the protonation of the nitrogen atom followed by rotation about the C-N bond. Since protonation could decrease the C=N bond order, the rotation around the C-N bond axis seems reasonable (Scheme 27.2).

Mechanism 1: Protonation-Rotation

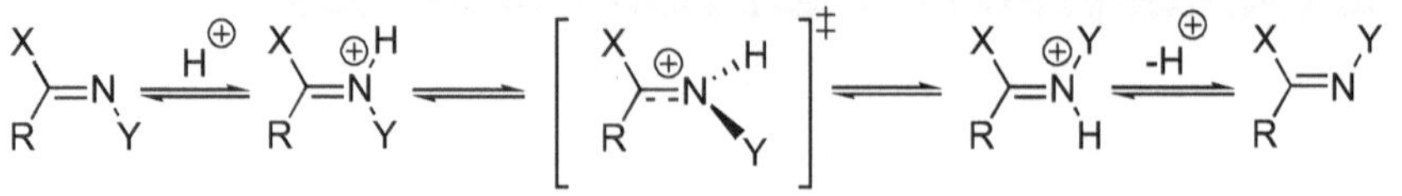

Scheme 27.2

In **Mechanism 2** (nucleophile catalysis), the protonated C=N bond undergoes nucleophilic attack by the acid counterion (Nu⁻) giving a tetrahedral intermediate **A** in equilibrium with conformation **B** by bond rotation. The loss of Nu⁻ in **B** leads to the other stereoisomer (Scheme 27.3).

Mechanism 2: Nucleophile Catalysis

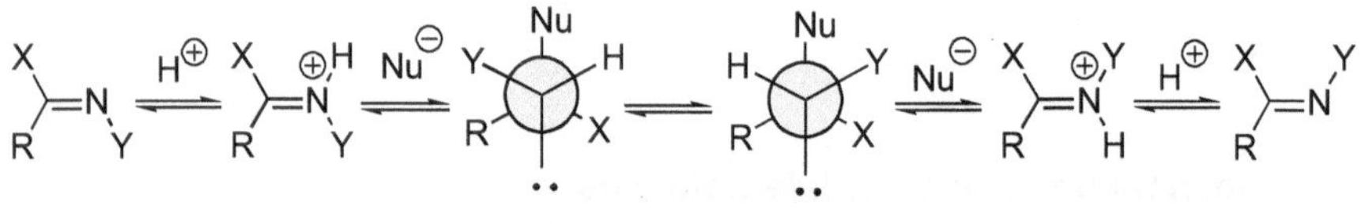

Scheme 27.3

Based on the results obtained in the isomerization of hydroximoyl chlorides 1 and 2 and hydroximates 3 in acidic media, discuss which of the proposed mechanisms is more reasonable.

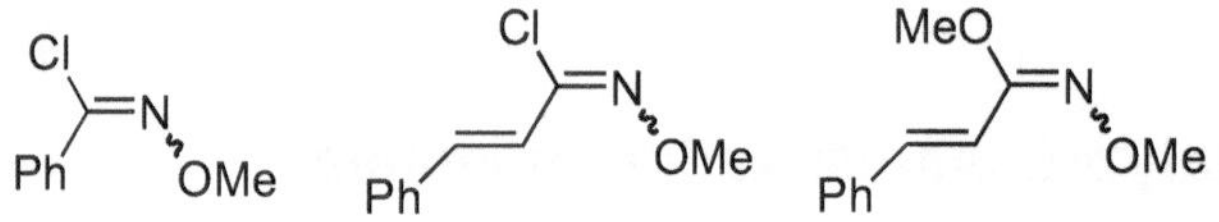

Experimental Data

The isomerization of compounds **1-3** was studied in dioxane solutions of HCl, trifluoromethanesulfonic acid (CF_3SO_3H, triflic acid) or tetrafluoroboric acid (HBF_4). Since the rates of isomerization depend on the concentration of the acid, all the measurements were carried out at the same concentration.

1. The pK_a values of the acids used are: CF_3SO_3H, (–13), HCl (–7), and HBF_4 (0.05). Although the values of pK_a are referred to water, it is reasonable to assume that triflic acid is also stronger than HCl in dioxane.

2. Hydroximoyl chlorides **1** and **2** isomerized exclusively in a HCl dioxane solution, whereas the isomerization of hydroximate **3** took place in dioxane solutions of all three acids.

Table 27. 1

kcal/mol	**4**	**5**	**6**
Z-isomer	47.2	30.3	20.4
E-isomer	47.6	31.9	21.5

3. The rates of isomerization for **3** followed the order of acidity of the three acids (triflic acid > HCl > HBF$_4$).

4. The rates of E/Z isomerization of **1** were compared to the rate of incorporation of radioactive chloride ion ([36]Cl$^-$) into the product during the isomerization. It was found that [36]Cl was incorporated into the hydroxymoyl chloride at a rate equal to one half the rate of isomerization.

5. Energies (kcal/mol) for rotation about the C=N bond of iminium ions **4-6** (E and Z isomers) were calculated using the Gaussian 94 and 98 series of programs (HF level, 6-31+G* basis set) and are collected in Table 27.1.

Discussion

The different behavior of compounds **1-2** (having a Cl substituent in the C=N bond) and **3** (having a OMe substituent in the C=N bond) under the same isomerization conditions (acid in dioxane) is possibly the most remarkable feature that arises from the analysis of the experimental data. The fact that hydroximoyl chlorides **1** and **2** isomerize exclusively with an acid (HCl) having a *nucleophilic counterion* (Cl$^-$) could indicate that the isomerization follows the **nucleophile catalysis pathway** in these cases. On the other hand, hydroximates **3** are capable of isomerizing in acids with *non-nucleophilic counterions*, as triflic acid (CF$_3$SO$_3^-$) and tetrafluoroboric acid (BF$_4^-$). This could suggest an isomerization through the **protonation-rotation pathway**. Nevertheless, all these preliminary conclusions are pointless if the rest of the experimental data does not support them. Next we will consider the two possibilities in detail.

Hydroximoyl chlorides 1 and 2. Do they really isomerize through a nucleophile catalysis mechanism?

An important data in favor of the nucleophile catalysis mechanism is the result obtained for **1** in the presence of radioactive chloride ion ([36]Cl$^-$). It was found that the label was incorporated into the hydroximoyl chloride at a rate equal of one half ot the isomerization rate, as indicated in Eq. 27.1.

$$v = 1/2 \; v_{isom} \qquad\qquad (27.1)$$

That is, the isomerization takes place through an intermediate that looses labeled and unlabeled Cl⁻ with equal probability. The nucleophile catalysis of **Mechanism 2** proposes a tetrahedral intermediate **7** that would justify this experimental result. The mechanism has been drawn starting with a mixture of E/Z imines **1**. After losing unlabeled chlorine, ion **7** evolves to the labeled imines **8** (as E/Z mixture) whereas the loss of labeled chlorine yields the starting imines **1** (Scheme 27.4, labeled atoms colored red). We should remark that if Mechanism 1 (rotation-isomerization) were operative in this case, the labeled chlorine would have never been incorporated in the products.

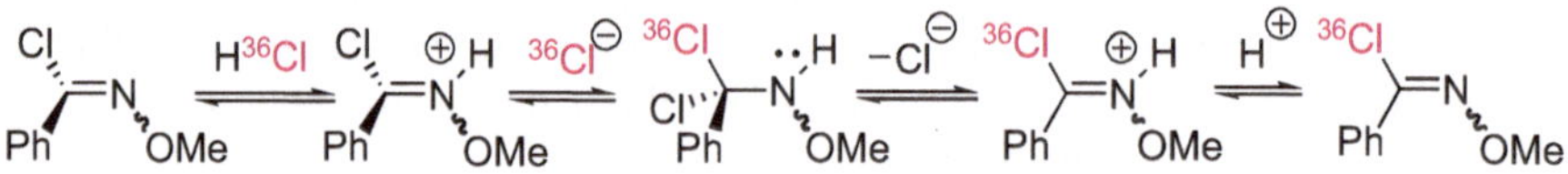

1   **7**   **8**

Scheme 27.4

Hydroximates 3. Do they isomerize through the protonation-rotation mechanism?

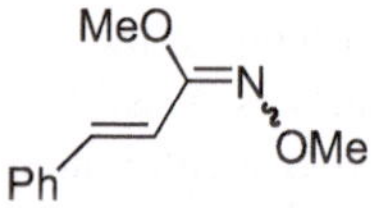

3

Compounds **3** isomerize in both nucleophilic and non-nucleophilic media and the isomerization rate increases with the strength of the acid employed. Both are solid arguments for a mechanism involving the rotation around a protonated iminium C=N bond.

The comparative study of the calculated energies for iminium C=N bonds could help us to rationalize all the previous considerations. The results indicated in Table 27.1 are referred to protonated hydroximoyl chlorides **4-5** and hydroximates **6**, all structurally referable to compounds **1-3** but in which a hydrogen atom has replaced the phenyl group. In all cases the energies have been calculated for both isomers **E** and **Z**, although the values are very similar in both cases. In Fig. 27.1 we have represented the average energy values for each case.

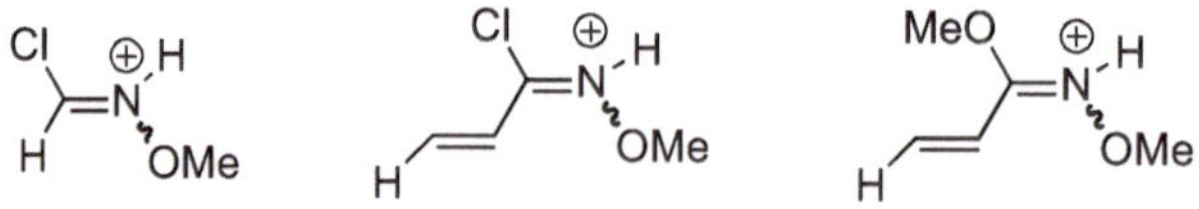

4   **5**   **6**

47.4 kcal/mol   31.1 kcal/mol   20.9 kcal/mol

Figure 27.1

The calculations show that protonated hydroximoyl chlorides **4** and **5** have a rotation barrier considerably higher than the corresponding hydroximates **6** (differences of about 17 and 10 kcal mol^{-1}, respectively). A higher energy rotation barrier in the iminium intermediate should disable the protonation-rotation pathway, favoring the alternative nucleophilic attack. This is in agreement with the nucleophile catalysis mechanism proposed previously for the *E/Z* isomerization of hydroximoyl chlorides **1** and **2** (Scheme 27.5).

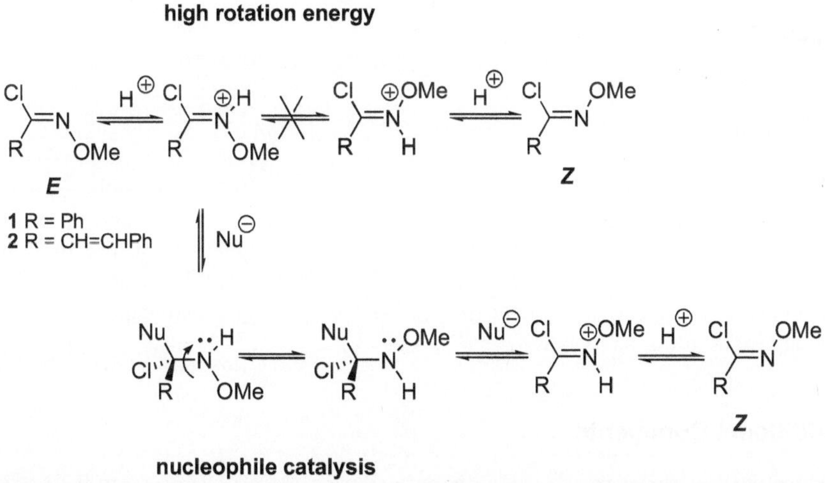

high rotation energy

E

1 R = Ph
2 R = CH=CHPh

Z

nucleophile catalysis

Scheme 27.5

The addition of substituents that stabilize the positive charge should lower the barrier to rotation in iminium ions. This is due to the contribution of the canonical form in which the positive charge is placed on the carbon atom (**D** in Fig. 27.2).

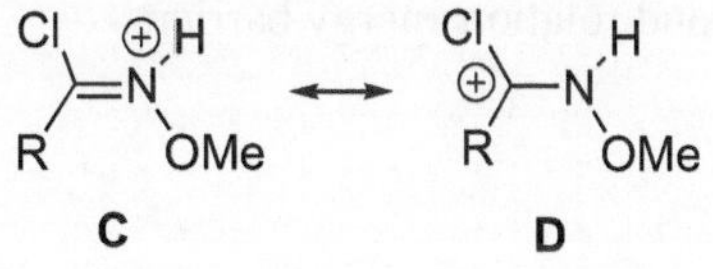

C   **D**

Figure 27.2

In fact, protonated hydroximates **6**, in which the Cl atom has been replaced by a MeO group, show considerably lower rotational barriers than their hydroximoyl chloride counterparts **4** and **5**. The lowest value corresponds to **6**, in which adding a vinyl group has increased the conjugation. Figure 27.3 represents the delocalization of the positive charge by conjugation in protonated hydroximates **3H$^+$**. The delocalization of the positive charge in **3H$^+$** has two effects in favor of the protonation-rotation pathway: a) lowers the iminium rotation energy barrier; b) complicates the attack of the nucleophile, as the magnitude of the positive charge in the C-N bond is diminished by conjugation. The result is that compounds **3** only isomerize by the protonation-rotation mechanism.

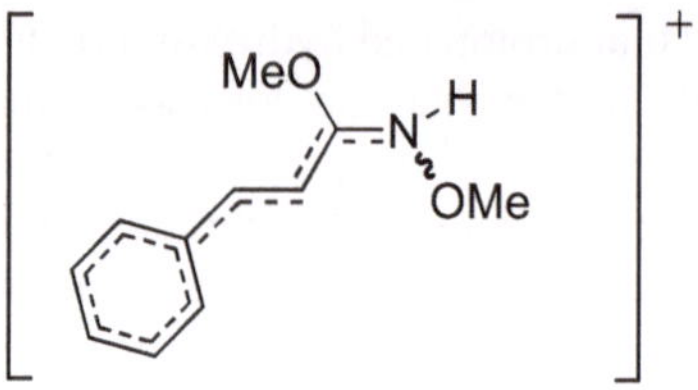

3H$^+$

Figure 27.3

In Summary

The *E/Z* isomerization of C=N bonds can occur either by a nucleophile catalysis mechanism or by iminium ion rotation. The electrophilicity and the barrier to rotation of the iminium ion formed during the first step of the reaction determine the isomerization pathway. Thus, hydroximoyl chlorides **1** and **2** isomerize only by the nucleophile catalysis mechanism whereas hydroximates **3** are capable of isomerizing by iminium ion rotation.

Additional Comments

This problem is based on the work by Johnson JE, Morales NM, Gorczyca AM, Dolliver DD, McAllister MA (2001) *J. Org. Chem.* 66:7979-7985.

Subjects of Revision

Nucleophile catalysis. Bond rotation energy barriers.

Level 2 – Case 28
A Dearomatizing Disrotatory Electrocyclic Ring Closure

Key point: *Electrocyclic reactions*

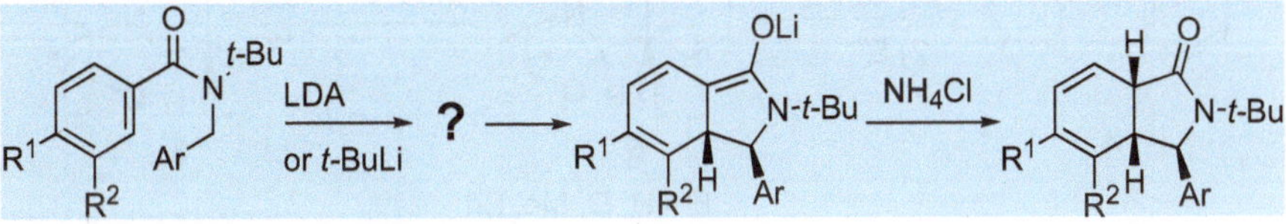

A Dearomatizing Disrotatory Electrocyclic Ring Closure

Aromatic amides like **1** (both benzamides and naphthamides) can be dearomatized to yield bi- and polycyclic amides **2** in a stereoselective cyclization reaction triggered by a benzylic lithiation α to the amide nitrogen to form organolithium intermediate **3**. The proposed mechanism of the reaction consists of the intramolecular conjugate addition of the benzylic anionic center in **3** into the electron-deficient *ortho* position of the aromatic ring (Scheme 28.1). In most cases, the addition of 1,3-dimethyltetrahydro-2(1*H*)-pyrimidinone (DMPU) to the reaction medium is required to promote the cyclization step. Considering the proposed mechanism, the high stereoselectivity observed in the cyclization is truly *remarkable*.

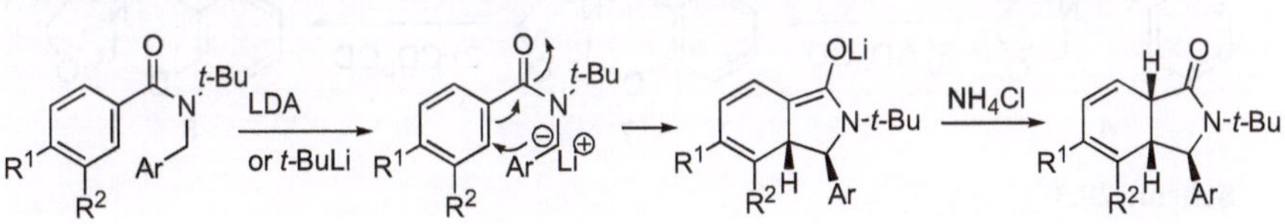

R^1 = H, OMe; R^2 = H

R^1, R^2 = benzo

Scheme 28.1

Explain the exquisite control of the stereochemistry of the reaction products with the help of the following experimental data?

Experimental Data

1. Lithiation of *N*-benzoyl oxazolidine **4** with either *t*-BuLi or *s*-BuLi in the presence of DMPU, at –78°C, followed by quenching with an electrophile, yielded tricyclic products **5** as single diastereomers.[1] On stirring **5** with acid (2M HCl in ether) a clean epimerization takes place to yield compounds **6** (Scheme 28.2).

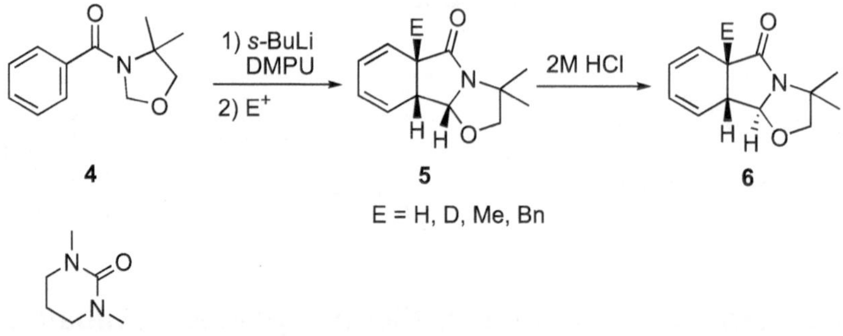

E = H, D, Me, Bn

DMPU

Scheme 28.2

2. Lithiation of **4** in absence of DMPU, followed by *rapid* quenching (after 30 min) at –78°C with CD_3OD only yielded *ortho*-deuterated product **7** (95%). By repeating this sequence, both *ortho* protons could be replaced by deuterium to give **8** (Scheme 28.3).

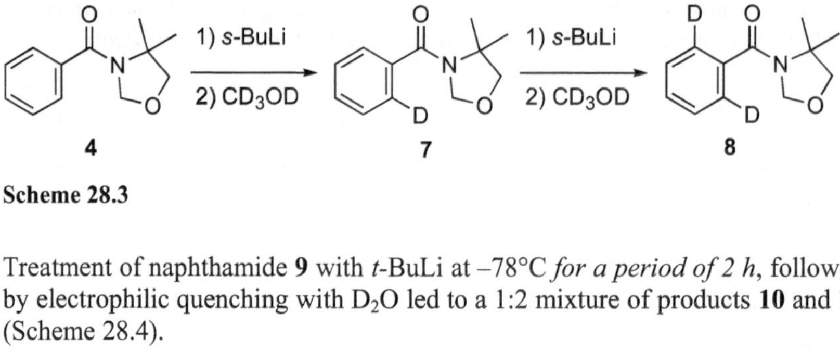

Scheme 28.3

3. Treatment of naphthamide **9** with *t*-BuLi at –78°C *for a period of 2 h*, followed by electrophilic quenching with D_2O led to a 1:2 mixture of products **10** and **11** (Scheme 28.4).

---

[1] The evidence of the stereochemistry of compounds **5** was gained from X-ray crystallography and nuclear Overhauser effect (NOE) studies.

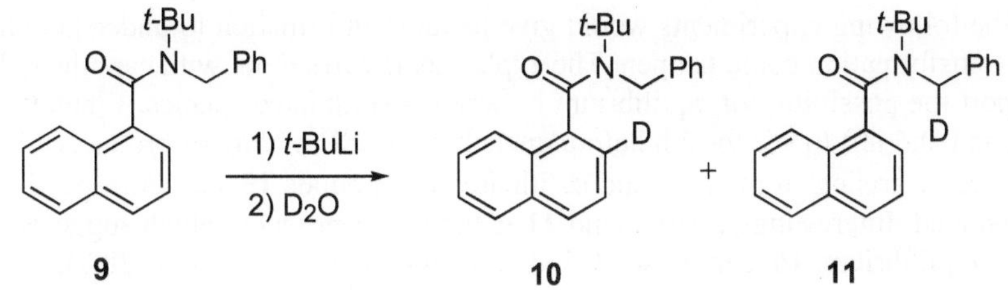

Scheme 28.4

When the reaction was performed under the same conditions (*t*-BuLi at –78°C) but followed by a *rapid* quenching with D_2O, only compound **10** was isolated.

4. The reaction of **9** with *t*-BuLi at –78°C *for 16 h*, in the presence of DMPU, followed by electrophilic quenching with D_2O yielded exclusively bicyclic amide **12** (Scheme 28.5).

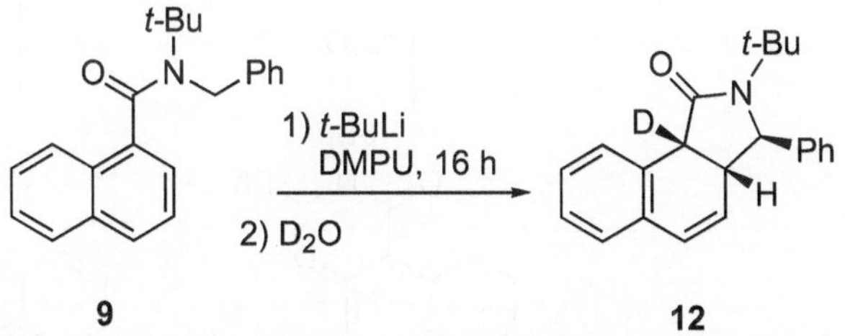

Scheme 28.5

Discussion

To start the discussion, we will try to explain the experimental data on the basis of the anionic cyclization mechanism proposed in Scheme 28.1. First, we will consider the lithiation step. Following the proposed mechanism, lithiation of *N*-benzoyl oxazolidine **4** should take place at the α position to the amide nitrogen. However, the deuterium-labeling experiments indicate that lithiation of **4** occurs exclusively in the *ortho* position(s) of the aromatic ring. The first step of the reaction should be then the formation of lithiated compound **13** which must be transformed *in some way*, into α-lithiated compound **14** in order to cyclize (Scheme 28.6).

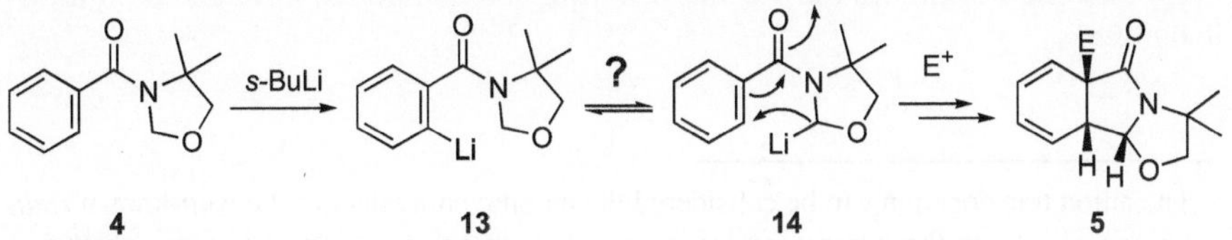

Scheme 28.6

The following experiments would give us more information to understand how this transformation could happen. The experiments carried out with naphthamide **9** support the possibility of equilibrium between two lithiated species. When **9** was lithiated and left to stir for 2 h before quenching, a 1:2 mixture of products **10** and **11** (clearly arising from *ortho* and α-lithiated compounds **15** and **16** respectively) is obtained. Interestingly, compound **11** is the major product, which suggests that in the equilibrium, **16** is more stable but **15** is more reactive (Scheme 28.7).

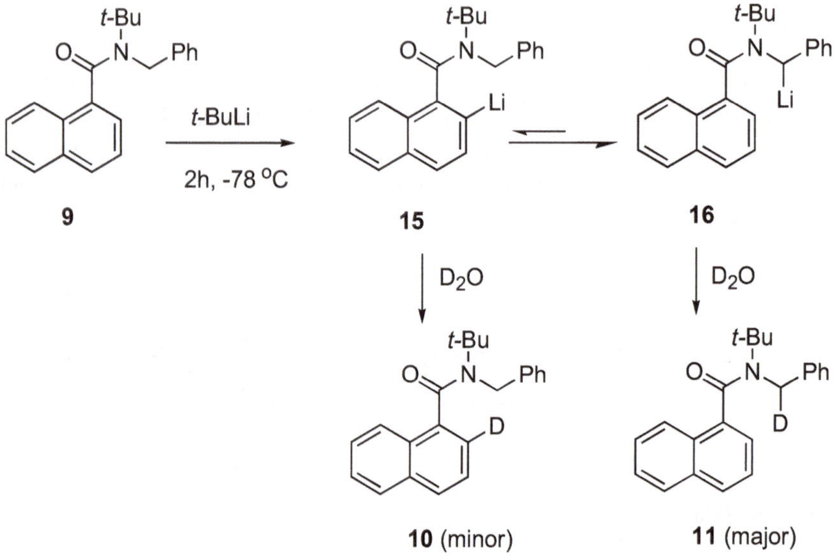

Scheme 28.7

We know from the experimental data that lithiation of **9** followed by subsequent quenching with D_2O yields exclusively compound **10**. Obviously, *ortho*-lithiated compound **15** must be formed at first instance. However, if **15** is left for 2h before quenching, it has time to equilibrate with the more stable α-lithiated species **16** and the mixture of products **10** and **11** is formed. The equilibrium between the anionic species **15** and **16** is called **anion translocation** and occurs when an anion formed under kinetic control, undergoes an intramolecular proton transfer to improve its stability.[2]

Considering all the experimental evidence, we can conclude that the lithiation step occurs initially at the ortho position of the aromatic ring, followed by rapid formation of an equilibrium mixture with the α-lithiated compound through anion translocation.

---

[2] The anion translocation can be considered the anionic equivalent of the well-known *radical translocation*: the intramolecular radical abstraction of a hydrogen atom, which is a key step in some important radical reactions.

Once we know that the α-lithiation of the amide **4** is a stepwise process, we can devote ourselves to the cyclization step. An ionic cyclization of α-lithiated **14** would be expected to generate the more stable *trans* cyclic enolate **17** (and hence **6**) rather than the less stable *cis*-enolate **18** precursor of **5** (Scheme 28.8). The facile epimerization of *cis*-**5** to *trans*-**6** in acidic medium is an argument in favor of the higher stability of the *trans*-isomer **6** with respect to **5**. However, *cis*-fused tricyclic system **5** is exclusively formed in the reaction. We should remark that the attack of the electrophile in the quenching step would be controlled in all cases by the preference for the ring 6,5-*cis*-fusion in the final products.

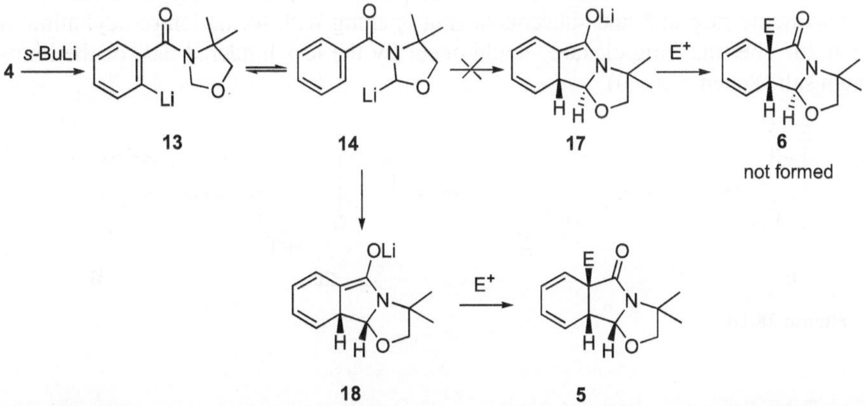

Scheme 28.8

If the stereochemistry of the reaction product **5** is exactly the opposite of what is expected by the mechanism in Scheme 28.8, it is evident that we should consider other alternatives for the cyclization step. An attractive option could be to suppose that the cyclization of lithiated compound **14** is a *pericyclic reaction*. In fact, if we discard the stepwise mechanism for the cyclization of **14** and consider a 6π disrotatory thermal electrocyclic ring closure in this intermediate (better indicated by the canonical form **18**), the result would be *cis*-tricyclic enolate **19**. Quenching with an electrophile will give the reaction product **5** with total stereoselectivity (Scheme 28.9).

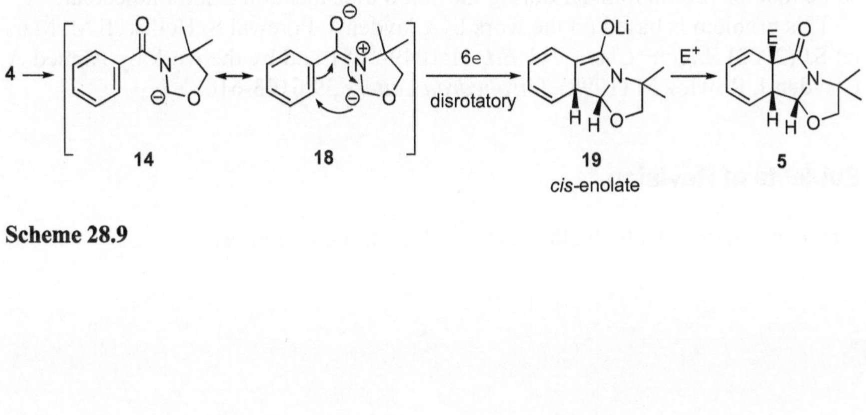

Scheme 28.9

It is worth noting that DMPU is added in all cases to obtain the cyclization products. DMPU (like HMPA) is a good cation-solvating solvent. In this case it is believed that DMPU forces *ortho* organolithium **13** to undergo anion translocation to give the α-lithiated intermediate **14** and favors the electrocyclic ring closure step. The effect of the removal of the counterion to accelerate a pericyclic reaction has been also observed in anionic *oxy*-Cope rearrangements.

A final question remains. The suggested electrocyclization produces a *cis*-fused tricyclic system **5**, significantly less stable than the corresponding *trans*-isomer **6**. The epimerization of **5** to **6** easily takes place by treatment with acid. A reasonable mechanism for this reaction could be the protonation of the oxygen atom of the oxazolidine ring in **5** and subsequent ring opening with formation of acyliminium ion **20**. The final ring closure should occur by the less-hindered face of the iminium salt (Scheme 28.10).

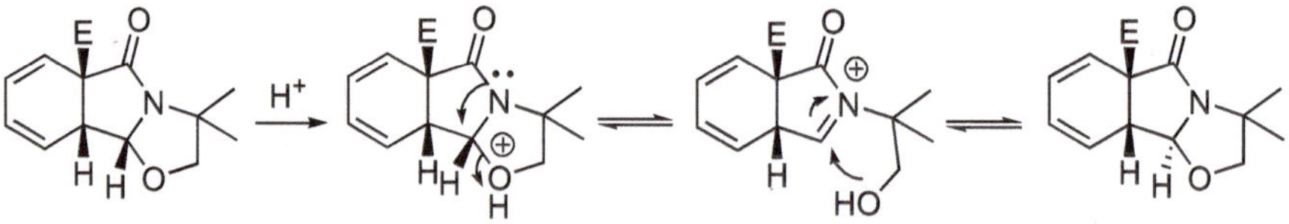

Scheme 28.10

In Summary

The mechanism of the dearomatization-cyclization of aromatic amides takes place by initial lithiation at the ortho position of the aromatic ring. This species is in equilibrium with the more stable α-amide lithiated compound, which forms the final product by means of a 6π electrocyclic ring closure.

Additional Comments

For the cyclization of naphthamide **9**, design a crossover experiment to demonstrate that the proton transfer during the anion translocation is intramolecular.

This problem is based on the work by Clayden J, Purewal S, Helliwell M, Mantel SJ (2002) *Angew. Chem. Int. Ed.* 41:1049-1051 and by the work by Ahmed A, Clayden J, Rowley M (1998) *Tetrahedron Letters* 39:6103-6106.

Subjects of Revision

Stereochemistry. Electrocyclic reactions. Reactivity of anions.

Level 2 – Case 29
Stereoselective Debromination of Vicinal Dibromides

Key point: *Stereochemistry*

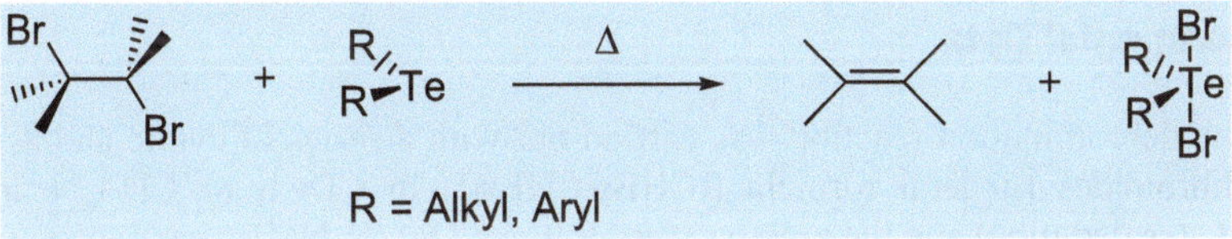

R = Alkyl, Aryl

Stereoselective Debromination of Vicinal Dibromides

Organotellurides **1** have been used in debromination of vicinal (*vic*) dibromides. The reaction takes place by heating the dibromide and the organotelluride in $CHCl_3$ or CH_3CN at 90–100°C as shown in Scheme 29.1.

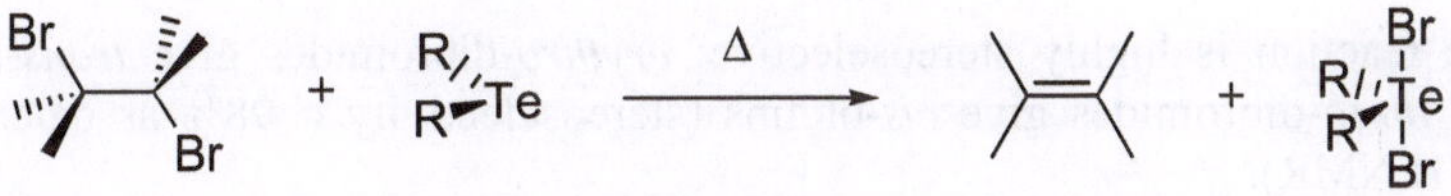

1 R = Alkyl, Aryl

Scheme 29.1

Mechanisms 1 and 2 have been proposed to explain the role of the tellurides in these reactions.

Comment both options and discuss the experimental data justifying which of the proposed mechanisms seems to be more likely.

MECHANISM 1

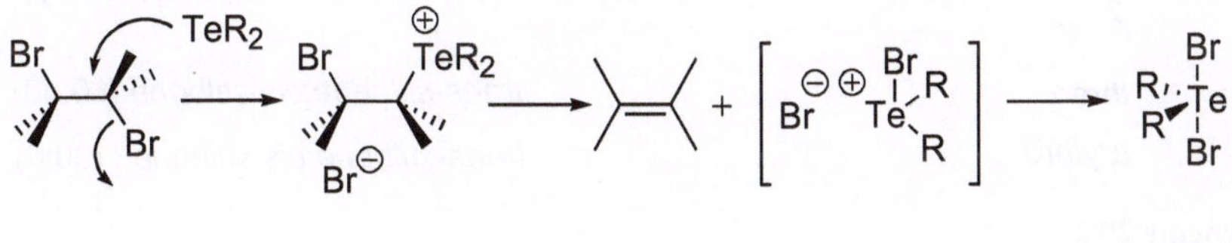

MECHANISM 2

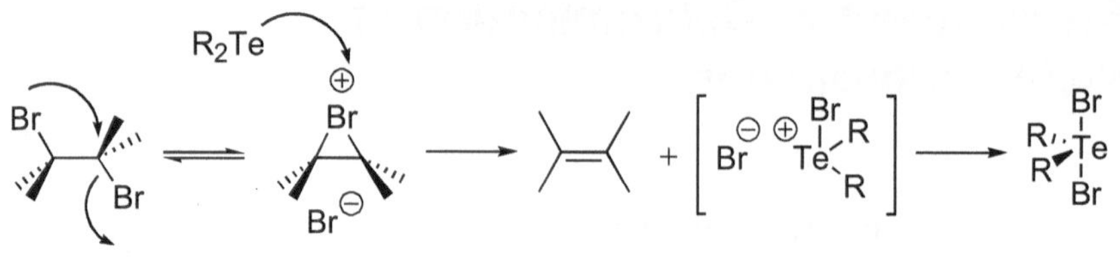

Experimental Data

The debromination reaction was carried out with a series of cyclic and acyclic vic-dibromides (general formula RCHBr-CHBrR) in $CDCl_3$ or CD_3CN at 90-100°C (sealed tube) and the kinetics were followed by ^{1}H-NMR spectroscopy.

1. The kinetic study suggests a first-order dependence in both telluride and vic-dibromide for the debromination reaction. The rate law can be expressed as:

$$-d[\text{dibromide}]/dt = k_{obs}[\text{dibromide}][\text{telluride}] \qquad (29.1)$$

2. The reaction is accelerated by more electron-rich diorganotellurides.

3. The reaction is faster in acetonitrile than in chloroform.

4. The reaction is highly stereoselective: erythro-dibromides give trans-olefins and threo-dibromides give cis-olefins (stereoselectivity > 98% as determined by ^{1}H NMR).

5. erythro-Dibromides are much more reactive than threo-dibromides.

6. The reaction is particularly slow in the case of trans-1,2-dibromocycloalkanes.

7. 1,2-Dibromo-2-methyl-1-phenylpropane reacts faster than 1,2-dibromodecane ($k_{rel} = 330$).

8. The debromination of 1,2-dibromo-1,2-diphenylethane 2 is only stereoselective for the erythro-isomer (Scheme 29.2).

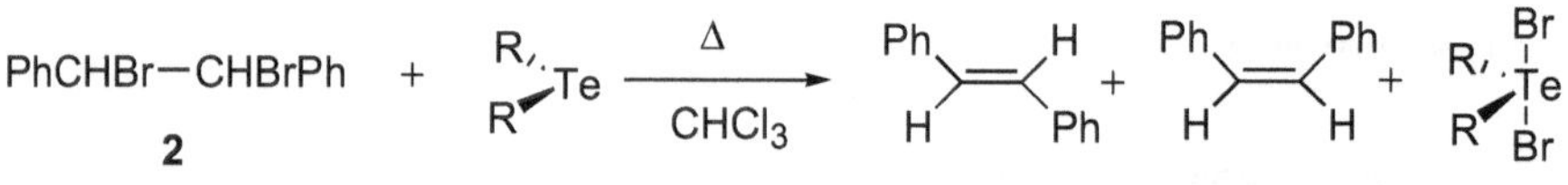

threo	trans-stilbene:cis-stilbene (60:40)
erythro	trans-stilbene:cis-stilbene (100:0)

Scheme 29.2

Discussion

Both mechanisms propose a debromination in two steps.

Mechanism 1 suggests a backside **nucleophilic** attack by the telluride at the C-Br bond of the dibromide, followed by **elimination** from the resulting telluronium salt **3** (Scheme 29.3).

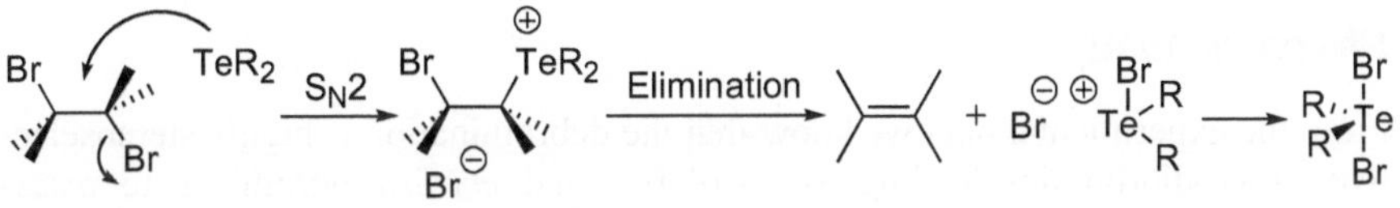

Br TeR₂ S_N2 Br TeR₂⊕ Elimination + Br⊖ ⊕Te$\overset{Br}{\underset{R}{\mid}}$R → R$\underset{R}{\overset{Br}{\mid}}Te\underset{Br}{\mid}$

3

Scheme 29.3

On the other hand, **Mechanism 2** proposes the displacement of a Br⁻ by nucleophilic attack of the neighboring bromine atom (**neighboring group participation**), and formation of a **bromonium ion intermediate 4**. In a second step, the telluride acts as a scavenger of the Br⁺ in the bromonium intermediate to yield the olefin (Scheme 29.4). In both cases, the bromotelluronium salts **5** formed at first instance, give the isolated neutral Te(IV) by-products **6** after bromide addition.

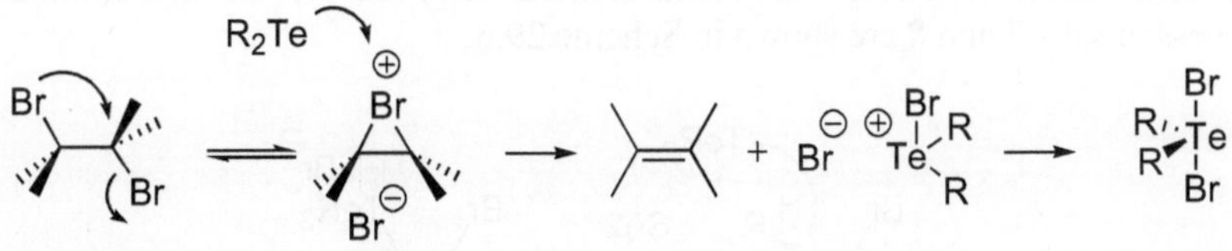

R₂Te → ⇌ → + Br⊖ ⊕Te$\overset{Br}{\underset{R}{\mid}}$R → R$\underset{R}{\overset{Br}{\mid}}Te\underset{Br}{\mid}$

4   **5**   **6**

Scheme 29.4

At first glance both mechanisms seem to be very reasonable. Nevertheless, before deciding which one is more likely, it will be necessary to carry out a careful analysis of all the experimental information we have in hand.

Kinetic Data

Both mechanisms are overall second-order processes involving organotelluride and dibromide, and could fit well with the experimental rate law.

$$-d[\text{dibromide}]/dt = k_{obs}[\text{dibromide}][\text{telluride}] \tag{29.1}$$

Accordingly to Eq. 29.1, any changes made in the concentration or reactivity of the reagents will affect the reaction rate. In consequence, more electron-rich tellurides (also more nucleophilic) will accelerate the debromination. This experimental result could be justified by both mechanisms. On the other hand, either Mechanism 1 or Mechanism 2 propose the formation of a polar intermediate and hence, both should be influenced by a change in solvent polarity. In fact, debromination rates are faster in acetonitrile than in chloroform.

After analyzing the kinetic data we are not able to decide which one is the best option. Both mechanisms are in good agreement with the rate law and both can justify why the reaction is accelerated in more polar solvents and with more electron-rich organotellurides.

It is time to discuss the stereochemistry of the reaction.

Stereochemistry

From the experimental data we know that the debromination is highly stereoselective, *threo*-dibromides leading to *cis*-olefins and *erythro*-dibromides to *trans*-olefins. Let us consider whether the two proposed mechanisms can explain these results.

In **Mechanism 1** the initial S_N2 attack by the Te atom at a C-Br bond of the dibromide, leads to **inversion of the configuration** at the carbon. In consequence, *threo*-dibromides will produce *erythro*-salts **7** and *erythro*-dibromides will produce *threo*-salts **8** exclusively (Scheme 29.5).

The salts will lead to the olefins by means of an elimination process and a **concerted *syn*-elimination** could explain the stereoselectivity of the reaction. An *anti*-elimination could not be considered, since it would lead to alkenes with the opposite stereoselectivity. The conformations required for the *syn*-elimination process in salts **7** and **8** are shown in Scheme 29.6.

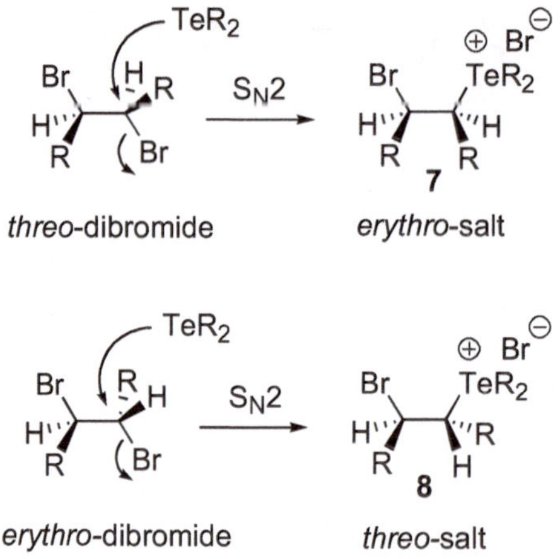

threo-dibromide   erythro-salt

erythro-dibromide   threo-salt

Scheme 29.5

Exercise: Represent the conformations required for the *anti*-elimination process and justify the stereochemistry of the olefin obtained for each case.

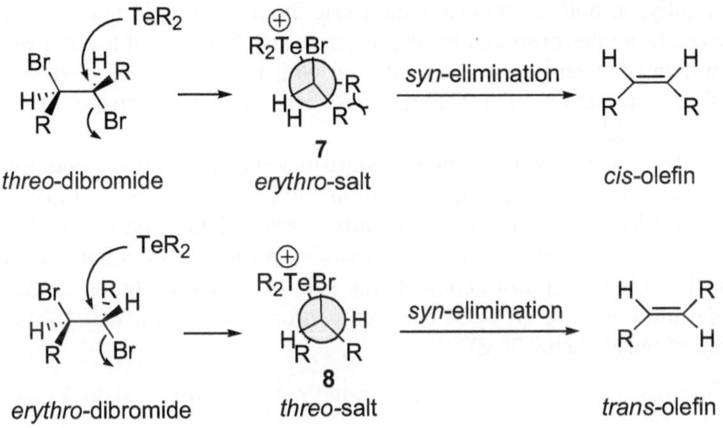

Scheme 29.6

A concerted *syn*-elimination requires the salt to adopt an **eclipsed conformation** that should be very sensitive to the steric interactions between the R groups. Then we should expect that a more crowded conformation (as in **7**) would lead to a slower elimination process. This is consistent with the experimental observation that *erythro*-dibromo derivatives react much faster than their *threo*-analogs. In addition, the increasing ring strain for *syn*-eliminations in cyclic molecules could justify the slow debrominations observed for *trans*-1,2-dibromocycloalkanes.

All of the experimental data commented above can be also explained by **Mechanism 2**. In this case, the stereochemistry of the reaction products would be determined by the structure of the **bromonium ion intermediate**. *erytho*-Dibromides would yield *trans*-olefins whereas *threo*-dibromides would yield *cis*-olefins, as it is shown in Scheme 29.7. The eclipsing interactions between the R groups in the intermediate ions would decrease their stability. Therefore, bromonium ions **9** derived from *erytho*-dibromides should be formed faster than those obtained from *threo*-dibromides **10**, which is in agreement with the experimental data (Scheme 29.7).

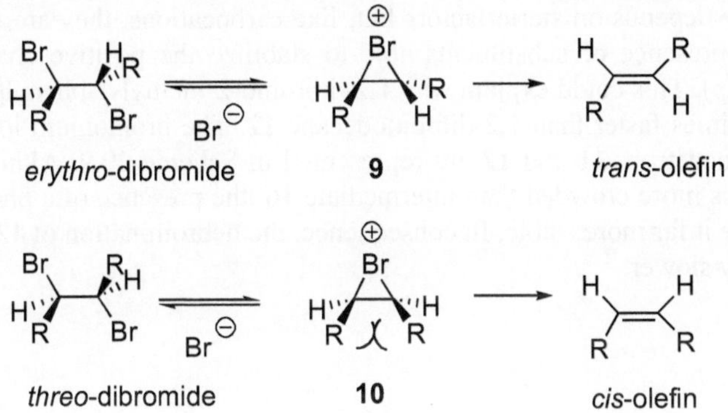

Scheme 29.7

Apparently, in both mechanisms the steric factors are important to explain the relative reaction rates observed for the different substrates. At this point we should study in detail the remaining data, starting with the differences observed in the rates of debromination of 1,2-dibromo-2-methyl-1-phenylpropane **11** and 1,2-dibromodecane **12**.

As we discussed previously, in **Mechanism 1** the steric effects determined the relative reaction rates during the *syn*-elimination step: *a more crowded eclipsed conformation leads to a slower elimination reaction*. Thus, one would expect that 1,2-dibromodecane **12** should be a very reactive substrate, (minimal steric interactions in the eclipsed conformation of the salt **14**). Instead, **11** reacts 330 times faster, despite the more crowded eclipsed conformation of the salt **13** during the elimination step (Scheme 29.8).

This is the first disagreement we have found between **Mechanism 1** and the experimental data.

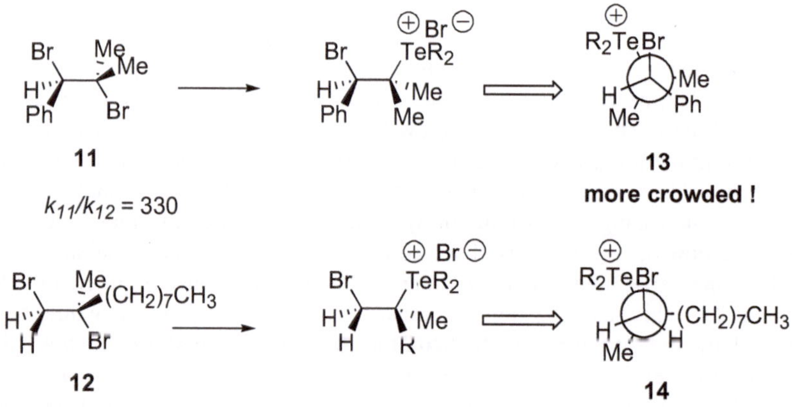

Scheme 29.8

Let us consider whether **Mechanism 2** could justify the differences in reaction rates between **11** and **12**. We should remember that the stability of a bromonium ion not only depends on steric factors but, like carbocations, they are very sensitive to the presence of substituents able to stabilize the positive charge (i.e. a phenyl group). This could explain why 1,2-dibromo-2-methyl-1-phenylpropane **11** reacts 330 times faster than 1,2-dibromodecane **12**. The bromonium ion intermediates obtained from **11** and **12** are represented in Scheme 29.9. Although intermediate **15** is more crowded than intermediate **16**, the presence of a phenyl group should make it far more stable. In consequence, the debromination of **12**, has to be considerably slower.

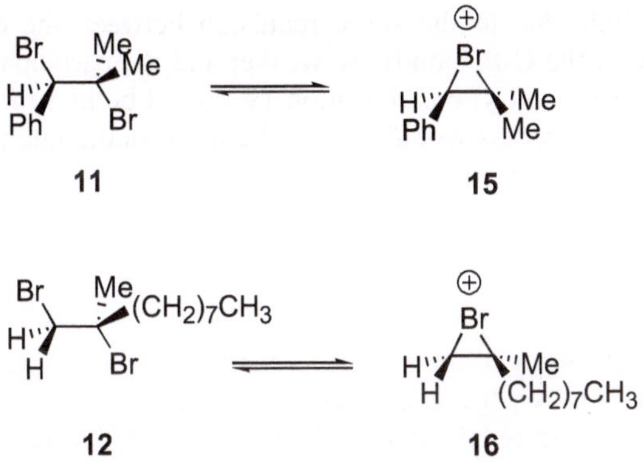

11 ⇌ **15**

12 ⇌ **16**

Scheme 29.9

At this point it is clear that **Mechanism 1** is unable to explain the differences in reaction rates for compounds **11** and **12** and in consequence it has to be ruled out.

Once Mechanism 1 has been discarded **Mechanism 2** is the alternative of choice. The formation of a bromonium ion intermediate seems to justify very well all the experimental data we have checked. However, we cannot decide that Mechanism 2 is the right option without discussing a final point of evidence: *The odd stereoselectivities obtained in the debromination of 1,2-dibromo-1,2-diphenylethane* **2**. The debromination is stereoselective in the case of the *erythro*-isomer but the *threo*- yields a mixture of *cis/trans*-olefins.

The bromonium ion intermediates that should be formed in each case are represented in Scheme 29.10. Intermediate **17**, formed from the *erythro*-isomer, benefits from the conjugation with the two phenyl groups, that are arranged to minimize the steric interactions. In this case, only *trans*-stilbene would be formed. In contrast, the bromonium ion intermediate **18**, obtained from the *threo*-isomer,

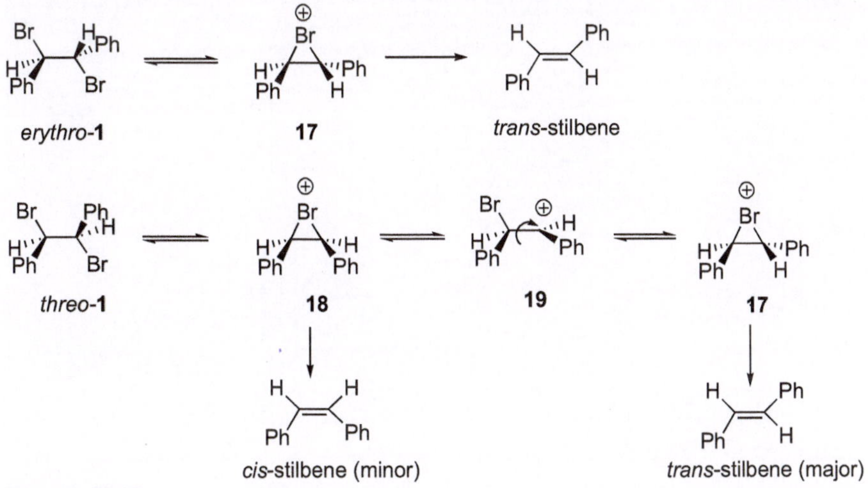

erythro-**1** ⇌ **17** → *trans*-stilbene

threo-**1** ⇌ **18** ⇌ **19** ⇌ **17**

cis-stilbene (minor)

trans-stilbene (major)

Scheme 29.10

should be less stable due to the steric repulsion between the eclipsing phenyl groups. In this case, the C-Br bonds are weaker and the participation in the equilibrium of a stabilized benzylic carbocation **19** should be more favored. The ring opening-free rotation process would lead to the more stable intermediate **17** from which *trans*-stilbene is formed.

In Summary

As we have seen, the study of the stereochemistry of the debromination reaction is the key choosing between the two mechanistic pathways. Both proposals could justify the kinetic data (rate law, nucleophilicity of the telluride, effects of solvent polarity) however, only Mechanism 2 could satisfactorily explain the stereoselectivity in all cases. The intermediacy of a bromonium ion and the role of the telluride as a scavenger of the Br^+ seem to be the best option with all the data in hand.

Additional Comments

Suggest some experiments that could bring additional evidences in support of the bromonium ion. Comment other alternatives to the intermediacy of *free* bromonium ions in the reaction.

This problem is based on the work by Butcher TS, Zhou F, Detty MR (1998) *J. Org. Chem.* 63:169-176.

Subjects of Revision

S_N2. Elimination reactions. Bromonium ion intermediates. Neighboring group participation.

Level 2 – Case 30
Diels-Alder Reactions of *N*-Acyl-1,2,4,5-tetrazines

Key point: *Inverse electron demand Diels-Alder reaction*

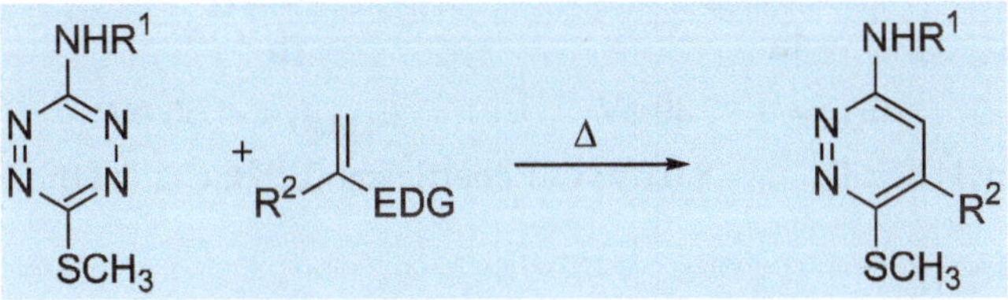

Diels-Alder Reactions of *N*-Acyl-1,2,4,5-tetrazines

The reaction between *N*-acyl-6-amino-3-(methylthio)-1,2,4,5-tetrazines **1** with electron-rich dienophiles **2** gives the corresponding 1,2-diazines **3** in excellent yields and with high regioselectivity (Scheme 30.1).

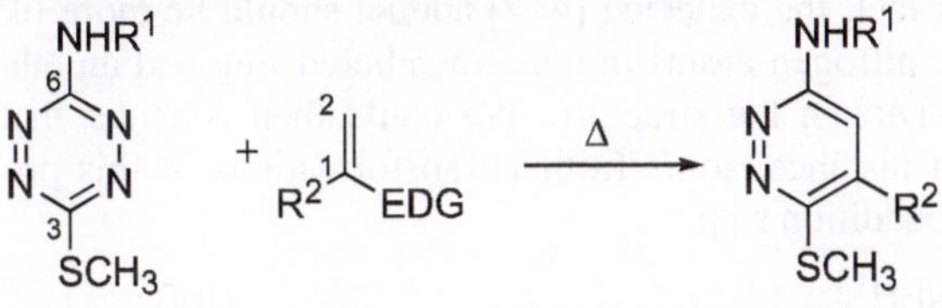

1a, R^1 = Ac
1b, R^1 = BOC; R^2 = H, OMe, alkyl

EDG = electron-donating group: OEt, OMe, OTs, pyrrolidino, morpholino

Scheme 30.1

Considering that the dienophile employed in the reaction is ethyl vinyl ether, *answer the following questions:*

1. *Propose a reasonable mechanism to interpret how diazines **3** are formed.*
2. *Predict which tetrazine would be more reactive towards this dienophile.*
3. *Draw the structure of the reaction products between tetrazines **1** and ethyl vinyl ether.*

Experimental Data

1. LUMO energy levels and atom coefficients in 1,2,4,5-tetrazines **1** (obtained from AM1 computational studies).

ELUMO = -1.39 eV   ELUMO = -1.36 eV

2. Ethyl vinyl eher: E_{HOMO} –8.0 eV; C-1 coefficient 0.280; C-2 coefficient 0.720).

Discussion

Mechanism of the reaction

If we consider the general reactivity between tetrazines **1** and electron-rich dienophiles indicated in Scheme 30.1, the reaction between ethyl vinyl ether (EDG = OEt and R^2 = H) and **1** should yield products with the structure **4** (Scheme 30.2).

However, **4** does not look like the expected cycloaddition adduct between **1** and ethyl vinyl ether. In fact, the expected [4+2] adduct should be more like bicyclic product **5**, with four nitrogen atoms in a six-membered ring and an ethoxy group incorporated somewhere in the structure. We could then consider that, if **5** has been formed at first instance, some further transformations of this product have arisen after the cycloaddition step.

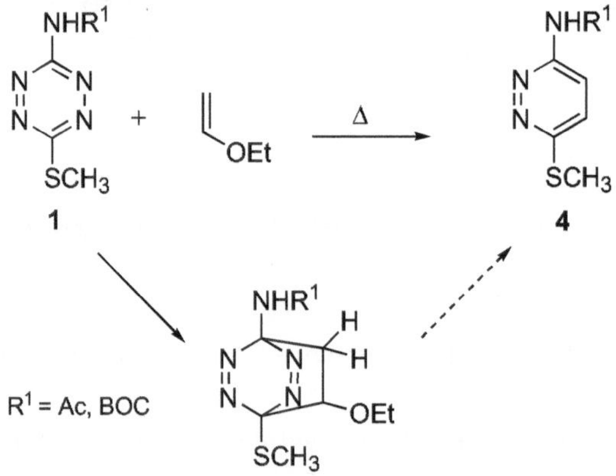

Scheme 30.2

From the experimental results summarized in Scheme 30.1 we know that the electron-donating (EDG) substituent on the dienophile is not present in any of the reaction products **3**. Then, it is possible that under the reaction conditions, adducts **5** eliminate ethanol to give **6** (Scheme 30.3). In consequence, the formation of the more stable (aromatic) diazines **4** from intermediates **6** could proceed by a **retro-Diels-Alder cycloaddition** with the expulsion of a N_2 molecule.

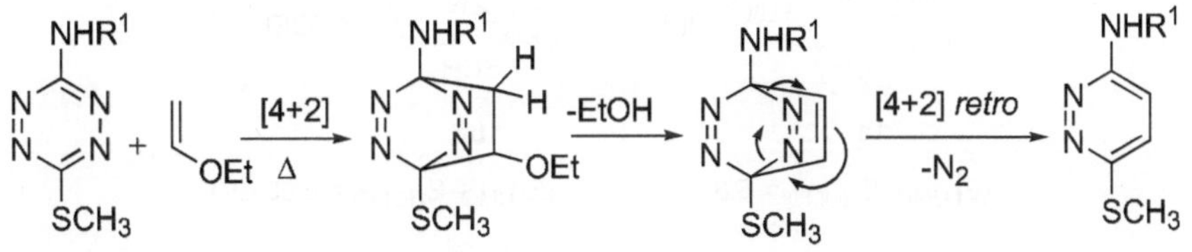

R¹ = BOC, Ac

Scheme 30.3

Reactivity

Contrary to the *normal* Diels-Alder processes, the FMO (Frontier Molecular Orbitals) involved in *inverse electron demand* Diels-Alder reactions are the LUMO of the diene (1,2,4,5-tetrazine) and the HOMO of the dienophile (electron-rich alkene). The reactivity of tetrazines **1** toward ethyl vinyl ether would depend on the LUMO diene-HOMO dienophile gap. Thus, the combination LUMO-HOMO having a smaller energy gap will lead to the faster reaction. In this case the LUMO for **1a** has a lower energy than that of **1b** and this will be reflected in their relative reactivities. Energy gaps between tetrazines **1** and ethyl vinyl ether are 6.61 eV for **1a** and 6.64 eV for **1b**. In consequence, tetrazine **1a** will be more reactive toward ethyl vinyl ether than **1b**.

Regioselectivity

An important point to discuss in any Diels-Alder reaction is the regioselectivity of the cycloaddition. This aspect requires considering the coefficients of the atomic orbitals in the FMO involved in the cycloaddition process (Scheme 30.4). Since the tetrazines are the dienes in the reaction, we should look at C-3 and C-6 coefficients of the LUMO in **1a** and **1b**. In both cases C-6 bears the largest LUMO orbital coefficient and this fact dominates the regioselectivity of the reaction. The C-6 center should preferentially combine with the dienophile C-2 center, which possesses the largest HOMO orbital coefficient. Then, overlapping between positions C–6 of the diene and C–2 of ethyl vinyl ether would yield adducts **5** from which diazines **4** are formed. We should notice from the general formula of diazines **3** in Scheme 30.1, that products derived from the other possible regioisomers **7** are not observed in the reaction.[1]

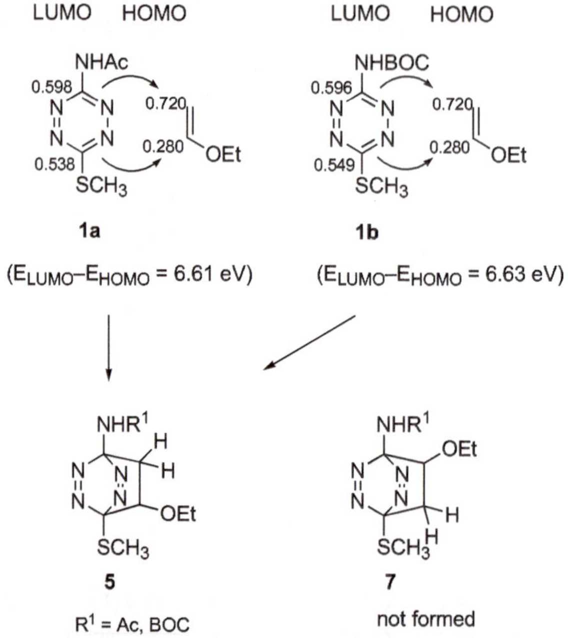

Scheme 30.4

In Summary

1,2,4,5-Tetrazines **1** undergo regioselective inverse electron demand Diels-Alder reactions with a variety of electron-rich dienophiles to yield 1,2-diazines in good yield. The process takes place in three steps: (a) [4+2] cycloaddition, (b) elimination of the electron-donating group, (c) extrusion of N_2 by means of a retro-Diels-Alder process.

Questions

Once we have understood the mechanism of the cycloaddition between tetrazines **1** and electron-rich alkenes, we would be able to interpret the following experimental results:

---

[1] The regioselectivity is consistent with the polarization of the diene and the ability of the methylthio group to stabilize a partial negative charge at C-3 and the *N*-acylamino group to stabilize a partial positive charge at C-6.

1. The reaction of **1a** with phenyl acetylene yields a mixture of regioisomers **8** and **9** in a product ratio 2:1 (Scheme 30.5).
   DATA: Phenyl acetylene: E_{HOMO} –9.5 eV; C-1coefficient 0.580; C-2 coefficient 0.650.

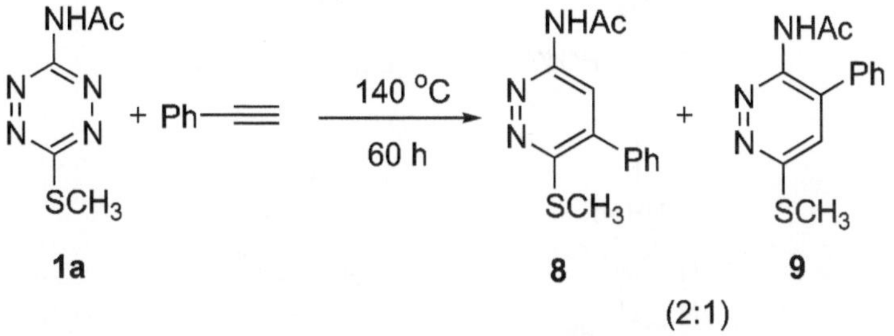

Scheme 30.5

2. Formulate the structure of the product obtained after heating compound **10** in dioxane.

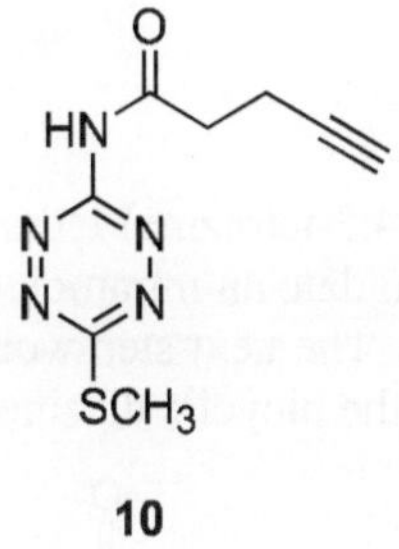

10

Answer to Question 1

It is clear that by using phenyl acetylene as dienophile the reaction is less regioselective than in the case of ethyl vinyl ether (a 2:1 mixture of regioisomers **8** and **9** is obtained in this case). By looking at the C-1 and C-2 coefficients in the HOMO of the phenyl acetylene we will realize that they have very similar values (Scheme 30.6). As discussed previously, the regioselectivity is governed by the overlapping between the positions bearing the largest coefficients in the LUMO of the diene and the HOMO of the dienophile (of course having the same sign). In this case, the difference between the coefficients of both acetylenic carbons is small and this would be the reason for the lower regioselectivity of the reaction in comparison to that of the dienophiles previously described. Again the major regioisomer **8** would result from the overlapping between the positions C-6 and C-2 having the largest coefficients, whilst the minor regioisomer **9** would result from the interaction between the positions C-6 and C-1.

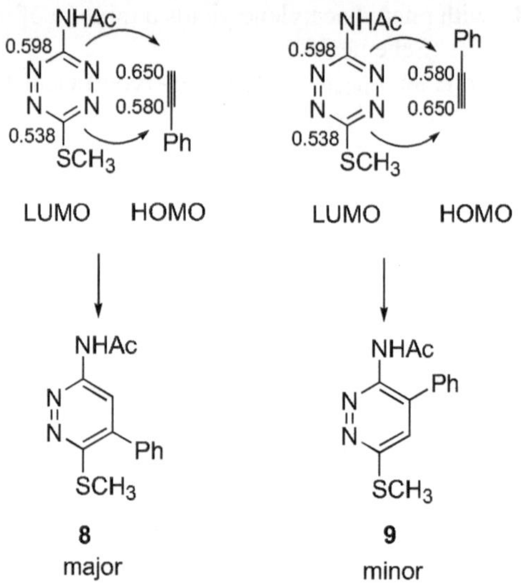

Scheme 30.6

Answer to Question 2

Compound **10** has a diene (1,2,4,5-tetrazine) tethered to a dienophile (C–C triple bond), so it is reasonable to formulate an intramolecular [4+2] cycloaddition in the first place, leading to adduct **11**. The next step would be a retro-Diels-Alder process with N_2 extrusion, to obtain the bicyclic diazine **12** (Scheme 30.7).

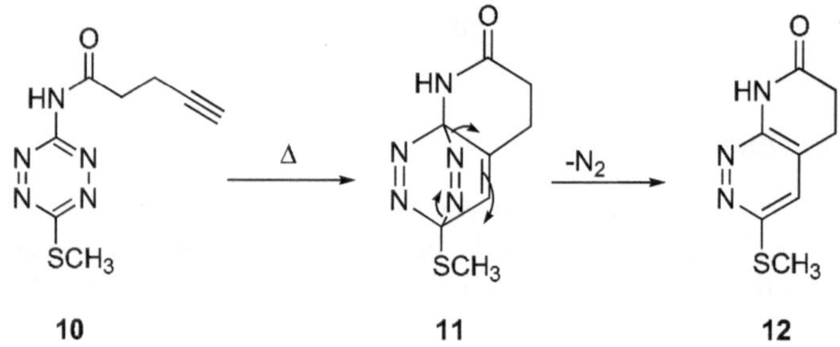

Scheme 30.7

Additional Comments

This problem is based on the work by Boger DL, Schaum RP, Garbaccio RM (1998) *J. Org. Chem.* 63:6329-6337.

Subjects of Revision

Diels-Alder reactions: regioselectivity, reactivity.

Level 2 – Case 31
Stereoselective Synthesis of 2-Acylaziridines

Key point: *Stereochemistry. Solvent effects*

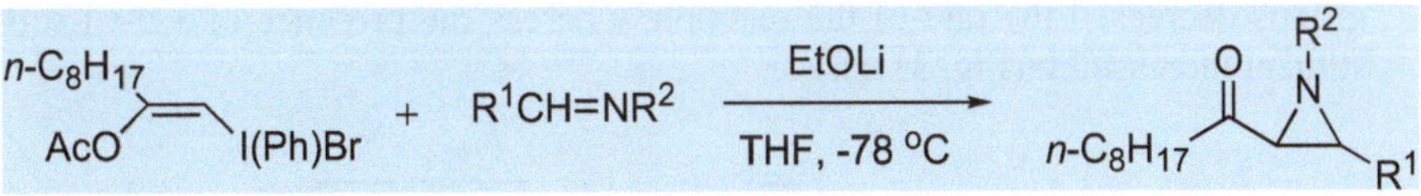

Stereoselective Synthesis of 2-Acylaziridines

The reaction between Z-(2-acetoxyvinyl)iodonium bromides **1** and imines **3** in EtOLi, provides a route for the synthesis of 2-acylaziridines **4** in good yields. It has been proposed that the first step of the reaction consists in the *in situ* formation of a monocarbonyl iodonium ylide **2**, that adds to the imine to give the corresponding aziridines as *cis:trans* mixtures (Scheme 31.1).

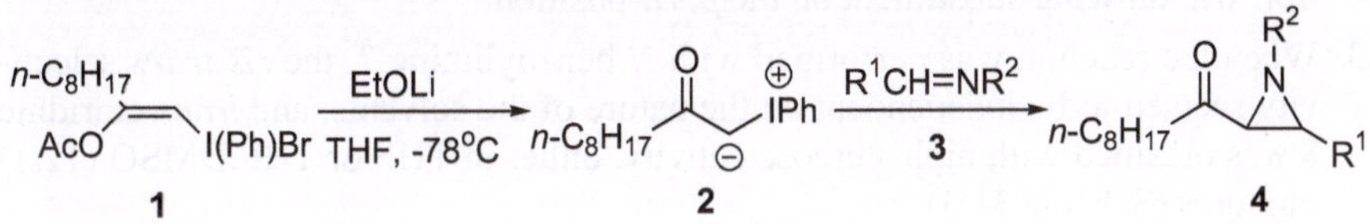

Scheme 31.1

A study of the stereoselectivity of the reaction has demonstrated that the stereochemical outcome of the aziridination depends on the solvent employed. Thus, when the reaction between **1** and *N*-(phenylsulfonyl)benzaldimine **5** was carried out in THF, *cis*-aziridine **6** was stereoselectively obtained. By contrast, *trans*-aziridine **6** was isolated as the major reaction product in THF:DMSO (12:1) (Scheme 31.2).

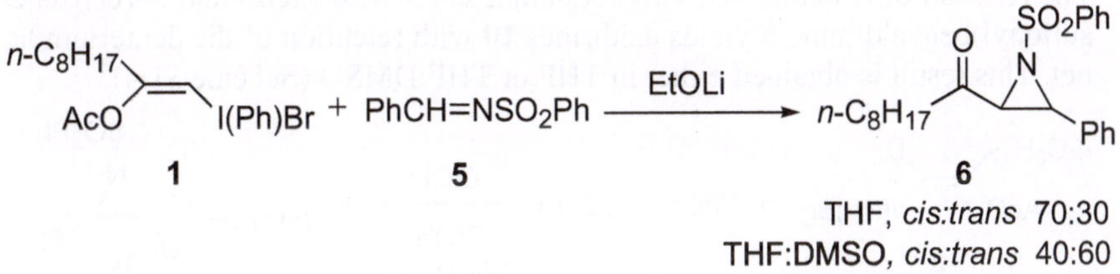

THF, *cis:trans* 70:30
THF:DMSO, *cis:trans* 40:60

Scheme 31.2

Justify the stereochemistry of the reaction discussing the influence of the solvent in the process.

Experimental Data

1. A kinetic study performed in a series of *para*-substituted *N*-(arenesulfonyl)-benzaldimines in THF indicated that the reaction rates were dependent on the substituent present in the aromatic ring. Electron-donating p-MeO or p-Me groups decreased the rate of the reaction, whereas the presence of a p-Cl substituent increased it (Fig. 31.1).

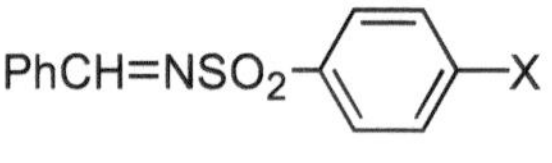

k_{rel}	X
0.56	OMe
0.84	Me
1	H
1.82	Cl

$$PhCH=NSO_2-\text{(aryl)}-X$$

Figure 31.1

2. In the experiments carried out in THF it has been shown that the electron-donating *para*-substituents of *N*-sulfonylimines increase the *cis*-selectivity of the reaction, whereas lower *cis*-selectivity results from the presence of an electron-withdrawing substituent on the *para*-position.

3. When the reaction was performed with *N*-benzoylimine 7, the *cis:trans* selectivity seemed to be independent on the nature of the solvents, and *trans*-aziridine 8 was obtained with high stereoselectivity, either in THF or THF:DMSO (12:1) mixtures (Scheme 31.3).

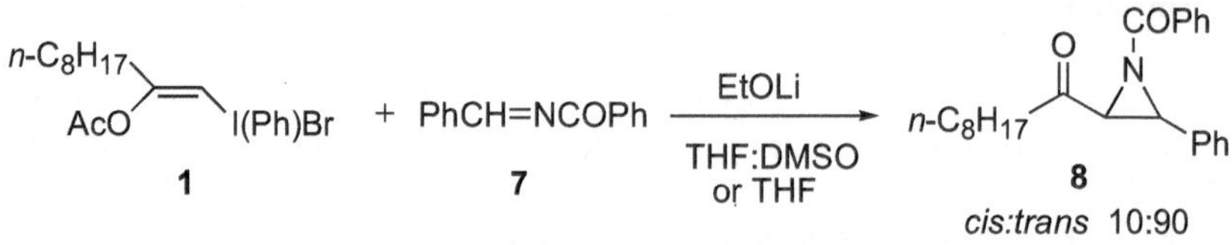

$$n\text{-}C_8H_{17}\text{—CH(AcO)}=\text{I(Ph)Br} \quad + \quad PhCH=NCOPh \xrightarrow[\text{THF:DMSO or THF}]{\text{EtOLi}} n\text{-}C_8H_{17}\text{-aziridine-Ph}$$

1   **7**   **8**

cis:trans 10:90

Scheme 31.3

4. The reaction of α-deuterated vinyliodonium salt **9** with EtOLi and *N*-(benzenesulfonyl)benzaldimine **5** yields azidirines **10** with retention of the deuterium label. This result is obtained either in THF or THF:DMSO (Scheme 31.4).

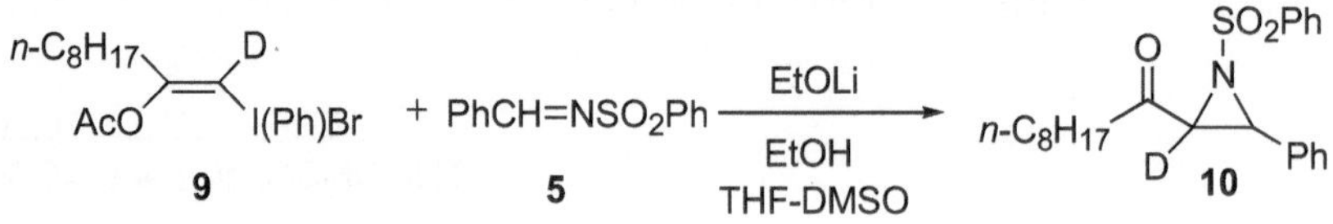

$$n\text{-}C_8H_{17}\text{—C(D)(AcO)}=\text{I(Ph)Br} \quad + \quad PhCH=NSO_2Ph \xrightarrow[\text{THF-DMSO}]{\text{EtOLi, EtOH}} n\text{-}C_8H_{17}\text{-aziridine(D)-Ph}$$

9   **5**   **10**

Scheme 31.4

Discussion

Any proposed mechanism for the aziridination of *Z*-(2-acetoxyvinyl)iodonium bromides **1** with imines **3** must give a reasonable explanation for the two steps of the reaction. In other words, it must be able to interpret the formation of ylides **2** and their further addition to imines **3** to form the aziridines **4**.

Do we have any information about the slow step of the process? Well, since we know that the reaction of iodonium salts **1** with *para*-substituted *N*-(arenesulfonyl)benzaldimines is sensitive to the nature of the substituents present on the sulfonyl aromatic ring, the imines must be necessarily involved in the rate-determining step of the reaction. In consequence, if the reaction between the ylides and the imines (second step) is rate-limiting, the formation of the ylides (first step) must be a fast pre-equilibrium.

Next we are going to discuss each of the two reaction steps in detail.

Step 1: Formation of ylides 2

During the formation of ylide **2** in the presence of EtOLi the acetate moiety has been lost. As the reagent is a nucleophile, it is reasonable to consider that the removal of the acetate group takes place by a classical addition-elimination process induced by ethoxide. Thus, nucleophilic attack of the EtOLi on the carboxylate in **1** would form intermediate **11** that by elimination of ethyl acetate gives ylide **2** (Scheme 31.5). This mechanistc proposal is fully in agreement with the retention of the deuterium label in the reaction products **10** when starting from the α-deuterated salt **9**.

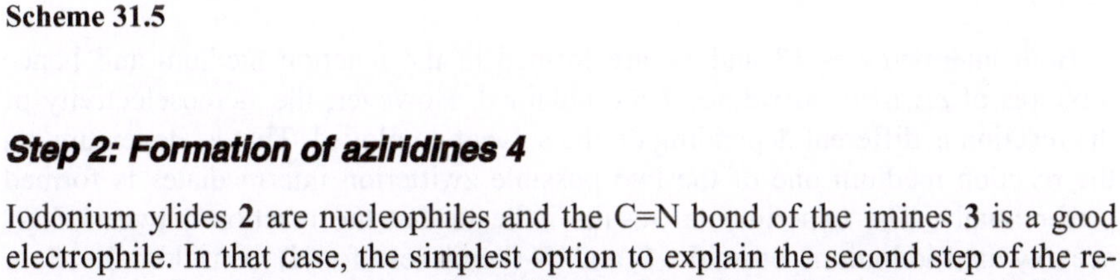

Scheme 31.5

Step 2: Formation of aziridines 4

Iodonium ylides **2** are nucleophiles and the C=N bond of the imines **3** is a good electrophile. In that case, the simplest option to explain the second step of the reaction would consist in the nucleophilic attack of the carbanion to the imine to form zwitterions **12**. Subsequent intramolecular nucleophilic attack (S_N2) by the nitrogen atom in **12** would yield aziridines **4** with concomitant liberation of iodobenzene (Scheme 31.6).

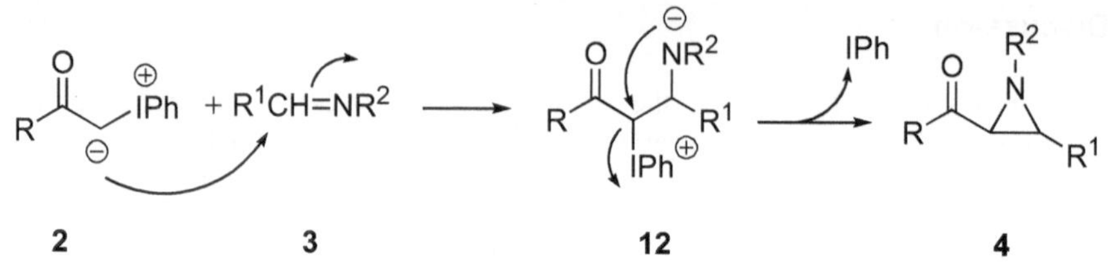

2   **3**   **12**   **4**

Scheme 31.6

During the nucleophilic addition of the ylide to the imine C=N bond two possible zwitterions could be formed: zwitterion **13** and zwitterion **14** (Scheme 31.7). As an intramolecular backside attack by the nucleophile is required to obtain the aziridines, the two groups involved in the S_N2 reaction must be antiperiplanar. Thus, intramolecular nucleophilic attack in zwitterion **13** will lead to aziridines with *cis*-stereochemistry, whereas *trans*-isomers will be obtained from zwitterion **14** (Scheme 31.7).

zwitterion 13   **cis-aziridine**

zwitterion 14   **trans-aziridine**

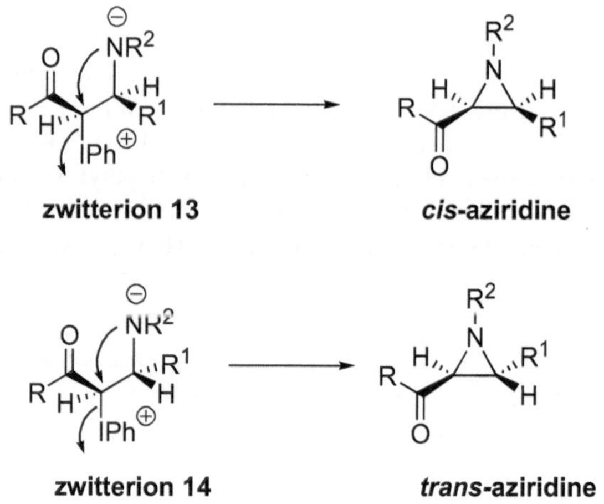

Scheme 31.7

Both intermediates **13** and **14** are formed in the reaction medium and hence mixtures of *cis:trans*-aziridines **4** are obtained. However, the stereoselectivity of the reaction is different depending on the solvent employed. That is, depending on the reaction medium one of the two possible zwitterion intermediates is formed preferentially. The experimental results indicate that the reaction between Z-(2-acetoxyvinyl)iodonium bromide **1** and N-(benzenesulfonyl) benzaldimine **5** in THF, leads preferentially to *cis*-aziridine **6**, whereas the *trans*-isomer is the major reaction product in THF:DMSO (12:1) mixtures. DMSO is an excellent coordinating solvent (DN = 29.8), much better than THF (DN = 20). Since the formation of both zwitterions is controlled by the addition of the polar ylide **2** to the imine **3**, it is reasonable to think that any changes in the solvent polarity should affect the nucleophilic addition step and hence the **13:14** ratios.

In a non-coordinating medium like THF, the intramolecular coordination between the sulfonyl oxygen atom of the imine **5** and the positively charged iodine (III) of the ylide **2** should be possible. This would lead to more tightened transition states during the nucleophilic addition step that could be represented as **15** and **16**. Undoubtely **15** is considerably less crowded than **16** and hence should be preferentially formed, finally leading to *cis*-aziridines **6** as major reaction products (Scheme 31.8).

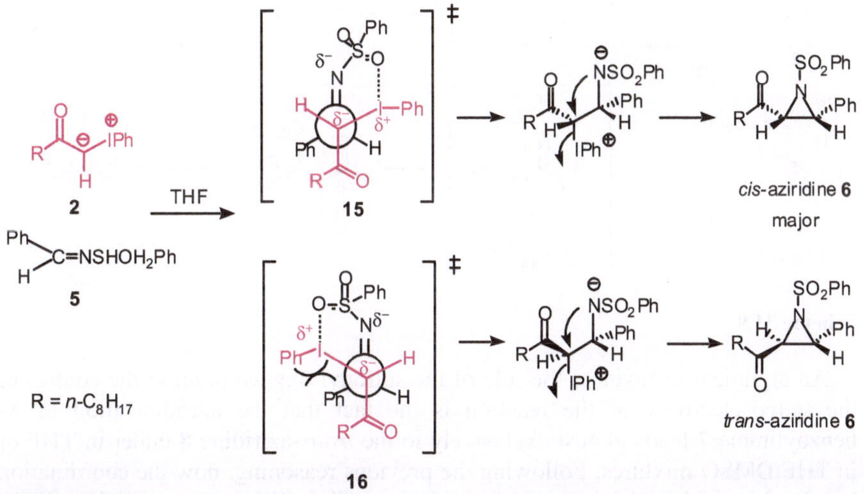

Scheme 31.8

The intramolecular coordination proposed in THF is supported by the fact that electron-donating substituents in the *para*-position of the aromatic ring in sulfonylimines **5**, favor the formation of the *cis*-aziridine isomers. Undoubtedly, electron-donating groups enhance the electron density of the sulfonyl oxygen atom favoring the coordination with the positively charged iodine (III) of the ylide **2** and hence the preferential formation of zwitterion intermediates like **13**.

The intramolecular coordination is not possible if a good solvating agent like DMSO is present in the reaction medium. DMSO is an excellent cation solvator and will coordinate to the positively charged iodine (III) of the ylide **2**, increasing the effective size of the PhI^+ group. In this case, transition states **17** and **18**, in which the bulky NSO_2Ph and PhI^+ groups are antiperiplanar will be formed in preference (Scheme 31.9). Now **18** is considerably less crowded than **17** and hence should be favored, leading finally to *trans*-aziridines **6** as major reaction products.

Comment

Remember that Donor Numbers (DN) are used as an empirical semiquantitative measure of the nucleophilic properties of solvents and are particularly used when discussing the influence of solvent polarity in reactions involving cations. They range from dichloromethane (DN = 0.0 kcal mol^{-1}, reference solvent) to HMPA

(DN = 38.8 kcal mol⁻¹). The higher the donor number, the stronger is the interaction between solvent and acceptor.

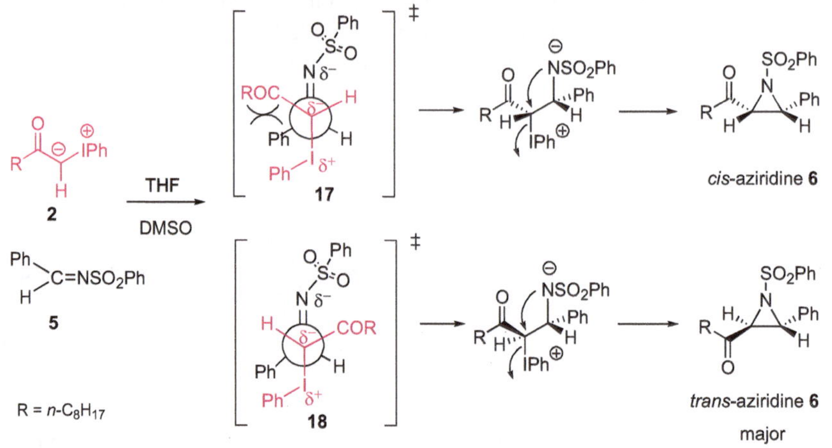

Scheme 31.9

An argument in favor of the role of the sulfonyl oxygen atom in the control of the stereoselectivity of the reaction is the fact that the aziridinization of *N*-benzoylimine **7** leads almost exclusively to the *trans*-aziridine **8** either in THF or in THF:DMSO mixtures. Following the previous reasoning, now the coordination between the carbonyl oxygen atom with the iodine (III) is not possible and the most favorable transition state for the nucleophilic addition step will be the less crowded, with bulky PhI⁺ and benzoyl groups placed antiperiplanar. Transition state **19** should be more stable than more crowded **20** leading preferentially to *trans*-aziridines **8** as major reaction products (Scheme 31.10).

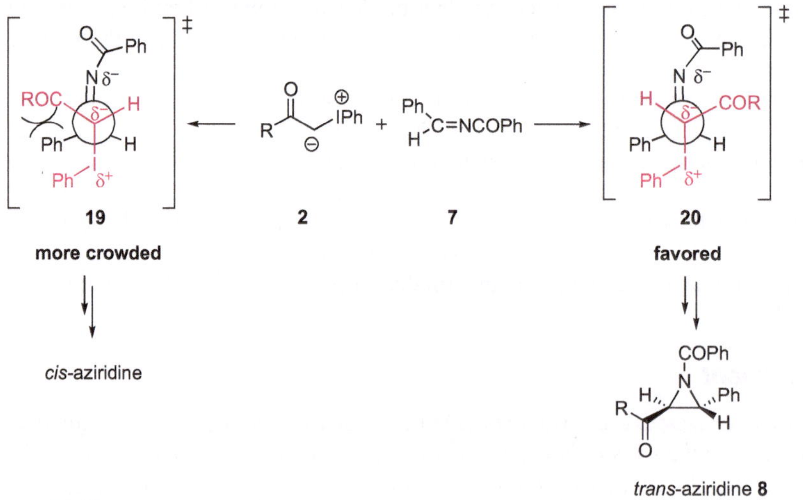

Scheme 31.10

In Summary

We can conclude that 2-acylaziridines **4** can be obtained from Z-(2-acetoxy-vinyl)iodonium bromides **1** and imines **3** in the presence of EtOLi. The reaction is a stepwise process. During the first step a monocarbonyl iodonium ylide **2** is formed by removal of the acetate group promoted by EtOLi. The second step consists in the nucleophilic attack of the ylide **2** to the imine C=N bond to yield a zwitterion intermediate (rate-determining step) that leads to the aziridine **4** by means of an intramolecular S_N2 reaction. The overall mechanism could be drawn as in Scheme 31.11. Although the aziridines **4** are obtained as *cis:trans* mixtures, the stereochemical outcome of the reaction can be controlled by the reaction solvent and by the substituent on the nitrogen atom in the imines.

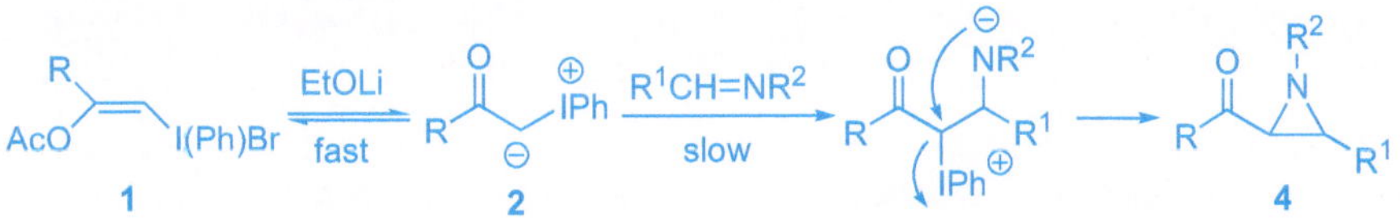

Scheme 31.11

Additional Comments

This problem is based on the work by Ochiai M, Kitagawa Y (1999) *J. Org. Chem.* 64:3181-3189.

Subjects of Revision

Solute-solvent interactions. Empirical parameters of solvent polarity.

Additional Comments

This problem is based on the work by ...

Subject of Revision:

Level 3 – Case 32
The Baylis-Hillman Reaction

Key point: *Catalysis. Volume of activation*

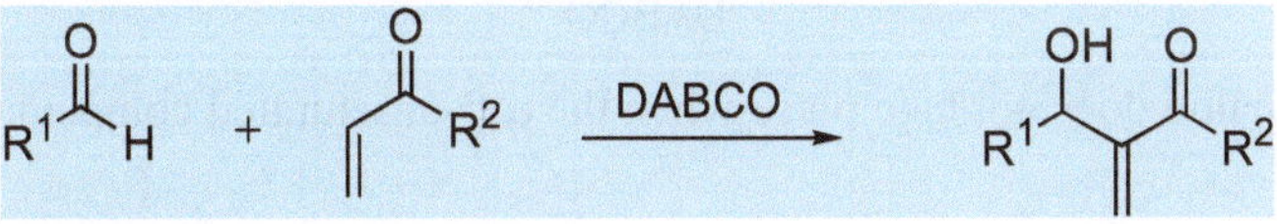

The Baylis-Hillman Reaction

The tertiary amine-catalyzed C-C bond-forming reaction of aldehydes with α,β-unsaturated compounds like acrylates or vinyl ketones is known as the Baylis-Hillman reaction (Scheme 32.1). The process occurs in the presence of catalytic amounts of base, generally DABCO (1,4-diazabicyclo[2.2.2]octane), although other tertiary amines have also been employed.

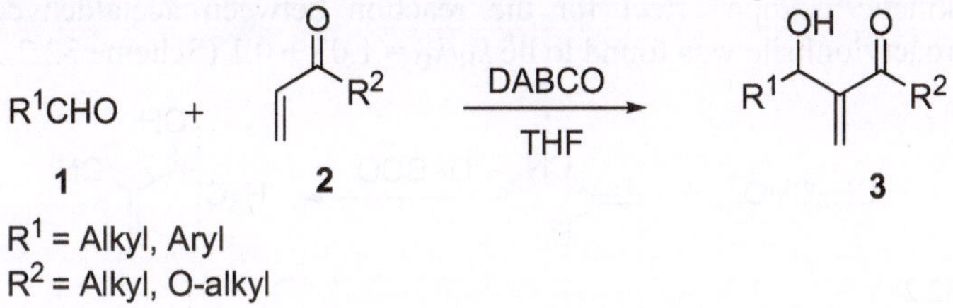

R^1 = Alkyl, Aryl
R^2 = Alkyl, O-alkyl

Scheme 32.1

Although the Baylis-Hillman reaction is notorious for having slow reaction rates (some reactions take as much as 10 days to take place at room temperature) a suitable combination of reagents, catalyst and reaction conditions, can remarkably increase the rate of the reaction.

Based on the following experimental data:

1. *Discuss the type of catalysis (base or nucleophile) of the process and propose a mechanism for the reaction.*

2. *Considering the volume of activation of the reaction, comment the effect of an increase in pressure on the reaction rate.*

Experimental Data

1. The reaction rate was significantly improved when bases such as DABCO were used rather than simply tertiary amines like Et_3N.

DABCO

2. The reaction does not take place when the α,β-unsaturated compound is β-substituted.

3. For aromatic aldehydes, the reaction is much faster when electron-withdrawing substituents (like nitro groups) are placed on the *para*-position of the ring.

4. The reaction has a very high negative volume of activation ($\Delta V^{\ddagger} = -70$ cm^3 mol^{-1}).

5. The reaction rate is increased in polar solvents such as N,N-dimethylformamide (DMF), methanol or acetonitrile.

6. The kinetic law for the reaction is:

   $$v = k[R^1CHO][CH_2{=}CHCOR^2] \qquad \text{where } k = k_{obs}[\text{tertiary amine}] \quad (32.1)$$

7. The kinetic isotope effect for the reaction between acetaldehyde and α-deuteroacrylonitrile was found to be $k_H/k_D = 1.03 \pm 0.1$ (Scheme 32.2).

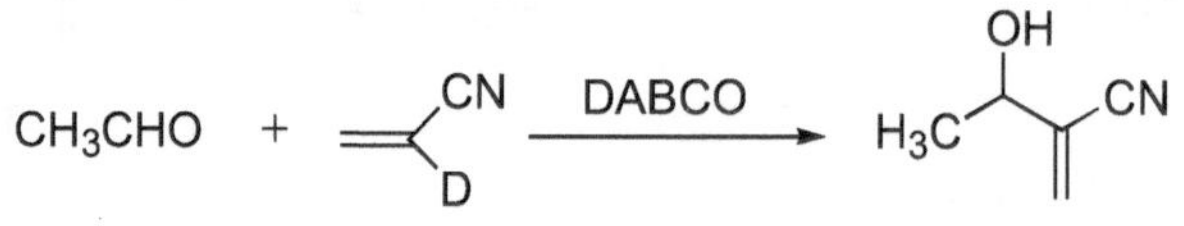

Scheme 32.2

Discussion

The Role of the Amine as Catalyst in the Reaction

Accordingly to the overall transformation represented in Scheme 32.1, product **3** results from the substitution of the α-H of the enone **2** by the aldehyde fragment. The rate expression (Eq. 32.1) shows a linear dependence on the first power of the concentration of aldehyde **1**, α,β-unsaturated compound **2** and base. This is suggesting a stepwise mechanism, as the *synchronous* coming together of three molecules in the activated complex of the slow step of the reaction is very unlikely. In addition, we know from the experimental data that the amine is the catalyst of the reaction but we have yet to decide whether the catalyst acts as a base or as a nucleophile.

Let us suppose that the amine is **acting as a base catalyst**. In this instance, the first step of the reaction could be the α-deprotonation of compound **2** to give enolate **4**. This species would then add to the aldehyde **1** to give intermediate **5**, which after protonation would yield the final product **3**. The catalyst will be recovered unaltered at the end of the process (Scheme 32.3). If step 1 were rate-determining, the reaction would be zero order in aldehyde and this is not in agreement with the kinetic law (Eq. 32.1). In consequence, the slow step should be the addition of enolate **4** to the aldehyde, and the deprotonation of **2** must be a fast pre-equilibrium.[1]

The main drawback of this route is that vinylic protons α- to carbonyls are not very acidic. Therefore, it is hard to believe that a weak base like a tertiary amine would be able to remove them.

Considering these arguments, the catalysis by base is very unlikely.

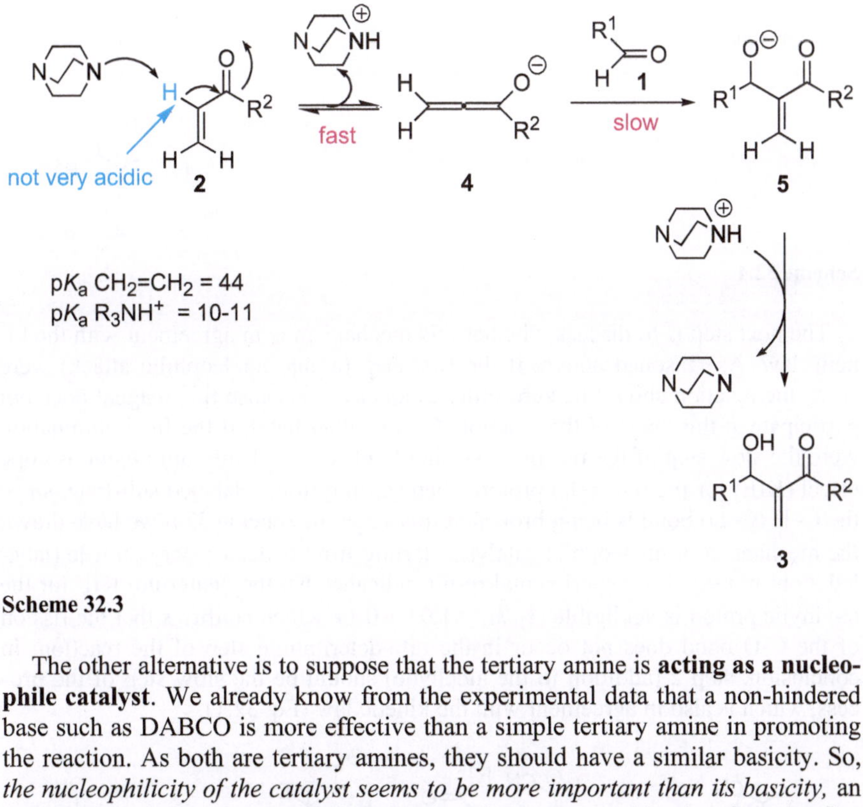

$$pK_a \ CH_2=CH_2 = 44$$
$$pK_a \ R_3NH^+ = 10\text{-}11$$

Scheme 32.3

The other alternative is to suppose that the tertiary amine is **acting as a nucleo-phile catalyst**. We already know from the experimental data that a non-hindered base such as DABCO is more effective than a simple tertiary amine in promoting the reaction. As both are tertiary amines, they should have a similar basicity. So, *the nucleophilicity of the catalyst seems to be more important than its basicity*, an argument in favor of a nucleophile-catalyzed process. In this situation, the first step of the reaction could be the conjugate addition of the amine to the enone **2** to give enolate **6** (Scheme 32.4). Supporting this proposal is the fact that the Baylis-

---

[1] The pK_a data given in the Scheme 32.3 have been taken from M. B. Smith and J. March, *Advanced Organic Chemistry*, Wiley, New York, 5th Ed., 2001, p 329.

Hillman reaction does not occur if **2** is β-substituted, probably because the addition of the amine to the α,β-unsaturated system **2** is inhibited by steric interactions. Once the enolate **6** is formed, it could add to the aldehyde **1** to give zwitterion **7**. The final product would be obtained from **7** by means of an elimination process. Again, as expected, the catalyst is recovered unaltered at the end of the reaction (Scheme 32.4).

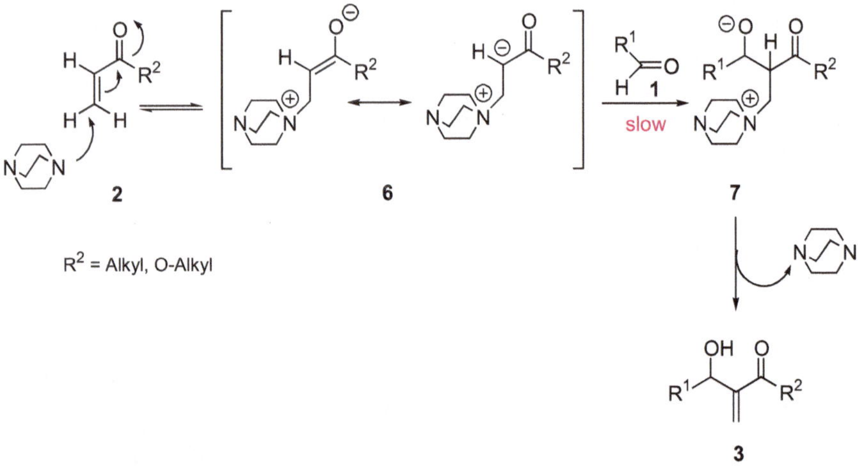

R^2 = Alkyl, O-Alkyl

Scheme 32.4

The next step is to discuss whether this mechanism is in agreement with the kinetic law. As reasoned above, if the first step (amine nucleophilic attack) were slow, the reaction should be zero order in aldehyde because this reagent does not participate at this stage of the reaction. On the other hand, if the final elimination were the slow step of the reaction, we should observe a deuterium kinetic isotope effect (KIE) for the α-vinylic proton when starting from a labeled substrate, since the C–H (C–D) bond is being broken at this stage. In Scheme 32.5 we have drawn the mechanism of nucleophile catalysis starting from α-deuteroacrylonitrile (labeled atom in red). The experimental result indicates that the deuterium KIE for the α-vinylic proton is negligible ($k_H/k_D = 1.03 \pm 0.1$), which confirms that the fission of the C-D bond does not occur in the rate-determining step of the reaction. In conclusion, step 2 (addition to the aldehyde) should be the slow step of the process, which is also in agreement with the kinetic law (Eq. 32.1).

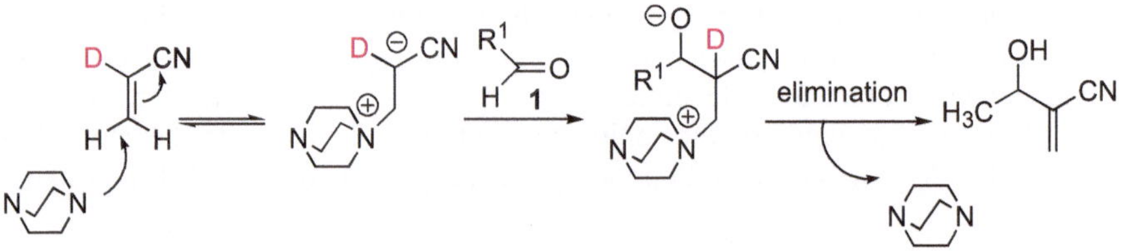

Scheme 32.5

Further experimental evidence corroborates the nucleophile catalysis mechanism previously discussed. The effect of the substitution in the aromatic ring is very significant when aromatic aldehydes are used: the reaction rate is increased when electron-withdrawing groups are placed in the *para*-position. Clearly, if the slow step consists on the nucleophilic attack of enolate **6** to the aldehyde **1**, the increase of the electrophilicity of the carbonyl group due to the presence of electron-withdrawing substituents should accelerate the reaction. Other data to be considered is the increase of the reaction rate in polar solvents. This is fully consistent with the formation of zwitterion **7** in the slow step, since an increase in the polarity of the solvent results in the stabilization of this polar intermediate by effect of solvation.

Apparently the hypothesis of a nucleophile-catalyzed reaction fits well with all the experimental data. However, there is still an additional question being the transformation of zwitterion **7** into the Baylis-Hillman product **3**. Intermediate **7** may evolve to compound **3** either by an E2 or by an E1cB elimination process (Scheme 32.6). Based on experimental evidence, we only know that the fission of the vinylic α-proton should occur at some stage after the rate-determining step, because no deuterium KIE is detected. Only with this data in hand, it is impossible to decide between the E2 and E1cB elimination processes (Scheme 32.6).

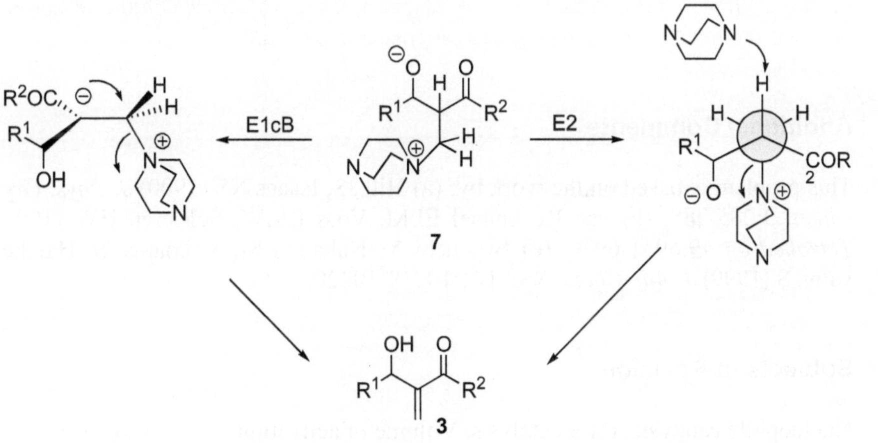

Scheme 32.6

Volume of Activation

The volume of activation $\Delta V^{\ddagger}$ is the difference between the partial molar volume of the transition state and that of the reagents. Individual values of $\Delta V^{\ddagger}$ and $\Delta S^{\ddagger}$ provide similar information on the structural changes that take place in the transition state of a reaction. In this case, the calculated volume of activation for the Baylis-Hillman reaction ($\Delta V^{\ddagger} = -70$ cm^3 mol^{-1}) is very high in magnitude and negative, indicating a volume reduction when passing from reagents to the transition state. This is in agreement with the mechanism proposed in Scheme 32.4, in which two successive bond formation steps take place from the reagents to the activated complex in the rate-determining step.[2]

The Baylis-Hillman reaction is accompanied by a considerable reduction in volume and it is evident that an increase in pressure would accelerate the reaction. For example, with an increase of pressure from 1 bar to 1000 bar, a 15-fold increase in the value of k_{obs} was observed. The effect of pressure has been used to extend the limited scope of the Baylis-Hillman reaction. In fact β-substituted acrylates (unreactive at atmospheric pressure) give the Baylis-Hillman products under high pressure conditions (8 kbar) in reasonable yields.

In Summary

The Baylis-Hillman reaction is generally defined as a base-catalyzed reaction between an α,β-unsaturated carbonyl compound and an aldehyde. However, the experimental evidence clearly demonstrates **that the catalyst is involved in the reaction as a nucleophile, not as a base.** The steps of the process include the conjugate addition of the catalyst to the α,β-unsaturated carbonyl compound, addition to the aldehyde, and elimination. The reaction has a very high volume of activation and the rate is very sensitive to an increase in pressure.

Additional Comments

This problem is based on the work by: (a) Hill JS, Isaacs NS (1990) *J. Phys. Org. Chem.* 3:285-288. (b) van Rozendaal ELM, Voss BMW, Scheeren HW (1993) *Tetrahedron* 49:6931-6936. (c) Iwabuchi Y, Nakatani M, Yokoama N, Hatakeyama S (1999) *J. Am. Chem. Soc.* 121:10219-10220.

Subjects of Revision

Nucleophile catalysis. Base catalysis. Volume of activation.

---

[2] Diels-Alder reactions are known to have high $\Delta V^{\ddagger}$ values as they are accompanied by a considerable reduction in volume. In such processes two molecules are combining to a single product, though no charges are involved. The typical $\Delta V^{\ddagger}$ values are about -35 cm^3 mol^{-1}. However, it is a fact that associative processes leading to ion formation, have much larger volumes of activation due to **solvent electrostriction** (that is, the tendency of a solvent to associate tighly with a charged center by electrostatic forces). The solvent electrostriction is responsible for a reduction of both, entropy and volume of activation and has been claimed to be responsible for the high magnitude of $\Delta V^{\ddagger}$ observed in this case.

Level 3 – Case 33
The Rate-Determining Step in the S_NAr Reaction

Key point: *Kinetic isotope effects. Base catalysis*

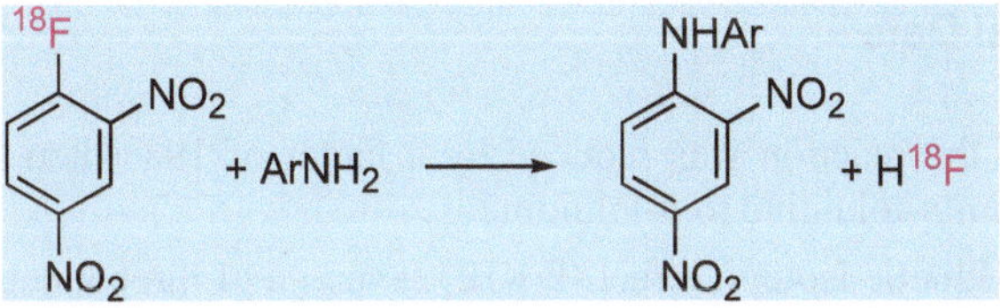

The Rate-Determinig Step in the S_NAr Reaction

In spite of the fact that the nucleophilic aromatic substitution in electron-deficient aromatic rings (S_NAr) is one of the most venerable organic reactions, it is still a subject of debate. The generally accepted mechanism for this reaction is the classical addition-elimination sequence that involves the formation of a Meisenheimer-type intermediate **1** (Scheme 33.1).

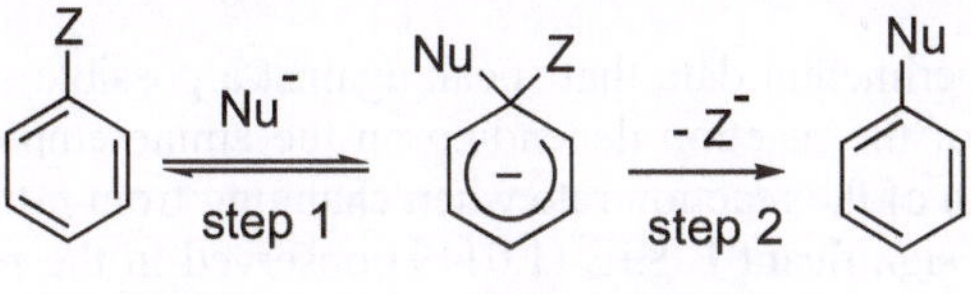

Z = electron-withdrawing group

Scheme 33.1

However, most of the data available concerns electronic effects and little is known about the influence of steric effects in determining whether the formation (step 1) or the decomposition of the intermediate complex **1** (step 2) is rate limiting. To obtain more information about this question the reaction between fluorine-labeled 2,4-dinitrofluorobenzene (DNFB) and *o*- and *p*-toluidines (2 and 4-methylanilines, respectively) has been studied (Scheme 33.2).

Based on the experimental data, discuss the influence of the steric effects on the rate-determining step (rds) of the reaction with o- and p-toluidine, formulating the kinetic law for each case.

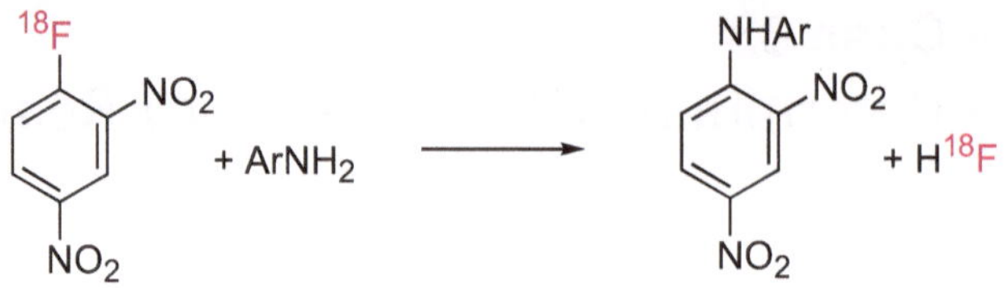

Scheme 33.2

Experimental Data

1. The rate of the reaction was reduced by a factor of 198 when the amine was changed from *p*-toluidine to *o*-toluidine.

2. The $^{18}F/^{19}F$ kinetic isotope effect (F KIE) (average of three kinetic experiments performed in DMSO at 30°C) was 1.0005 ± 0.0030 for *p*-toluidine and 1.0119 ± 0.0037 for *o*-toluidine. The estimated maximal $^{18}F/^{19}F$ KIE value is 1.032.[1]

3. The overall rate law in all cases can be written as $v = k_{obs}$ [DNFB] [amine]. It has been observed that k_{obs} shows a *nonlinear dependence* (small, but distinct curvilinear dependence) with the amine concentration in the case of *o*-toluidine. This effect has not been detected in the case of *p*-toluidine (k_{obs} is not dependent on the amine concentration).

Discussion

There are two experimental data that speak against a possible change in the rate-determining step of the reaction depending on the amine employed. First the remarkable reduction of the reaction rate when changing from *p*-toluidine to *o*-toluidine. Second, the significant F KIE (1.0119) observed in the reaction with sterically more-hindered *o*-toluidine compared with the negligible F KIE (1.0005) of the reaction with unhindered *p*-toluidine. Before starting a full discussion on these topics it is convenient to have a detailed picture of the two steps of the S_NAr mechanism for the reaction. The first step involves the nucleophilic attack of the amine to the DNFB and the subsequent formation of the Meisenheimer complex **2**. The second step consists of the decomposition of the intermediate to yield the substitution product. This step can or can not be catalyzed by base, as we will comment later on (Scheme 33.3).

---

[1] Despite their great value in mechanistic and bioorganic chemistry, isotope effects for the element fluorine have been scarcely used. This is due to the fact that natural fluorine consists entirely of the isotope ^{19}F. However, natural fluorine may be used in combination with the accelerator-produced short-lived radionuclide ^{18}F ($t_{1/2} = 110$ min) to determine KIEs. See Matsson O, Persson J, Axelsson BS, Långström B (1993) *J. Am. Chem. Soc.* 115:5288-5289.

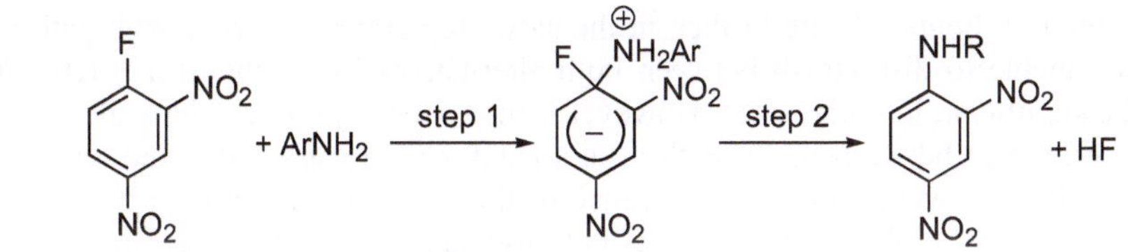

Scheme 33.3

What is evident from the two steps represented in Scheme 33.3 is that the expected F KIEs will be different depending on which step is rate-limiting. The F KIE is observed if the **C-F bond is being broken in the transition state of the slow step of the reaction**, and that occurs only during the decomposition of intermediate **2**. In other words, we can consider that the absence of F KIE is an argument in favor of step 1 being rate-determining. Based on the measured $^{18}F/^{19}F$ KIE for *o*-toluidine (1.0119) and $^{18}F/^{19}F$ KIE for *p*-toluidine (1.0005) we can affirm that the slow step of the reaction is different depending on the amine employed. *With p-toluidine the rate-determining step is the formation of the Meisenheimer intermediate* **2**, *whereas with o-toluidine the decomposition of* **2** *is the slow step of the process.*

To assert that the decomposition of the Meisenheimer intermediate **2** (step 2) is the rds in the case of *o*-toluidine requires being more precise, as this step is a little more complicated than what has been described in Scheme 33.3. The decomposition of **2** can follow two alternative pathways that are indicated in Scheme 33.4. The first alternative (path a) is a base-catalyzed decomposition process, which consists of the fast deprotonation of intermediate **2** by the amine to form intermediate **3**, followed by slow departure of the fluorine ion. The other, is an uncatalyzed decomposition pathway (path b) that consists of the initial elimination of fluorine ion leading to aromatic ammonium salt **4**, that is subsequently deprotonated by the base present in the medium in a fast acid-base exchange (Scheme 33.4).

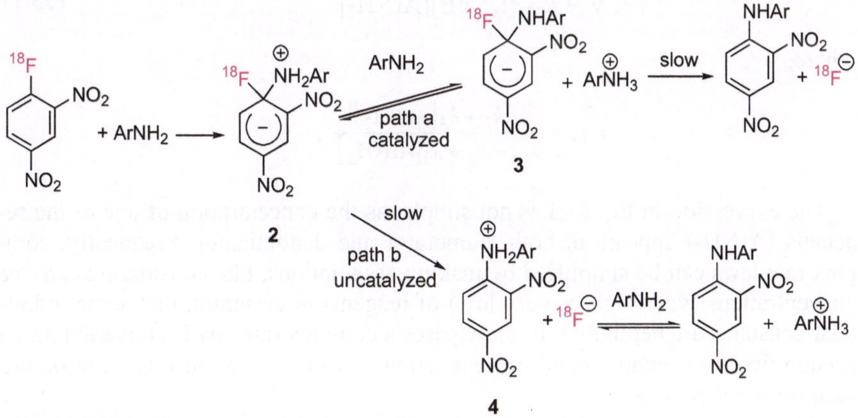

Scheme 33.4

As the C-F bond is being broken in the slow step either in path a or in path b, we are unable to distinguish between both alternatives by means of a F KIE (it will be significant in both cases). However from a kinetic point of view, there is a main difference between them. Path a, is a catalyzed process and if the reaction follows this pathway, a linear dependence of the rate constant and the base concentration must be observed. Path b, is not base-catalyzed and in consequence, if the reaction follows this pathway, the rate constant should be independent of the concentration of base. The experimental data indicate that the reaction of DNFB with o-toluidine shows a dependence of k_{obs} with the concentration of base. This should point to the base-catalyzed mechanism. The odd fact is that the dependence of k_{obs} with the concentration of base is not linear, as should be generally expected for the catalysis process. To understand this apparent incongruity we should make a kinetic analysis of the overall process indicated in Scheme 33.5.

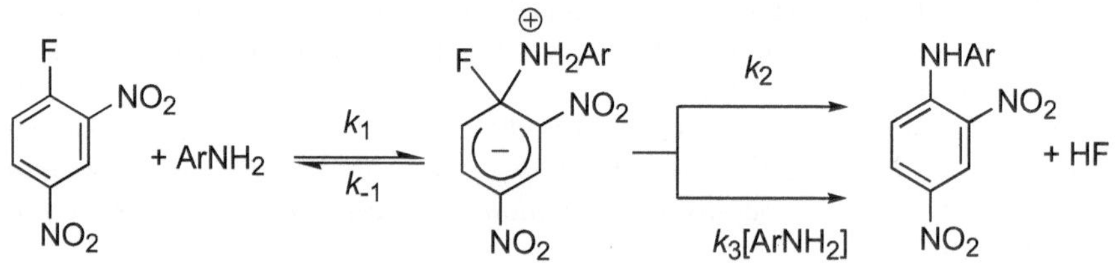

Scheme 33.5

Following the standard methodology, the analysis of a *complex reaction* like this one, can be simplified by the application of the *steady-state approximation*. To apply this approximation we assume that, after an initial brief period, the concentration of an intermediate species (the Meisenheimer intermediate **2** in this case) achieves a steady-state in which its rate of formation is equal to its decomposition rate. The resulting equation will be (Eq. 33.1):

$$v = k_{obs}[\text{DNFB}][\text{ArNH}_2]$$ (33.1)

where,

$$k_{obs} = \frac{k_1(k_2 + k_3[\text{ArNH}_2])}{k_{-1} + k_2 + k_3[\text{ArNH}_2]}$$

The expression in Eq. 33.1 is not simple, as the concentration of one of the reactants [ArNH₂] appears in both, numerator and denominator. Frequently, complex rate laws can be simplified by making assumptions, like considering *extreme* concentrations (very high or very low) of reagents or assuming that some individual constants are negligible. In these cases a complex rate law is converted into a pseudo first- or pseudo second-order reaction. *In other cases, however, the simplification is not possible.*

Let us suppose that the slow step of the reaction is the addition of the amine to the DNFB to form the Meisenheimer intermediate (step 1). As we have discussed before, this is the situation when p-toluidine is used as the nucleophile. Under these circumstances we could consider that k_{-1} is negligible, ($k_{-1} <<< k_2 + k_3[ArNH_2]$) and hence, $k_{obs} \sim k_1$.

The kinetic law of the reaction will be $v = k_1$ [DNFB] [$ArNH_2$] and the complex expression of k_{obs} in Eq. 33.1 has been transformed into a simple pseudo second-order constant.

If we assume that for o-toluidine, the base-catalysed process is rate determining (k_3) then, we could consider that in Eq. 33.1, $k_{-1} >>> k_2 + k_3[ArNH_2]$. That is, the reversion to the intermediate complex to reactants must be very fast. The observed rate constant in this case would have the form:

$$k_{obs} = \frac{k_1\left(k_2 + k_3[ArNH_2]\right)}{k_{-1}}$$
(33.2)

In other words, making these assumptions k_{obs} depends **linearly** on the concentration of base.

This is not the experimental result. How could we explain the nonlinear dependence of k_{obs} with the concentration of base when the reaction is carried out with o-toluidine? When we simplify a rate law we are always considering limiting situations that can or cannot be true. In the case of o-toluidine, the nonlinear dependence with the base concentration is indicating that no simplification of Eq. 33.1 should be possible and hence that $k_{-1} \sim k_2 + k_3$ [$ArNH_2$]. That means that despite the decomposition of the intermediate complex to the final products ($k_2 + k_3[ArNH_2]$ is the slow step of the reaction) it competes to some extent with the partial reversion to the reactants (k_{-1}).

What is the cause of the differences observed in the S_NAr reaction with o- and p-toluidines? It seems that moving the methyl substituent from the *para*- to the *ortho*-position affects the rate of decomposition of the Meisenheimer intermediate. The nucleophilic attack (k_1) is more difficult in the case of o-toluidine and leads to a more congested Meisenheimer intermediate. Steric compression in the intermediate may be relieved by partial reversion to the reactants, enhancing k_{-1} and converting this route into a competitive alternative to the decomposition pathway. The rate expression for o-toluidine (Eq. 33.1) reflects that the two decomposition pathways of the intermediate (reversion to reactants k_{-1} and formation of the reaction products ($k_2 + k_3$[amine]) have to be considered.

The significant F KIE observed for o-toluidine as compared to p-toluidine clearly demonstrates a change from rate-limiting nucleophilic addition to rate-limiting nucleofuge detachment for the sterically more hindered nucleophile (Scheme 33.6).

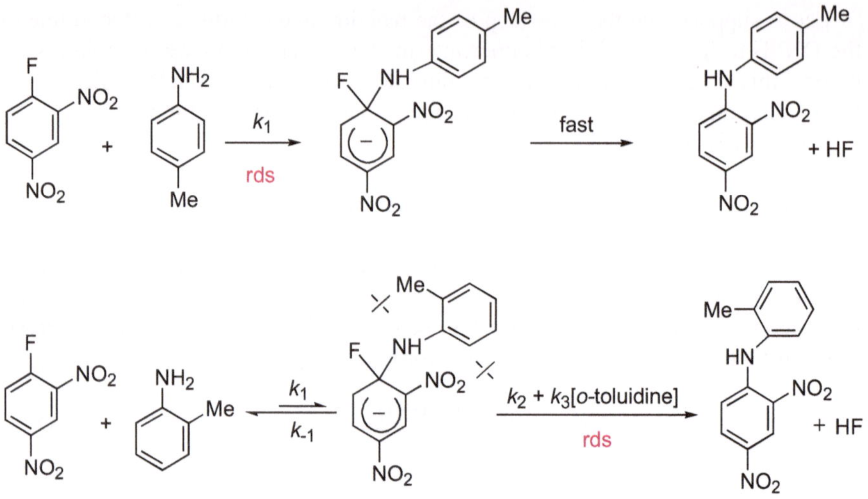

Scheme 33.6

In Summary

The influence of steric effects on the S_NAr mechanism has been studied by means of leaving group fluorine kinetic isotope effects. The reaction of DNFB with *o*- and *p*-toluidines shows a change in the rate-determining step depending on the steric hindrance of the amine employed.

Additional Comments

To obtain Eq 33.1 we have applied the steady-state approximation to the intermediate **2**. If **2** achieves a steady-state, then we can write:

$$\frac{-d[2]}{dt} = k_2[2] + k_{-1}[2] - k_1[DNFB][ArNH_2] + k_3[2][ArNH_2] = 0 \tag{33.3}$$

Hence,

$$[2] = \frac{k_1[DNFB][ArNH_2]}{k_2 + k_{-1} + k_3[ArNH_2]} \tag{33.4}$$

If we consider that

$$v = \frac{d[product]}{dt} = k_2[2] + k_3[2][ArNH_2] \tag{33.5}$$

the rate expression will be:

$$v = \frac{k_1 (k_2 + k_3[\text{ArNH}_2])}{k_2 + k_{-1} + k_3[\text{ArNH}_2]}[\text{DNFB}][\text{ArNH}_2] \tag{33.1}$$

were,

$$k_{\text{obs}} = \frac{k_1(k_2 + k_3[\text{ArNH}_2])}{k_{-1} + k_2 + k_3[\text{ArNH}_2]}$$

This problem is based on the work by Persson J, Matsson O (1998) *J. Org. Chem.* 63:9348-9350 and the work by Onydo I, Hirst J (1991) *J. Phys. Org. Chem.* 4:367-371.

Subjects of Revision

S_NAr mechanism. Rate-determining step. Steady state approximation. Kinetic isotope effects.

Level 3 – Case 34
Helicenophanes and their Racemization

Key point: *Activation parameters*

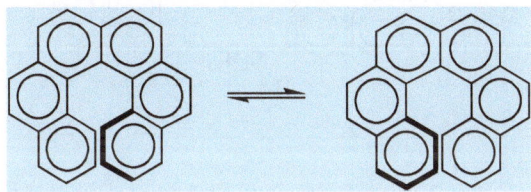

Helicenophanes and their Racemization

Helicenes are a fascinating class of compounds. They are prototypes of over-crowded molecules adopting helical configurations, and hence they are chiral. As a group, they share a very high specific rotation (for example $[\alpha]_D^{25} = 9620$ for [13]helicene), and have been found to racemize thermally (Scheme 34.1). These facts make them archetypical compounds to study the physico-chemistry of the racemization process by polarimetry. The accepted mechanism to account for this racemization takes place through a conformational change, with the necessary bond deformations being spread over much of the molecular skeleton.

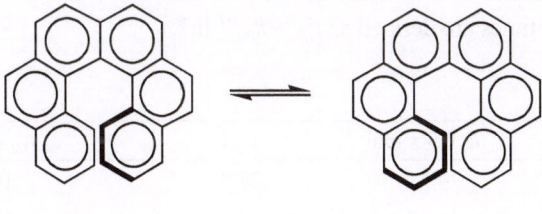

1

Scheme 34.1

It is known that alkoxy-substituted helicenes racemize more slowly than their unsubstituted congeners. For example helicene **2** has a $t_{1/2}$ more than twice as long as that of the unsubstituted parent helicene **1** (Figure 34.1).[1] To study the influence

---

[1] $t_{1/2}$ is the *half-life* of the reaction, that is, the time required for the amount or concentration of reactant to decrease by 50%.

of substituents on the process, the thermal racemization of [6]helicene **1** and its (bridged) substituted derivatives **3** and **4** has been studied. The first order racemization constants were measured in a temperature range between 120°C and 200°C, by following the changes of the optical rotation $[\alpha]_D$ in 1,2,4-trichlorobenzene.

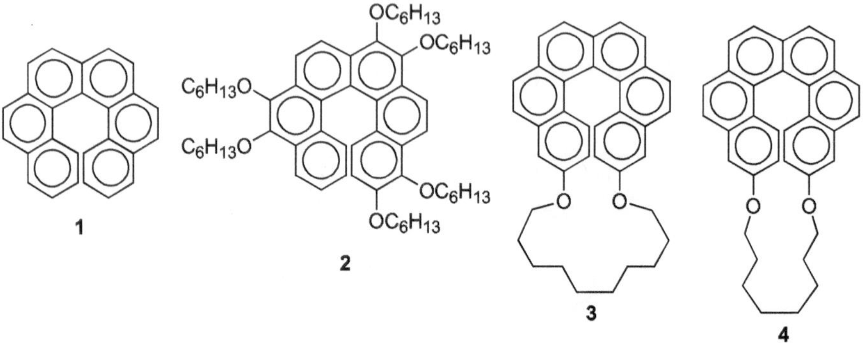

Figure 34.1

*Using the Arrhenius parameters calculate the activation parameters for the racemization of [6]helicenes **1**, **3** and **4** at 200°C. Explain the behavior towards racemization of unbridged[6]-helicene **1** and its bridged congeners **3** and **4**.*

Experimental Data

1. The activation energy (E_a) and the pre-exponential factor $(\ln A)$ were obtained graphically by application of the Arrhenius equation,

   $\ln k_{\mathrm{rac}} = \ln A - E_a/RT$ and are collected in Table 34.1.

   The half-life times are defined as $t_{1/2} = k_{\mathrm{rac}}^{-1} \ln 2$.

Table 34.1

Compound	E_a (kJ x mol^{-1})	$\ln A$	$t_{1/2}$ (min)
1	148.1	27.9	106
3	143.7	29.0	12
4	116.7	24.7	0.8

2. During the racemization process it is proposed that there is a change from a C_2-symmetry (ground state) to a C_s-symmetry (transition state).

Discussion

According to the transition state theory (Eyring equation) enthalpy ($\Delta H^{\ddagger}$), entropy ($\Delta S^{\ddagger}$) and free energy ($\Delta G^{\ddagger}$) of activation can be calculated from the Arrhenius parameters E_a and A obtained experimentally. The activation energy E_a is related to the enthalpy of activation through Eq. 34.1. On the other hand, the pre-exponential factor A can be related to the entropy of activation through Eq. 34.2. Once both $\Delta H^{\ddagger}$ and $\Delta S^{\ddagger}$ are known, it is easy to calculate $\Delta G^{\ddagger}$ by means of Eq. 34.3 (k_B and h are the Boltzmann and Planck constants respectively, T the absolute temperature and the gas constant R is given in J K^{-1}mol^{-1}).

$$Ea = \Delta H^{\ddagger} + RT \tag{34.1}$$

$$A = \frac{k_B Te}{h} e^{\Delta S^{\ddagger}/R} \tag{34.2}$$

$$\Delta G^{\ddagger} = \Delta H^{\ddagger} - T\Delta S^{\ddagger} \tag{34.3}$$

The calculated values for the activation parameters at 200°C are collected in Table 34.2.

Table 34.2

Compound	$\Delta G^{\ddagger}$ (kJ mol^{-1})	$\Delta H^{\ddagger}$ (kJ mol^{-1})	$\Delta S^{\ddagger}$ (J k^{-1} mol^{-1})
1	156.3	144.2	−25.6
3	147.7	139.8	−16.7
4	137.3	112.8	−51.8

If we compare the $t_{1/2}$ values in Table 34.1 it is clear that bridged compounds **3** and **4** racemize considerably faster than compound **1**. In addition, the shorter the bridge, the faster the racemization occurs. We have to remember that lower values of $t_{1/2}$ mean lower reaction times, faster reaction rates and greater k_{rac}. Checking the structural differences between compounds **1**, **3** and **4** the increase in the racemization rate must be related with:

- The presence of oxygen substituents (an electronic effect).
- The bridge that is present in 3 and 4 but not in 1.
- A combination of the two factors.

The experimental data state that [6]helicene **2** having **six alkoxy substituents!!** racemizes considerably slower than the parent [6]helicene **1**. Hence, we can conclude that the electron-donating ability of oxygen substituents should not be responsible for the increase in the racemization rates observed for compounds **3** and **4** regarding to **1**. We can discard the electronic effect of the substituents as the origin of the acceleration in the racemization process.

Let us check how the presence of a bridge in the helicene system influences the rate of racemization. From data in Table 34.1 it is clear that the shorter is the bridge the faster is the racemization (**4** racemizes faster than **3**). The influence of the bridge length is also manifested in the activation parameters. The entropy of activation ($\Delta S^{\ddagger}$) gives information about the increase (decrease) of order when passing from the reagents to the activated complex, whereas the enthalpy of activation $\Delta H^{\ddagger}$ measures all energy changes that occur between reagents and the activated complex. Finally, the $\Delta G^{\ddagger}$ values have no physical significance and give the same information as the rate constant to which they are directly related (transition state theory).

The $\Delta H^{\ddagger}$ values in Table 34.2 indicate that the energy balance of the racemization is more favorable for bridged compounds **3** and **4** than for the parent **1**, and even better for the helicene with the shorter bridge **4**. For this compound we observe the highest $\Delta S^{\ddagger}$ value, which indicates the most ordered transition state in the racemization process. These results are somehow surprising. One should have expected that the racemization would be easier in the compound with the longer bridge.

The polymethylenedioxy bridge clips the terminal benzene rings of the helicene structure together. This clipping places the aromatic rings close to the methylene chain causing *strong steric interactions in the ground state* that should be released during the transition state (Fig. 34.2). This is true due to the fact that this transition state has a C_s-symmetry, according to the data, and the evolution from a C_2 congested ground state to a C_s transition state should separate the aromatic termini. Therefore, compound **4** having the shorter chain should have also the higher ground state energy because the aromatic rings should be closer than in compounds **1** and **3** (evidently, these interactions are not present in [6]helicene **1**).

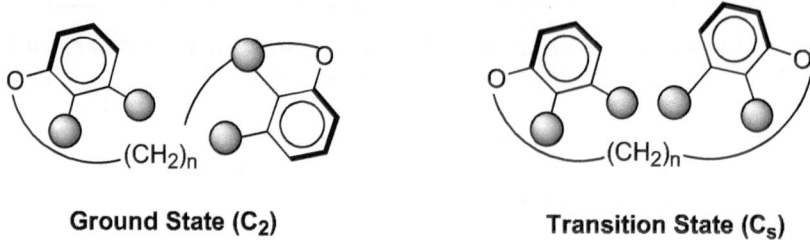

Ground State (C₂)   **Transition State (Cₛ)**

Figure 34.2

$\Delta H^{\ddagger}$ reflects the differences between the steric energies of the ground and the transition states of the helicenes. The actual situation for compounds **1, 3** and **4** is represented schematically in Fig. 34.3. The unbridged helicene **1** has the lowest ground state (steric) energy and therefore should go for the highest $\Delta H^{\ddagger}$ as in fact occurs. Because of the longer chain, bridged helicene **3** is sterically less congested than **4**. Therefore compound **4** should go for the smallest value of $\Delta H^{\ddagger}$. Furthermore, the high negative activation entropy of **4** is an indication of the lower probability for the turn in this compound compared to **3**, without doubt due to the shorter bridge chain. Thus, in these compounds both $\Delta H^{\ddagger}$ and $\Delta S^{\ddagger}$ cooperate to produce the highest racemization rate.

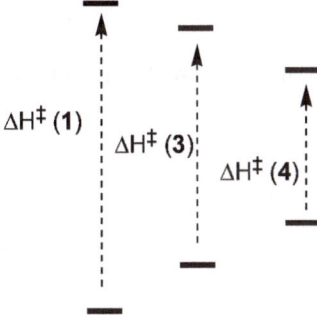

Figure 34.3

Comment

*Frequently when discussing the physical significance of the activation parameters we forget the reagents because we are too aware of the transition state, either in regard to the bonds that are formed or broken or in connection with the gain or loss of degrees of freedom. We should remember that the activation parameters refer to the changes from the **reagents** to the **activated complex in the transition state**.*

In Summary

The racemization of helicenes **1, 3** and **4** has been studied considering the values of the activation parameters. The easiness in the isomerization process is related with the release of steric energy that occurs during the change of a congested C_2-symmetric ground state to a C_s-transition state. The effect of the tethering chain of **3** and **4** into the entropy of activation can also be explained, since shorter chains provoke more ordered transition states.

Additional Comments

It was suggested that racemization of helicenes might take place by an internal double Diels-Alder reaction. Starting from any of the stereoisomers of [6]helicene formulate its racemization by this mechanism. Further work using isotope labeling discarded this suggestion. Formulate an experiment using isotope labeling to discard the double Diels-Alder mechanism.

This problem is based on the work by Meier H, Schwertel M, Schollmeyer D (1998) *Angew. Chem. Int. Ed.* 37:2110-2113.

Subjects of Revision

Activation parameters. Symmetry. Chirality in molecules devoid of chiral centers. Helicoidal chirality.

Level 3 – Case 35
Solvolysis of Vinyl Iodonium Salts

Key point: *Neighboring group participation.*
Stereochemistry. Reactive intermediates

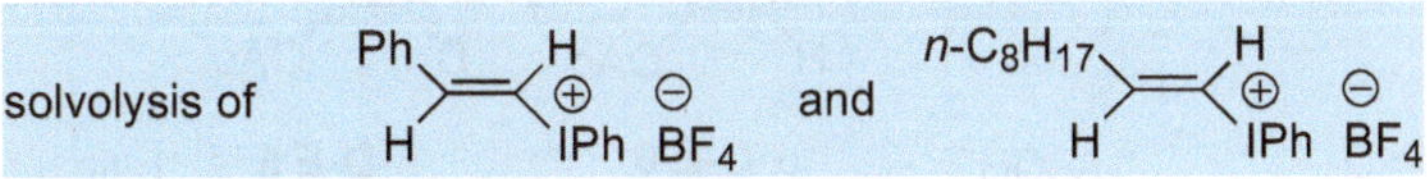

solvolysis of

$$Ph \qquad H \qquad \overset{\oplus}{IPh} \quad \overset{\ominus}{BF_4} \qquad and \qquad n\text{-}C_8H_{17} \qquad H \qquad \overset{\oplus}{IPh} \quad \overset{\ominus}{BF_4}$$

Solvolysis of Vinyl Iodonium Salts

The solvolysis in acetic acid (acetolysis) of *E*-styryl(phenyl)iodonium and *E*-1-decenyl(phenyl)iodonium tetrafluoroborates **1** and **2** has resulted in a particular interest from a mechanistic point of view, as both substrates behave in a different manner under the same conditions.

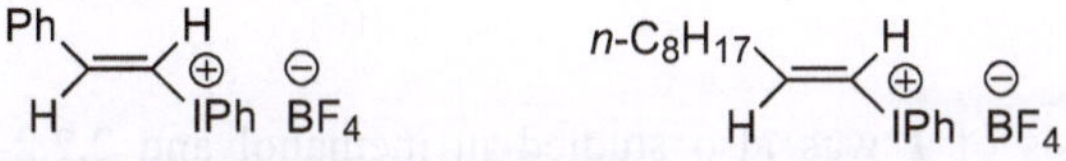

$$Ph \qquad H \qquad \overset{\oplus}{IPh} \quad \overset{\ominus}{BF_4} \qquad\qquad n\text{-}C_8H_{17} \qquad H \qquad \overset{\oplus}{IPh} \quad \overset{\ominus}{BF_4}$$

$$\mathbf{1} \qquad\qquad\qquad \mathbf{2}$$

Based on the following experimental data, discuss the mechanism of solvolysis of **1** *and* **2**.

Experimental Data

Acetolysis of **1** yielded an 85:15 mixture of *E/Z* acetates **3**, together with iodobenzene and phenylacetylene, whereas acetolysis of **2** exclusively gave *Z*-alkene **4** and iodobenzene (Scheme 35.1). In addition, the acetolysis of **1** was two orders of magnitude slower than that of **2**.

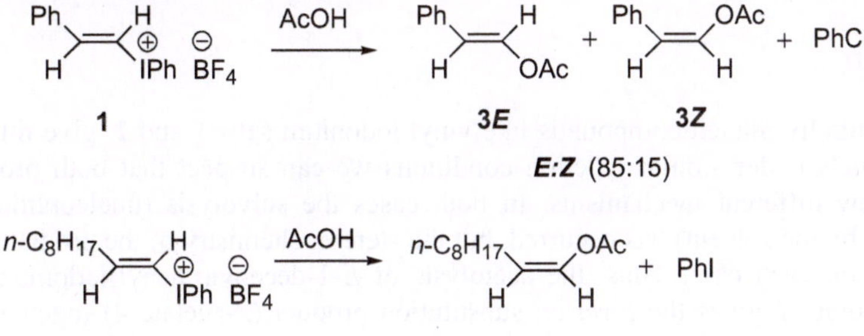

$$Ph \quad H \quad \overset{\oplus}{IPh} \overset{\ominus}{BF_4} \xrightarrow{AcOH} Ph \quad H \quad OAc \ + \ Ph \quad OAc \ + \ PhC{\equiv}CH + PhI$$

$$\mathbf{1} \qquad\qquad\qquad \mathbf{3E} \qquad\qquad \mathbf{3Z}$$

$$\mathbf{E{:}Z}\ (85{:}15)$$

$$n\text{-}C_8H_{17} \quad H \quad \overset{\oplus}{IPh}\overset{\ominus}{BF_4} \xrightarrow{AcOH} n\text{-}C_8H_{17} \quad OAc \ + \ PhI$$

$$\mathbf{2} \qquad\qquad\qquad\qquad \mathbf{4}$$

Scheme 35.1

Labeling experiments carried out with α-deuterated *E*-styryl(phenyl)iodonium tetrafluoroborate **5** resulted in the same 85:15 *E/Z* isomeric mixture of alkenes **6**. However, whereas the *Z*-isomer **6** retained the label at the original α-position, scrambling of deuterium between both the α and the β positions was detected in the *E*-isomer (Scheme 35.2). The analysis of the ^{1}H-NMR spectrum of the product mixture showed that the deuterium was distributed equally both at the α and β positions of the *E*-isomer **6**. Unlabeled phenylacetylene was also formed as by-product in these experiments.

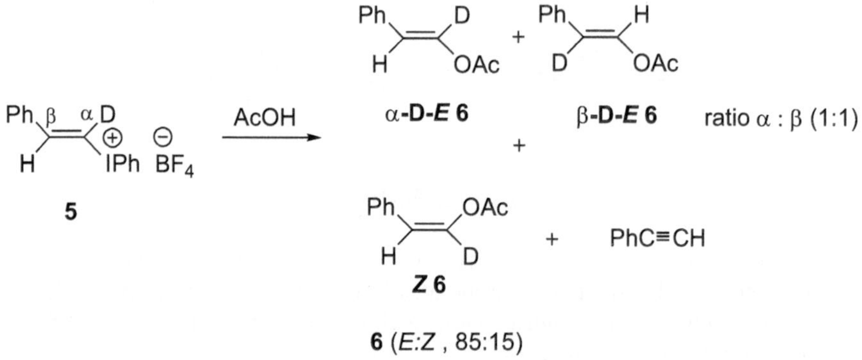

Scheme 35.2

The solvolysis of **1** was also studied in methanol and 2,2,2-trifluoroethanol (TFE). Whereas the results in methanol were similar to those obtained in acetic acid (see Scheme 35.1), the solvolysis of **1** in TFE exclusively yielded *E*-(2,2,2-trifluoroethoxy)styrene **7** (Scheme 35.3).

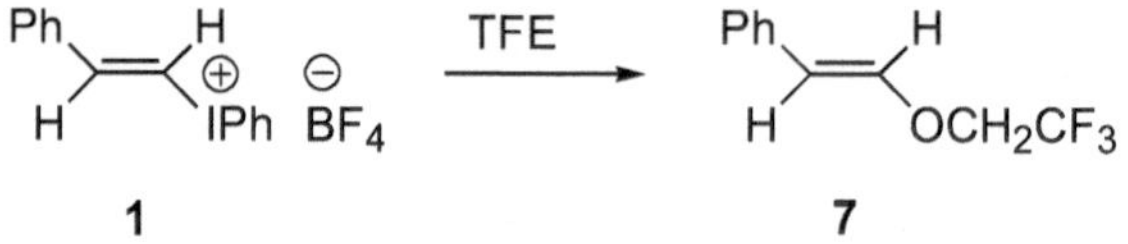

Scheme 35.3

Discussion

When structurally related compounds like vinyl iodonium salts **1** and **2**, give different products under similar reaction conditions we can suspect that both processes follow different mechanisms. In both cases the **solvolysis** (nucleophilic substitution by the solvent) has occurred, but the stereochemistry of the products is different on each case. Thus, the acetolysis of *E*-1-decenyl(phenyl)iodonium tetrafluoroborate **2** gives the *inverted* substitution product (*Z*-acetate **4**) together with iodobencene (the leaving group). This result suggests a nucleophilic substitution at a vinylic carbon with inversion of the configuration, which is known as *in-*

plane vinylic S$_N$2 mechanism. Like in the aliphatic bimolecular nucleophilic mechanism (S$_N$2), the reaction occurs by direct attack of the nucleophile (acetic acid) to the substrate. The nucleophile approaches the vinylic carbon opposite to the leaving group and both, nucleophile and substrate, take part in the transition state **8** (Scheme 35.4). The name *in-plane* derives from the flat topology of the transition state (represented by **8**).

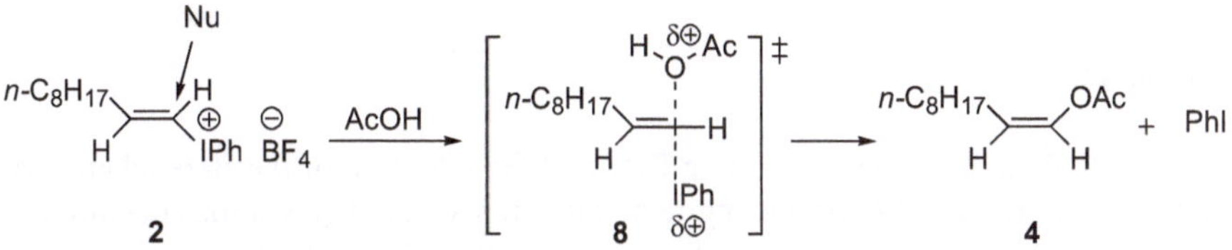

Scheme 35.4

To understand the results obtained in the acetolysis of *E*-styryl(phenyl)iodonium tetrafluoroborate **1** is a bit more complicated. As we can see in Scheme 35.1, the reaction affords a mixture of *retained* (**3E**) and *inverted* (**3Z**) solvolysis (substitution) products, together with iodobencene (the leaving group) and phenylacetylene (clearly an elimination product). It is almost impossible to formulate a single mechanism that could explain the formation of *all the reaction products*. In consequence, several mechanistic pathways should be considered in this case.

Let us first concentrate on the mixture of solvolysis products **3**. Isomer **3E** (the major isomer) *retains* the stereochemistry of the starting material, whereas isomer **3Z** (the minor isomer) *inverts* the stereochemistry of the starting material. In addition, the labeling experiments carried out with substrate **5** indicate scrambling of D/H in the *retained E*-isomers **6**, whereas the position of the label is not modified in the *inverted Z*-isomer (Scheme 35.5).

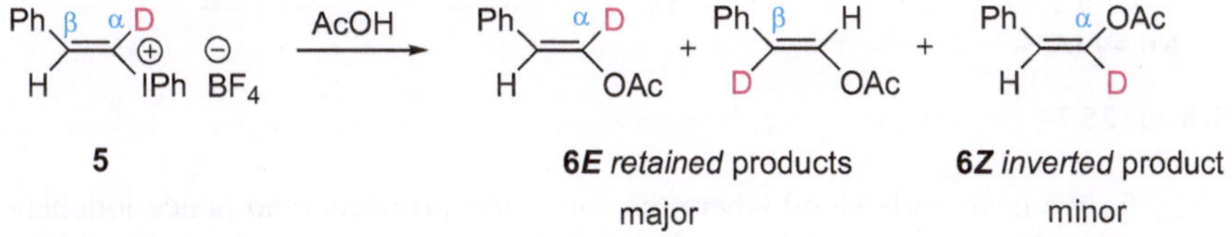

Scheme 35.5

Previously, we have discussed that the inversion of the stereochemistry in the solvolysis product of vinyl iodonium salt **2** could be explained by an *in-plane S$_N$2 vinylic substitution.* The same arguments can be employed now to explain how the minor isomer **6Z** is formed (Scheme 35.6). Direct attack of acetic acid to the α-carbon should lead to the observed reaction product in a one-step process through transition state **9**. The position of the deuterium atom should not change (no scrambling), only the stereochemistry of the product has been reversed during the reaction.

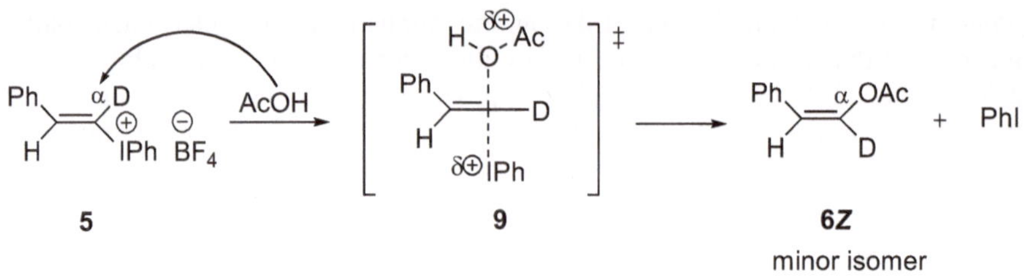

Scheme 35.6

Nevertheless, both the scrambling D/H and the retention of the stereochemistry in the major products **6E** are incongruent with an in-plane S$_N$2 vinylic substitution. It is known however, that an aromatic ring in the β-position of a leaving group can actively participate in a substitution reaction as *neighboring group*. The aryl behaving as neighboring group pushes out the leaving group to give a bridged ion called *phenonium ion*. Now consider this alternative for labeled *E*-styryl(phenyl)-iodonium tetrafluoroborate **5** (Scheme 35.7). Departure of iodobenzene with assistance of the phenyl group, would lead to the formation of the symmetric vinylene-phenonium ion **10**, which by nucleophilic attack of the solvent at the α- and β-positions, gives products **6E** with net retention of configuration and complete scrambling of deuterium. Phenonium ion **10** is symmetric and would undergo nucleophilic attack at the α- and β-carbons with an equal probability.

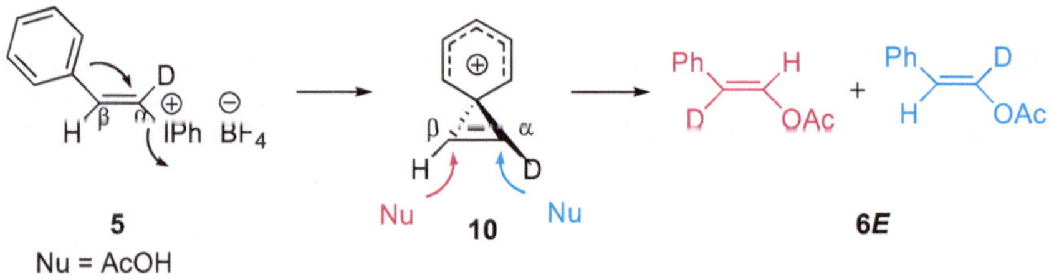

Scheme 35.7

So far we have understood where the solvolysis products (and hence iodobenzene) come from. However, the mechanisms discussed above cannot explain how phenylacetylene (the remaining reaction byproduct) is formed. A triple bond can result from an elimination process. Furthermore, when the reaction is carried out with deuterated vinyl iodonium **5**, **unlabeled** phenylacetylene is produced (see Scheme 35.2). Considering these arguments, a reasonable pathway to explain the formation of phenylacetylene could be a solvent-assisted α-elimination (Scheme 35.8). The removal of the H/D by the solvent and simultaneous departure of the iodobenzene would yield a vinylidene carbene **11**, which subsequently rearranges to the alkyne by migration of hydrogen. The complete loss of the label would be fully compatible with this route.

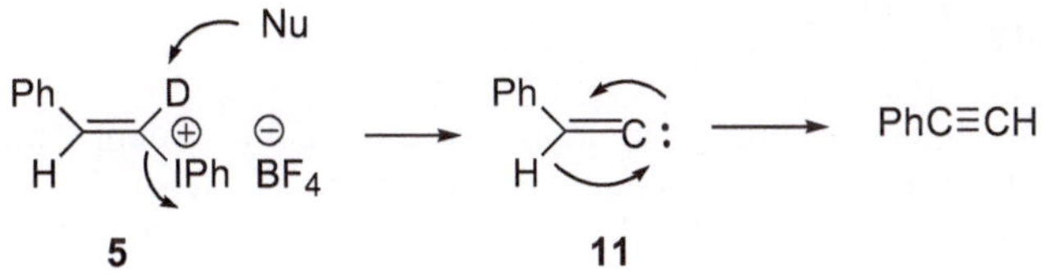

Scheme 35.8

In conclusion: The products obtained in the acetolysis of vinyl iodonium salt **1** are formed by three different mechanisms: a) direct in-plane S_N2 vinylic substitution, b) substitution with neighboring group participation of the phenyl group and c) solvent-assisted α-elimination.

The three routes are summarized in Scheme 35.9.

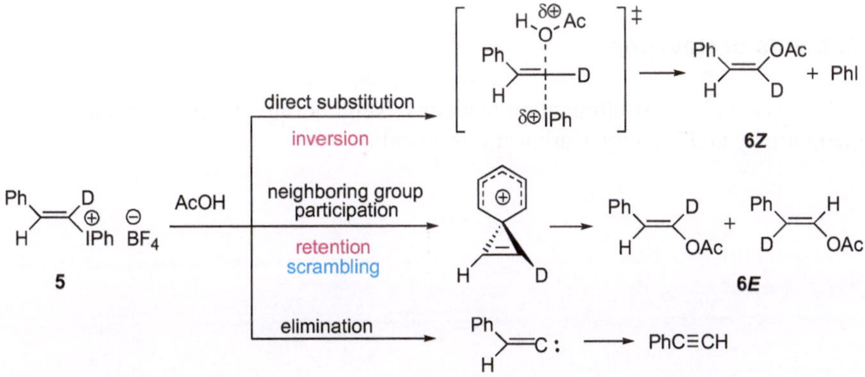

Scheme 35.9

The single question left is to comment the influence of the solvent in the process. The solvolysis of **1** in AcOH or MeOH yields a similar distribution of products, however, in TFE only the substitution product with retention of the configuration **7** is formed. TFE is a much weaker base and poorer nucleophile than AcOH and MeOH. In consequence, elimination and direct vinylic S_N2 substitution are much more difficult in TFE than in the other two solvents. The only isolated product in this case should come from the fenonium route (Scheme 35.10).

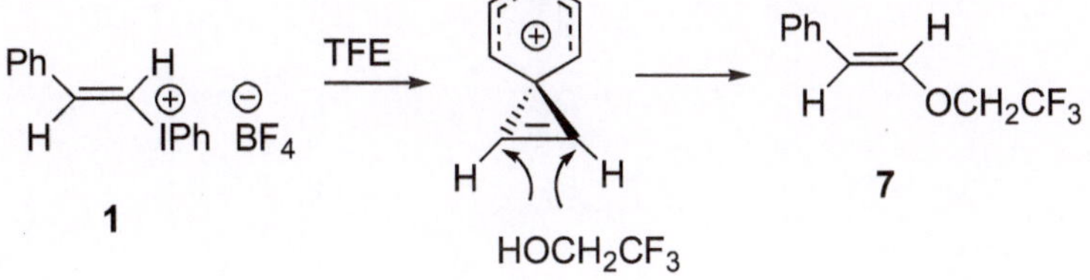

Scheme 35.10

In Summary

Owing to the superb leaving group ability of the iodonio group, vinyl iododium salts give different products under solvolysis conditions. Depending on the structure of the salt, in-plane vinylic S_N2 substitution, nucleophilic substitution with neighboring group participation, and even elimination processes can be observed.

Additional Comments

This problem is based on the work by Okuyama T, Ochiai M (1997) *J. Am. Chem. Soc.* 119:4785-4786 and by Gronheid R, Lodder G, Okuyama T (2002) *J. Org. Chem.* 67:693-702.

Subjects of Revision

Solvolysis. S_N2 and α-elimination reactions. Neighboring group participation. Nucleophilicity and basicity. Carbocations: stability.

Level 3 – Case 36
Vicarious Nucleophilic Substitution

Key point: *Change in the rate-determining step*

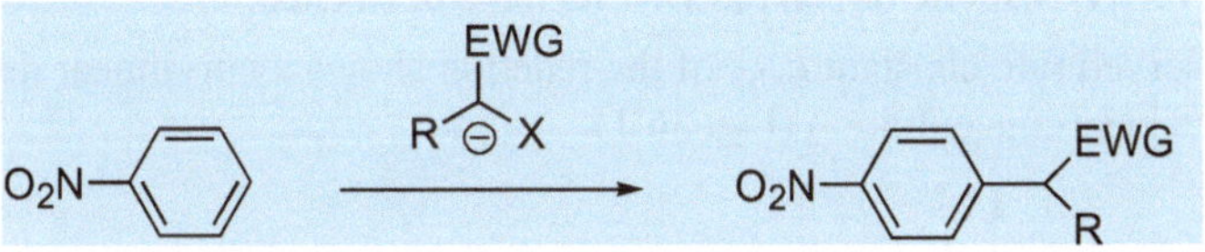

Vicarious Nucleophilic Substitution

The alkylation of aromatic nitro compounds by carbanions having a leaving group at the nucleophilic center is called *Vicarious Nucleophilic Substitution of Hydrogen* (VNS) (Scheme 36.1). This reaction is one of the scarce general processes that result in the formal nucleophilic aromatic substitution of hydrogen.

EWG = electron-withdrawing group
X = leaving group

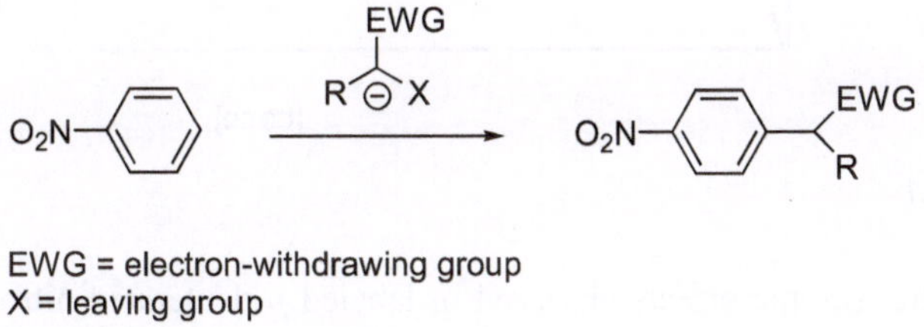

Scheme 36.1

The reaction proceeds by addition of the nucleophile to the *ortho*- or *para*-position of the nitroarene and subsequent base-promoted β-elimination of XH from the intermediate σ-adduct **1**. The VNS reaction is very general and the *ortho/para* regioselectivity is controlled by steric factors. Tertiary carbanions replace hydrogen in the VNS reaction exclusively in the *para*-position (Scheme 36.2).

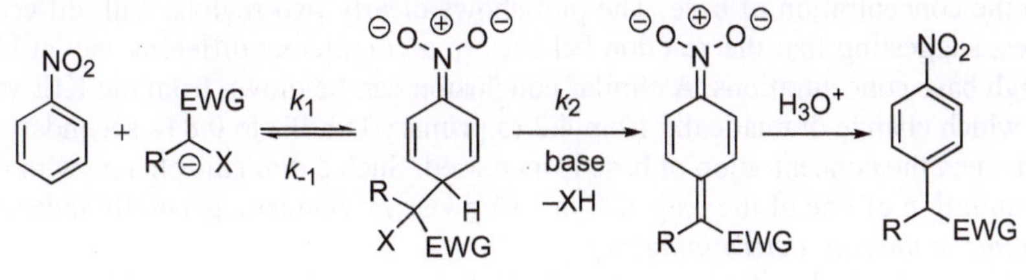

Scheme 36.2

Determine which is the slow step of the reaction with the help of the following kinetic data.

Experimental Data

1. The reaction proceeds with a variety of nucleophiles such as carbanions containing halogen, alkoxy, thioalkoxy, and other leaving groups. Among the bases, KOH, NaOH, *t*-BuOK or NaH in DMSO have been used.
2. The observed rate constant k_{VNS} of the reaction shows a curvilinear dependence upon the base concentration (Fig. 36.1).

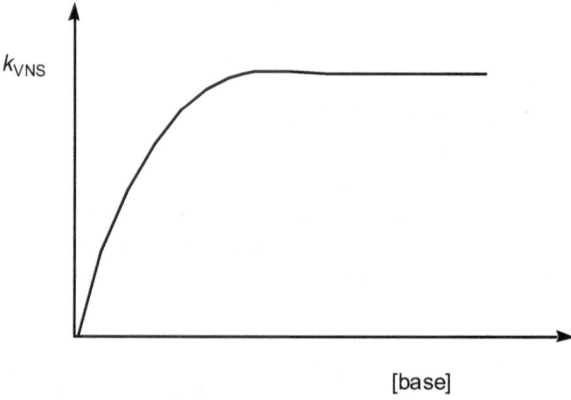

[base]

Figure 36.1

3. The kinetic isotopic effects observed in labeled *p*-deuterated nitrobenzene decrease with increasing base concentration. As an example KIE values from 4.2 ([base] < 0.1mol/L) to 0.8 ([base] in the range 0.1-0.6 mol/L) have been obtained.

Discussion

Figure 36.1 indicates a nonlinear dependence of the observed rate constant k_{VNS} with the concentration of base. The plot shows clearly two regions with different slopes, suggesting that the reaction behaves in a completely different way at low or high base concentrations. A similar conclusion can be drawn from the KIE values, which change dramatically from 4.2 (a primary D KIE) to 0.8 (a secondary D KIE) when the concentration of base is increased. Such a dramatic change with the concentration of one of the reagents, in a stepwise mechanism, generally indicates a *change in the rate-determining step*.

Let us study the kinetic law of the reaction. For a start it is reasonable to apply the steady-state approximation to the intermediate σ-adduct **1**. We assume that **1** is a reactive short-lived species and that after an initial brief period, its rate of formation and rate of disappearance are just balanced.

Considering the sequence of reactions in Scheme 36.2 and applying the steady-state approximation, the rate law for the VNS reaction will be given by Eq. 36.1:

$$v = \frac{k_1 \, k_2 [\text{base}]}{k_{-1} + k_2 [\text{base}]} [\text{nitrobencene}][\text{carbanion}] \qquad (36.1)$$

where

$$k_{\text{VNS}} = \frac{k_1 \, k_2 [\text{base}]}{k_{-1} + k_2 [\text{base}]}$$

At this point the reader is encouraged to deduce Eq. 36.1 step by step using the standard mathematic methodology of the steady-state approximation.

Equation 36.1 suggests a complex kinetic process and explains the observed non-linear dependence of k_{VNS} upon the base concentration. However, as shown in the plot in Fig. 36.1 we can consider two different limiting situations for the reaction, at low or high base concentration respectively.

At low base concentrations, we can assume that $k_{-1} \gg k_2$ [base] and then,

$$k_{\text{VNS}} = \frac{k_1 \, k_2 [\text{base}]}{k_{-1}}$$

Considering the equilibrium constant $K_1 = k_1 \, / \, k_{-1}$, the rate law will be reduced to Eq. 36.2:

$$v = K_1 k_2 [\text{base}][\text{nitrobencene}][\text{carbanion}] \qquad (36.2)$$

Equation 36.2 refers to a process that prevails when the intermediate σ-adduct **1** is formed in a fast pre-equilibrium (K_1). In this instance, the β-elimination step (k_2) is rate-determining and the rate of the overall VNS reaction will depend linearly on the concentration (and strength) of the base.

The other limiting situation corresponds to high concentration of base. Under these circumstances we can consider that k_2 [base] $\gg k_{-1}$ and Eq. 36.1 will be simplified to Eq. 36.3:

$$v = k_1 [\text{nitrobencene}][\text{carbanion}] \qquad (36.3)$$

As indicated in Eq. 36.3, at high base concentration the nucleophilic addition step (k_1) is rate limiting. This also means that the β-elimination step (k_2) is so fast that the formation of the intermediate σ-adduct **1** can be considered an irreversible process. There is not dependence on the base concentration. From the kinetic study we can conclude that the VNS reaction has a different rate law (and hence a different rate-determining step) depending on the concentration of base.

The change in the rate-determining step with the concentration of base is also in agreement with the observed deuterium KIEs. The experimental data indicate a decrease from 4.2 (a value typical of a primary KIE) to 0.8 (a secondary KIE value) with increasing base concentration. If we consider the previous kinetic discussion, at **low concentration of base** the transformation of σ-adduct **3** into intermediate **2** is rate-determining (Scheme 36.3). In consequence, as the labeled atom is directly involved in the elimination process, a primary KIE should be expected. This step probably proceeds via an E2-like mechanism.

low concentration of base

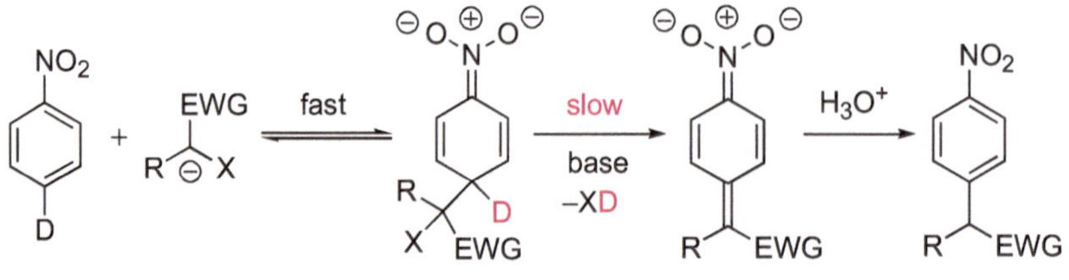

Scheme 36.3

At higher concentrations of base, the observed KIE values decrease up to 0.8. This is usual for an *inverse secondary deuterium isotope effect*, appropiate to the change from trigonal to tetrahedral carbon. This hybridization change occurs during the formation of the σ-adduct **3** (Scheme 36.4). At **high concentration of base**, the nucleophilic attack of the carbanion to the aromatic ring is the slow step of the reaction. The C-D bond is not broken during the process, but the aromatic Csp^2 changes to Csp^3 in the σ-adduct **3**, leading to the observed secondary KIE.

high concentration of base

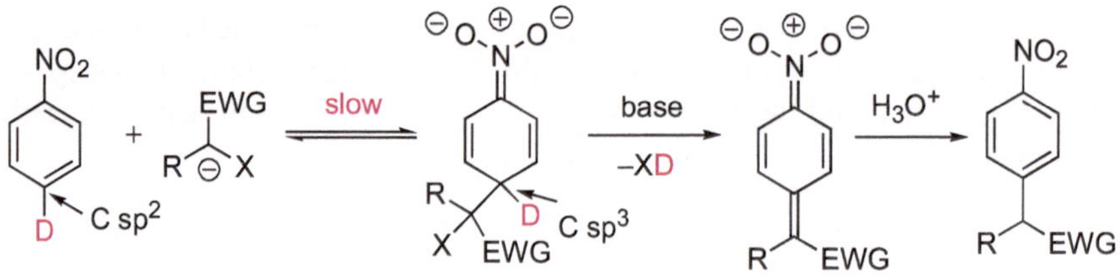

Scheme 36.4

In Summary

The vicarious nucleophilic substitution of hydrogen (VNS) occurs in two steps: a) nucleophilic addition of the carbanion to the nitroarene to form a σ-adduct and b) base induced β–elimination. Both of these processes can be rate-limiting: at low base concentration the β-elimination is the slow step whereas at high base concentration the nucleophilic addition is rate-determining.

Questions

1. Most of the kinetic data obtained to determine the influence of the base concentration in the VNS reaction have been obtained from competitive experiments between VNS and S_NAr reactions in fluoro nitrobenzenes like **4** with carbanions derived from sulfones **5**. Could you explain why? (R is a substituent that cannot be replaced by S_NAr).

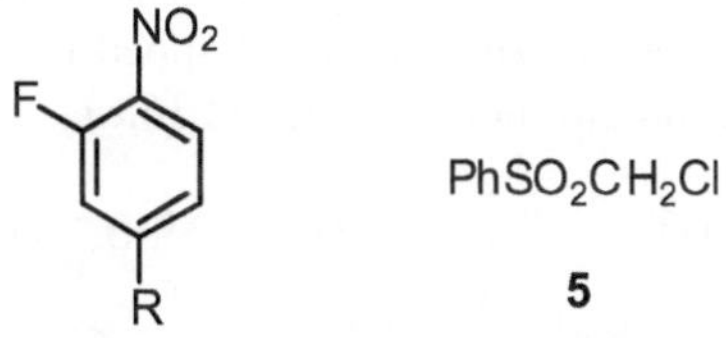

$PhSO_2CH_2Cl$

5

4

2. Draw a mechanism for the reaction between nitrobenzene and ethyl 2-chloropropionate in the presence of sodium hydride (DMSO as solvent) followed by addition of benzyl bromide. Justify the regiochemistry of the reaction.

Answer to Question 1

Substrates **4** have a fluorine and a hydrogen atom located in similarly activated positions of the same nitroarene molecule. The reaction of substrates **4** with nucleophiles would result in the substitution of either the hydrogen or the fluorine atom, leading to a mixture of S_NAr (**6**) and VNS (**7**) products (Scheme 36.5). In consequence, compounds **4** are ideal to evaluate the effect of base concentrations on the VNS mechanism. As the S_NAr mechanism is not influenced by base concentration (a reasonable assumption in most of the cases), the study of the VNS/S_NAr product ratio gives information about the effect of the base on the VNS reaction rate.

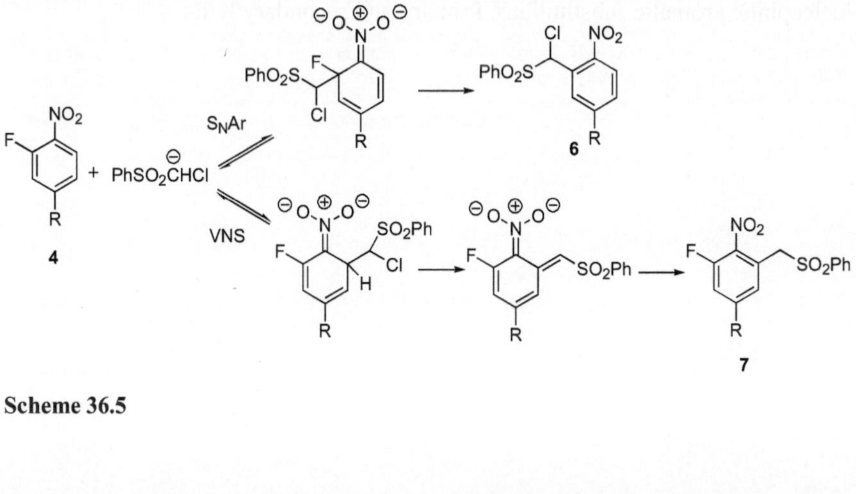

Scheme 36.5

Answer to Question 2

Following the VNS mechanism, the first step of the reaction is the nucleophilic addition of the enolate of ethyl 2-chloropropionate to the nitrobenzene ring (the ester forms the enolate in the reaction medium in the presence of NaH). As stated above, the *ortho/para* selectivity in VNS reactions is controlled by steric factors. Hence, a tertiary carbanion like **8** would attack exclusively in the *para*-position leading to σ-adduct **9**. The next step will be the base-induced elimination of HCl in **9** to yield anion intermediate **10**, which is trapped by the electrophile (benzyl bromide) giving the reaction product **11** (Scheme 36.6).

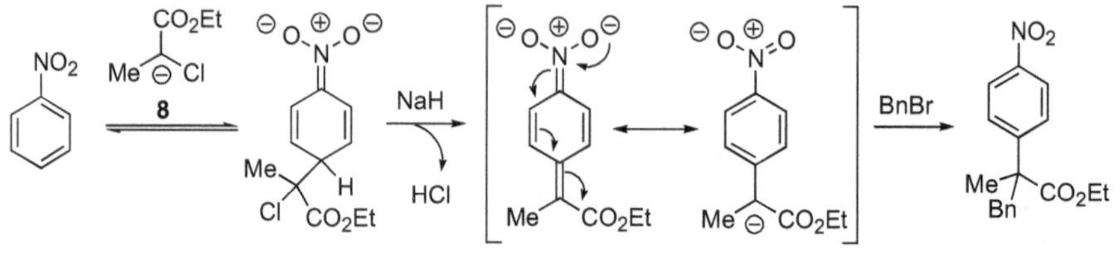

Scheme 36.6

Additional Comments

This problem is based on the work by Makosza M, Lemek T, Kwast A, Terrier F (2002) *J. Org. Chem.* 67:394-400 and by Lawrence NJ, Liddle J, Bushell SM, Jackson DA (2002) *J. Org. Chem.* 67:457-464.

Subjects of Revision

Nucleophilic aromatic substitutions. Primary and secondary KIEs.

Level 3 – Case 37
Base-promoted HF Elimination from 4-Fluoro-4-(4'-nitrophenyl)butan-2-one

Key point: *KIEs. Elimination reactions*

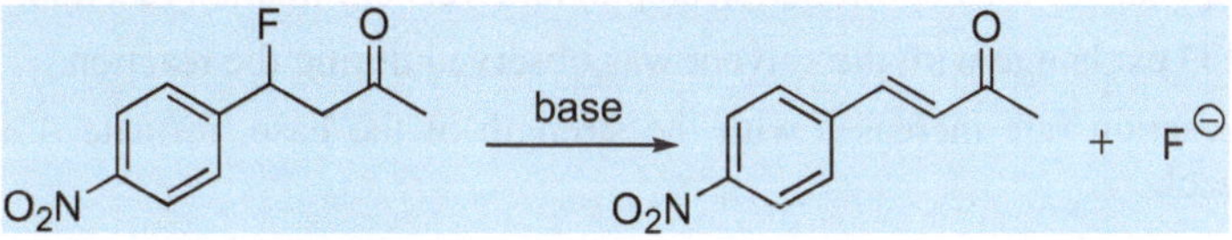

Base-promoted HF Elimination from 4-Fluoro-4-(4'-nitrophenyl)butan-2-one.

Three possible mechanisms of β-elimination (E1, E2 and E1cB) could be considered to explain the base-promoted HF elimination from 4-fluoro-4-(4'-nitrophenyl)butan-2-one **1** to give the *E*-arylbutenone **2** (Scheme 37.1).

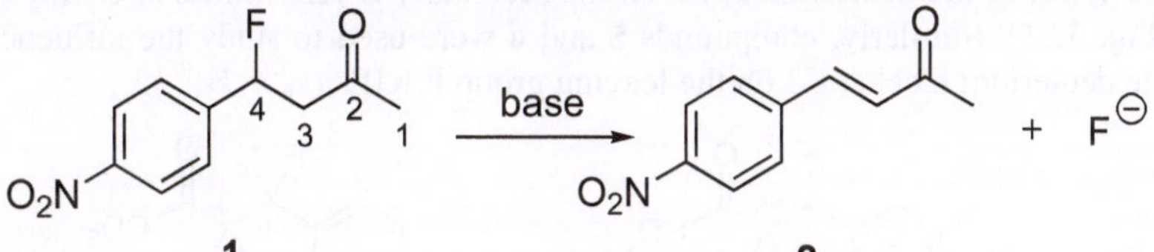

Scheme 37.1

Formulate the three mechanisms for the reaction and discuss which one seems to be more likely.

Experimental Data

1. Experiments carried out with labeled substrate **1** provided a set of primary (deuterium label placed at C3 position), secondary (deuterium label placed at C4 position) and leaving group (labeled F) kinetic isotope effects (KIEs). These data were determined for a set of bases with different strengths (formate, acetate and imidazole) in MeOH/H_2O as solvent and are collected in Table 37.1.

Table 37.1

pK_a	base	primary k^H/k^D	secondary k^H/k^D	[a] leaving group k^{18}/k^{19}
3.7	formate	3.2 ± 0.1	1.038 ± 0.013	1.0037 ± 0.0020
4.7	acetate	3.7 ± 0.1	1.050 ± 0.014	1.0047 ± 0.0012
6.95	imidazol	7.45 ± 0.1	1.014 ± 0.017	1.0013 ± 0.0012

[a] These values correspond to 5-15% of the estimated maximum F KIE of 1.03 (this value is the one expected for a complete C-F bond breakage).

2. Computational studies have indicated that secondary F KIEs as large as 1.0058 may appear for the deprotonation step in an E1cB elimination reaction.

3. No H/D exchange with the solvent was observed during the reaction.

4. The reaction rate increases with the strength of the base: formate < acetate < imidazol.

5. The *double-isotopic fractionation method* was employed in this study. This procedure consists of the use of deuterium substitution to selectively slow down the rate of one step in a reaction and observing the changes in a second kinetic isotope effect. In this study the aim is to obtain information from the secondary D KIE at C4 and the fluorine isotope effects, respectively. For that reason, substrates on which the deuterium has been placed at C3 to slow down the step in which the C3-H bond is being broken, have been employed. Then, the extra labels were deuterium at C4 (compound **4**) and labeled fluorine (compound **6**) (atoms colored red in Fig. 37.1). Compounds **3** and **4** were used to determine the effect of the deuterium at C3 on the secondary D KIE values at C4 ($k_3{}^H/k_4{}^D$) (Fig. 37.1). Similarly, compounds **5** and **6** were used to study the influence of the deuterium label at C3 on the leaving group F KIEs ($k_6{}^{18F}/k_5{}^{19F}$).[1]

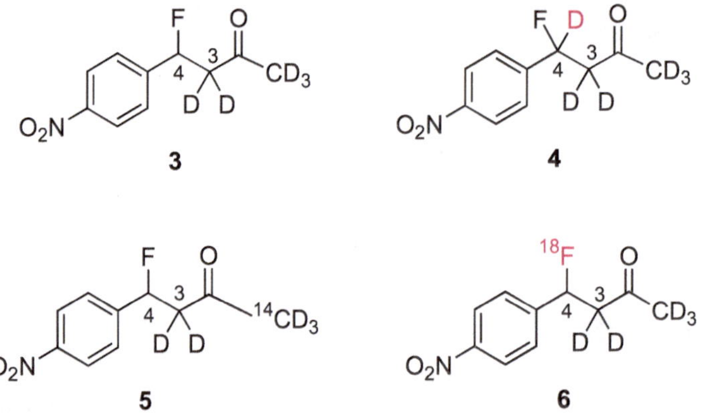

Figure 37.1

---

[1] The F KIEs are calculated from the $^{18}F/^{19}F$ isotopic ratios. As ^{18}F is radioactive while naturally abundant ^{19}F is not, a remote radioactive ^{14}C label was introduced in substrate **5** to enable radioactivity measurements.

If the two labeled positions of the substrates **4** and **6** are involved in **different steps of the mechanism**, the deuteration at C3 slows down its own step but also makes the subsequent steps slower, **decreasing the observed size of the evaluated KIE**. However, if the two isotopes **are involved in the same step of the reaction** the **observed KIE can be enhanced** (this step has become slower and thus more rate-limiting) **or** alternatively **shows no change** (if the step was in fact rate-limiting).

6. The C4-secondary deuterium KIEs obtained from the double-isotopic fractionation method in substrates **3** and **4** were 1.009, 1.000 and 1.010, for formate, acetate and imidazole, respectively.

The F KIE in substrates **5** and **6** was exclusively determined in presence of acetate as base. The observed value was 1.0009.

Discussion

Base-promoted β-eliminations can be E1, E2 and E1cB. In our case, the **E1** mechanism would consist of the rate-limiting detachment of the fluorine leaving group followed by a base-promoted fast proton-transfer step (equation A in Scheme 37.2). In the **E2** mechanism the bond breakage of the proton being transferred to the base and the detachment of the fluorine-leaving group are concerted (equation B in Scheme 37.2). Finally, the **E1cB** mechanism consists of a proton transfer step that leads to the formation of a carbanion intermediate, followed by the departure of the leaving group (equation C in Scheme 37.2). Depending on the rate-determining step of the reaction, we can distinguish between different types of E1cB mechanisms. If the slow step is the formation of the carbanion intermediate (step 1 in equation C) the elimination is called **(E1cB)$_{irr}$**. However, if the rate-determining step of the reaction is the departure of the leaving group (step 2 on equation C) the reaction is called **(E1cB)$_R$**.

With the help of the kinetic isotope data we should be able to distinguish between the different options. Deuterium KIEs at C3 indicate the degree of bond breakage of the C3-H bond, deuterium KIEs at C4 reflect the degree of rehybridization at the according position and the F KIEs are a measure of the breakage of the C-F bond. All are obviously referred to the transition state of the slow step of the reaction.

The experimental data in Table 37.1 indicate a primary deuterium KIE at C3 in all cases studied. This result definitively rules out the E1 mechanism, as the breakage of the C3-H bond does not occur during the rate-determining step. As we can see in the E1 transition state represented in Fig. 37.2, the C-F bond is being broken but the C3-H bond remains unaltered. A noticeable F KIE (F atom colored red) and a secondary D KIE at C4 (D colored blue) should be however expected in an E1 process.

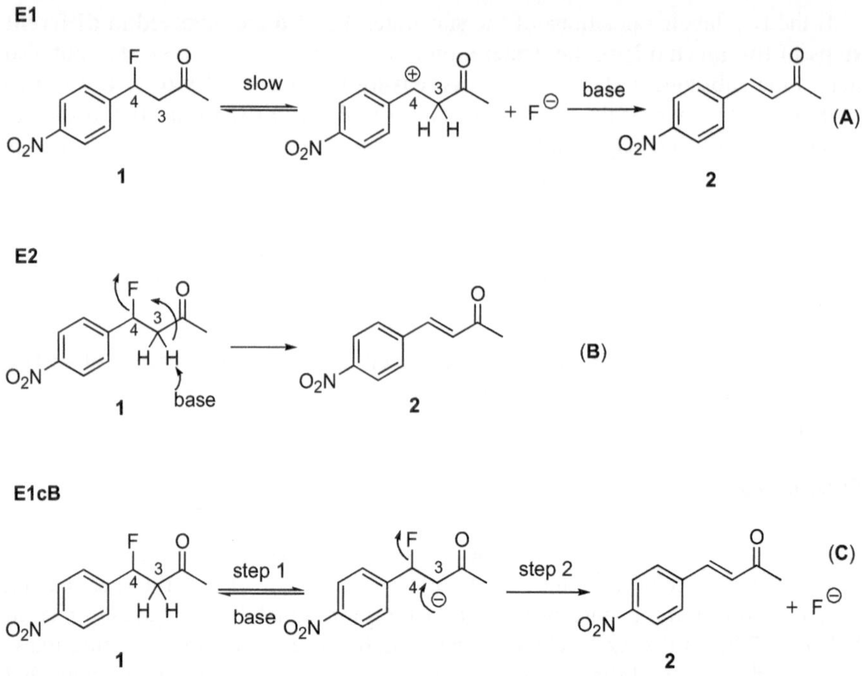

E1

E2

E1cB

Scheme 37.2

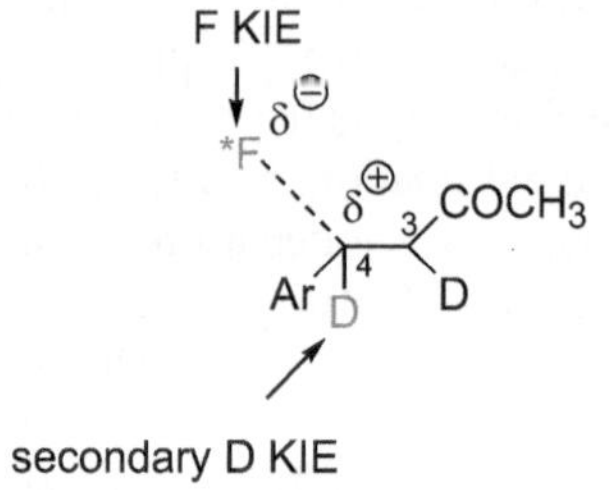

F KIE

secondary D KIE

E1 transition state

Figure 37.2

To distinguish between the two remaining alternatives is a bit more difficult. E2 and E1cB mechanisms have similar kinetic characteristics. They are bimolecular processes and, in principle, all will be affected in the same way by an increase in the base strength (the experimental data indicate that the elimination rate increases with the pK_a of the base employed). As the data of primary and secondary D KIEs provide information on the transition state of a reaction, a study of the different transition states could help us to better distinguish among the different options.

The transition states of the E2 and E1cB mechanisms are represented in Fig. 37.3 together with the KIEs that should be observed in each case. Just by comparing the three transition states, it becomes clear that no primary D KIE should be observed in the $(E1cB)_R$ mechanism, as the proton has been already removed. As the experimental fact is a clear primary D KIE at C3, the $E1cB_R$ mechanism must be discarded. Additionally it is an experimental fact that no H/D exchange with the solvent has been observed in the elimination of substrates 1. Solvent H/D exchange is indicative for an $E1cB_R$ mechanism where the carbanion is reprotonated by the solvent in the fast initial step (see equation C in Scheme 37.2).

All these arguments leave us with the E2 and $(E1cB)_{irr}$ alternatives, both in good agreement with the observed primary D KIEs at C3 (D colored red) and secondary D KIEs at C4 (D colored blue) (Fig. 37.3).

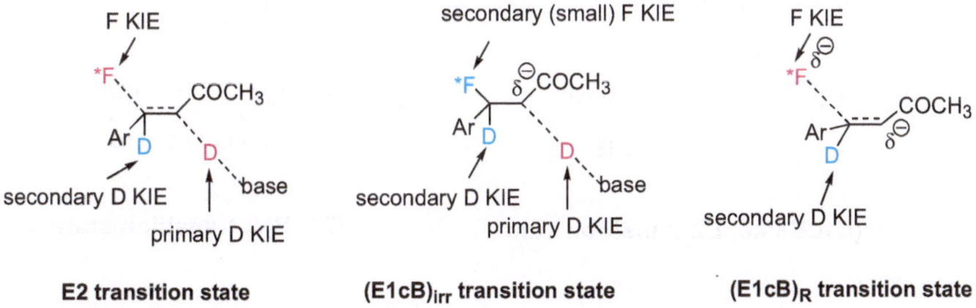

Figure 37.3

Let us now discuss the F KIE values. Leaving group KIEs (LG KIEs) indicate the degree of C-F bond breakage **in the transition state of the rate-limiting step**. The question now is: shall we then observe an F KIE in an $(E1cB)_{irr}$ mechanism? In principle, no F KIE should be detected, as the departure of the leaving group in $(E1cB)_{irr}$ processes takes place during the **fast** second step. However, we know from the given data that computational studies have pointed to *noticeable* secondary F KIEs in the transition state of the E1cB deprotonation step. This is remarkable because (primary) heavy atom effects are characterized by their very low values (they require to be determined with precisions higher than ± 0.0005) and secondary heavy atom effects are so small that they are almost undetectable and consequently of little use in mechanistic studies. The experimental F KIEs for the elimination of 1 range from 1.0013 to 1.0047 and correspond to 5-15% of the estimated maximum F KIE value of 1.03. This indicates that the degree of breakage of the C-F bond in the transition state is very small. This is not very compatible with an E2 reaction in which the C-F bond has to be broken in the transition state.

Is this reasoning solid enough to exclude an E2 mechanism? Well, possibly a fully synchronous E2 mechanism must be rejected, but such *central E2* mechanisms are rare. We could consider instead an unsymmetrical E2 process, that is, a **concerted** mechanism in which the degree of proton transfer to the base (C3-H bond breakage) is considerably higher than the breakage of the C-F bond in the transition state. This will be called an **(E1cB-like) E2** mechanism.

Either an **stepwise** (E1cB)$_{irr}$ or a **concerted** (E1cB-like) E2 mechanism could fit the results obtained in the labeling experiments: large primary D KIE at C3, secondary D KIE at C4 and a small F KIE, that in the case of the (E1cB)$_{irr}$ should correspond to a secondary effect but for the unsymmetrical E2 would be a small primary effect. Unfortunately, the experimental labeling data alone are not able to distinguish between the two possible alternatives. The transition states for both processes are represented in Fig. 37.4. The atoms involved in primary KIEs are colored red and the ones involved in secondary KIEs are colored blue.

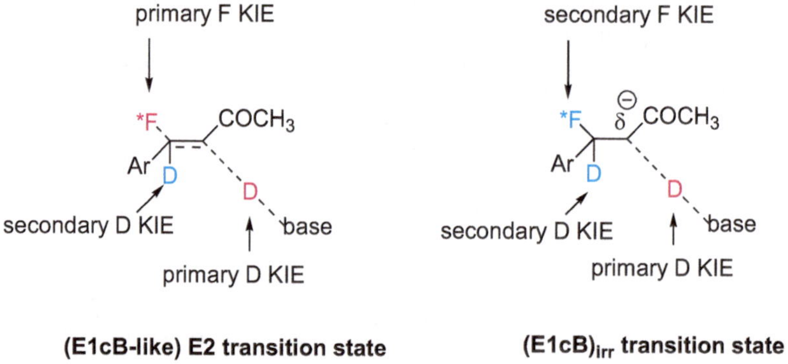

(E1cB-like) E2 transition state   **(E1cB)$_{irr}$ transition state**

Figure 37.4

The double-isotopic fractionation method is a good tool to distinguish between stepwise and concerted mechanisms and in our case it could be the ideal procedure to establish the subtle differences between the alternatives proposed. As it has been indicated in the introduction, the double-isotopic fractionation method evaluates the influence of the introduction of an extra isotope in a molecule already labeled and whose KIEs have been previously evaluated. An enhancement or no change of the previous KIE values is interpreted considering that the two isotopes are involved on the same reaction step. On the other hand, a decrease of the former KIE values indicates that the two isotopes are involved in different steps (Figure 37.5).

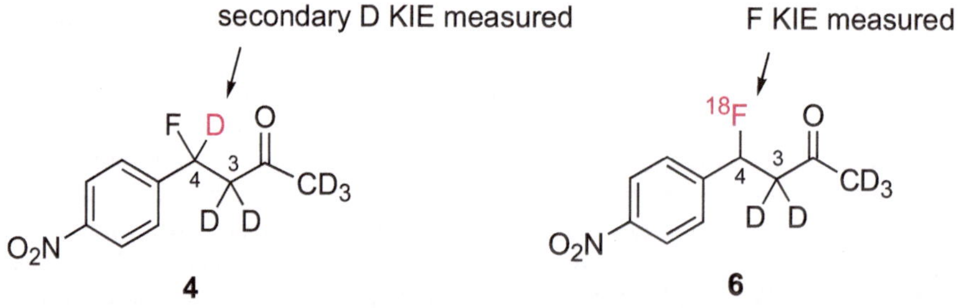

Figure 37.5

The secondary D KIEs for substrate **4** are 1.009, 1.000 and 1.010 for formate, acetate and imidazole, respectively. Except for imidazole, these values are smaller than those obtained by the standard method (1.038, 1.050 and 1.014). This would indicate that the breakage of the C3-H bond and the hybridization change at C4 (from sp^3 to sp^2) occur in different steps. Alternatively, the leaving group F KIE for deuterated substrate **6** (1.0009) is considerable smaller than the one previously recorded (1.0047). Again, this result could indicate that the breakage of the C3-H bond and the departure of the fluorine leaving group take place in different steps of the reaction. Both arguments are definitively pointing to a stepwise mechanism and not to a concerted one, even if this is not very symmetrical. In consequence, the stepwise $(E1cB)_{irr}$ mechanism seems to be the most likely for the elimination of substrates **1** (Scheme 37.3).

$(E1cB)_{irr}$

Scheme 37.3

As usual, before making a final conclusion we should test if the proposed mechanism fits well with all the experimental data. First, the elimination rate is affected by an increase in the base character. In the $(E1cB)_{irr}$ mechanism the first step is rate-limiting and the stronger the base the easier the removal of the acidic C3 hydrogen (k_1). In addition, a noticeable primary KIE must be observed at this position. The size of primary deuterium KIEs for proton transfer processes are considered as a measure of the symmetry of the transition state for the transfer. In this case, the highest value (7.5, the maximum calculated value for a primary KIE) corresponds to the stronger base (imidazol), which indicates a *central* transition state having the proton bounded with equal strength to C3 and base (A in Fig. 37.5). Weaker bases (formate, acetate) lead to less symmetrical transition states and hence to smaller KIEs (B in Fig. 37.5). Finally, secondary D KIEs at C4 and F KIEs are very small, as these bonds are not broken in the slow step of the reaction (k_2).

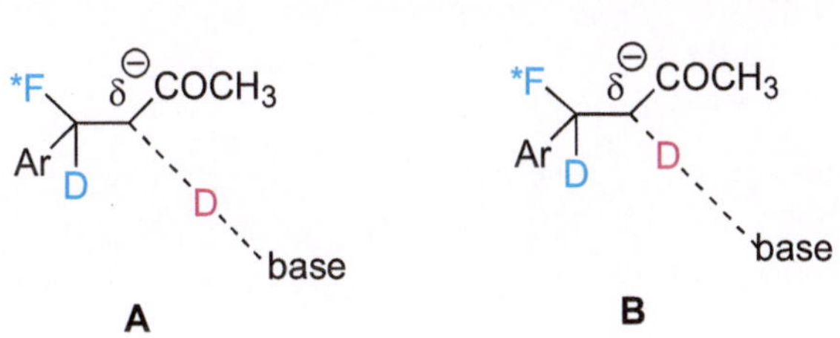

Figure 37.5

In Summary

Only by means of primary, secondary and leaving group F KIEs it is not possible to decide whether the base-promoted HF elimination of ketone **1** follows a concerted or a stepwise mechanism. The double-isotopic fractionation method provides strong evidence for a stepwise process, more likely via an $E1cB_{irr}$ mechanism.

Additional Comments

This problem is based on the work by Ryberg P, Matsson O (2001) *J. Am. Chem. Soc.* 123:2712-2718 and by the same authors in (2002) *J. Org. Chem.* 67:811-814.

Subjects of Revision

Elimination reactions. Primary, secondary and heavy atom kinetic isotope effects.

Level 3 – Case 38
Substitution of β-Halostyrenes by MeS⁻

Key point: *Concerted versus stepwise mechanism*

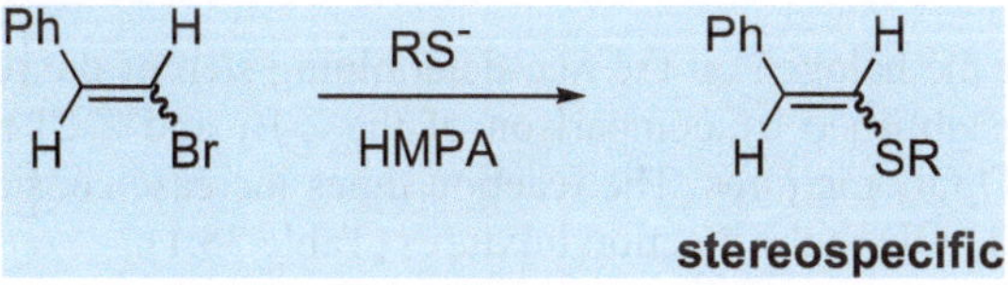

stereospecific

Substitution of β-Halostyrenes by MeS⁻

β-Bromostyrenes **1** react rapidly with MeS⁻ or *i*-PrS⁻ in HMPA, at room temperature, to give the vinylic substitution products **2** with high yields (> 95%) and almost complete retention of the configuration (> 98%) (Scheme 38.1).

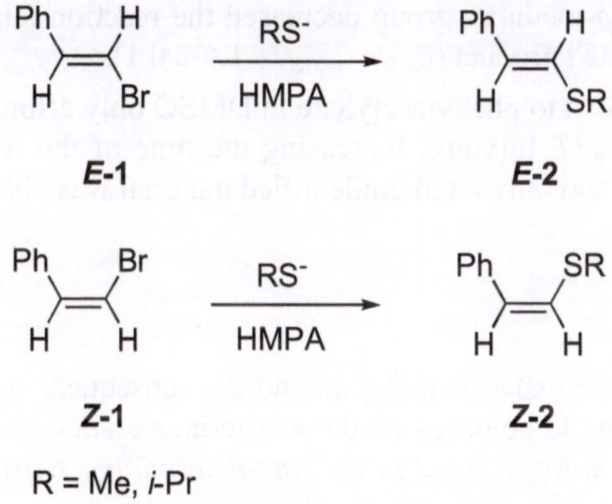

E-1   E-2

Z-1   Z-2

R = Me, *i*-Pr

Scheme 38.1

Among the possible concerted mechanisms for such systems, the in-plane S_N2-type substitution was discarded, but a perpendicular nucleophilic attack on the π* orbital was seriously considered. On the other hand, among the stepwise alternatives, an elimination-addition via phenylacetylene and a vinylic substitution via addition-elimination, may be operative.

Formulate and comment all options considered and discuss why the in-plane S_N2-type substitution was discarded. Decide which is the most reasonable mechanism for the reaction.

Experimental Data

1. The reactions in HMPA (hexamethylphosphoramide) were very fast (only took a few minutes). Similar stereoselectivities and yields were obtained when the reactions were carried out in DMSO or DMF. However, the reaction rates decreased when mixtures of $DMSO:CHCl_3$ were used.
2. Reactions carried out with mixtures Z-1/E-1 demonstrated that the Z-isomer reacted faster than the E-isomer: $k (Z-1)/k (E-1) = 2.6 \pm 0.1$. Similar results were observed when mixtures of the corresponding Z- and E-chlorides were employed.
3. The effect of the halogen on the rate-determining step of the reaction (*element effect*) was established by comparison of the Z-Br and Z-Cl reaction rates in $DMSO\text{-}d_6:CD_3OD$ mixtures. The reaction times increase considerably with the addition of CD_3OD to the reaction mixtures (Table 38.1).

Table 38.1

$DMSO\text{-}d_6:CD_3OD$	Reaction time, min	$k_{Z\text{-}Br}/k_{Z\text{-}Cl}$
95:5	17	1.0 ± 0.1
5:1	97	1.4 ± 0.1

4. The reaction was also fast and stereospecific with Z- and E-p-methoxy-β-bromostyrenes. Again, the Z-isomer reacted faster than the E-isomer, but the presence of the p-methoxy group decreased the reaction rate compared to that of the unsubstituted product ($k_{Z\text{-}1}/k_{Z\text{-}p\text{-MeO}} = 1.6 \pm 0.1$).
5. Addition of MeSNa to phenylacetylene in DMSO only afforded products **2** (2% yield) and as a Z+E mixture. Increasing the time of the reaction did not improve the yields and only a red unidentified material was obtained.

Discussion

The discussion of the experimental data and the subsequent study of the reaction mechanism are going to be based on three criteria: *stereochemistry*, *element effect* (*cleavage of the C-halogen bond in the rate-determining step*) and influence of a *p-MeO substituent on the aromatic ring*.

Stereochemistry

Possibly, the most striking aspect of the vinylic substitution of bromides **1** is the stereospecificity of the reaction. In fact, the retention of the stereochemistry during the process is decisive in deciding among the diverse mechanisms proposed. Let us use vinyl bromide E-**1** as a substrate to discuss the different options.

We will consider the concerted pathways first. **Why should an *in-plane concerted substitution* mechanism be rejected?** As it is shown in Scheme 38.2, a direct nucleophilic attack (similar to a S_N2 attack) would lead to transition state **3** and finally to product **Z-2** with total inversion of the stereochemistry during the process. Obviously, the retention of the stereochemistry of the reaction products cannot be explained through this route and in consequence, this mechanism must be discarded.

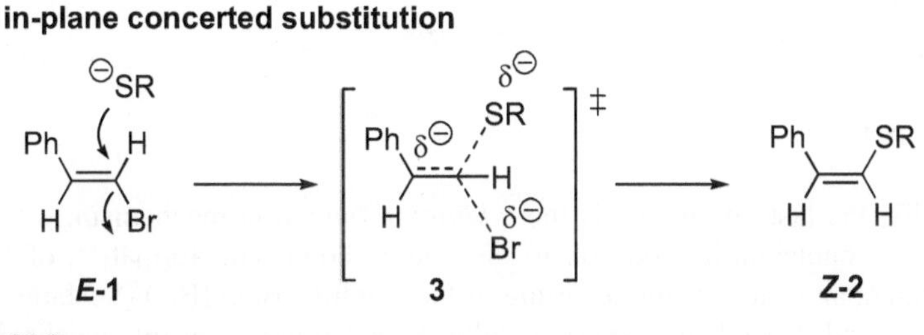

in-plane concerted substitution

Scheme 38.2

Alternatively, the perpendicular attack of the nucleophile on the π^* orbital seems to be much more reasonable. In that case, the transition state of the reaction can be represented as **4** with simultaneous formation and breakage of the C-nucleophile and C-Br bonds, respectively. The final product **E-2** would retain the original configuration. The stereochemistry of the starting compound is not altered during the process and hence, this pathway must be seriously considered (Scheme 38.3).

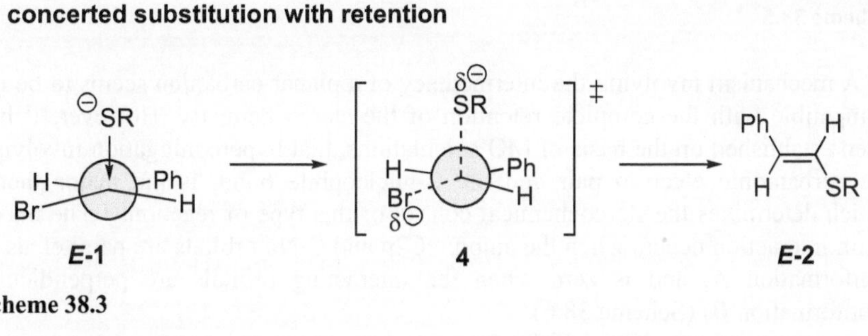

concerted substitution with retention

Scheme 38.3

If we think about the stepwise alternatives, the more likely for a nucleophilic vinylic substitution reaction are either an *elimination-addition* or an *addition-elimination* process. In the *elimination-addition* sequence (Scheme 38.4), the losts of HBr during the first step of the reaction would form phenylacetylene, which after nucleophilic attack would lead to anion **5**. The protonation of this species should yield the substitution products **2**, but as a *Z+E* mixture. It is evident that the stereospecificity of the reaction cannot be justified by this route. Furthermore, we know from the experimental data that phenylacetylene was virtually unreactive when treated with MeSNa under the reaction conditions employed for the vinylic substrates **1**. In view of these results, the *elimination-addition* mechanism should not be considered.

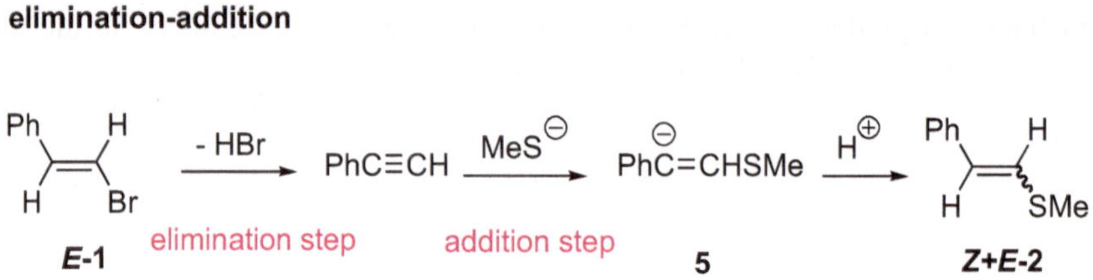

elimination-addition

Scheme 38.4

Finally, the last alternative is the *addition-elimination* mechanism, which consists of the nucleophilic addition to the double bond with formation of a planar carbanion **6**, and subsequent departure of the leaving group (Br⁻) (Scheme 38.5). It is known (and it has been experimentally demonstrated) that this mechanism occurs with retention of the configuration. Next we will explain why.

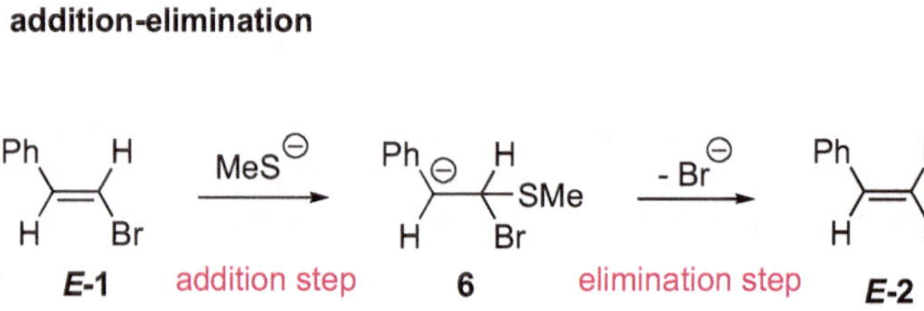

addition-elimination

Scheme 38.5

A mechanism involving the intermediacy of a planar carbanion seems to be incompatible with the complete retention of the stereochemistry. However, it has been established on the basis of MO calculations, that hyperconjugation involving the carbanionic electron pair and the C-nucleophile bond, is the major factor which determines the stereochemical control of this type of reactions.[1] The maximum interaction occurs when the anionic C2p and C-Nu orbitals are parallel, as in conformation **A**, and is zero when the interacting orbitals are perpendicular (conformation **B**) (Scheme 38.6).

---

[1] In fact, the hyperconjugative stabilizing effect is the net result of the destabilizing interaction of the occupied anionic C2p and the filled σ_{C-Nu} orbitals and the stabilizing interaction of the C2p with the σ^*_{C-Nu} orbital. A complete study of the stereochemistry in nucleophilic vinylic substitution reactions has been reported by Apeloig Y, Rappoport Z (1979) *J. Am. Chem. Soc.* 101:5095-5098.

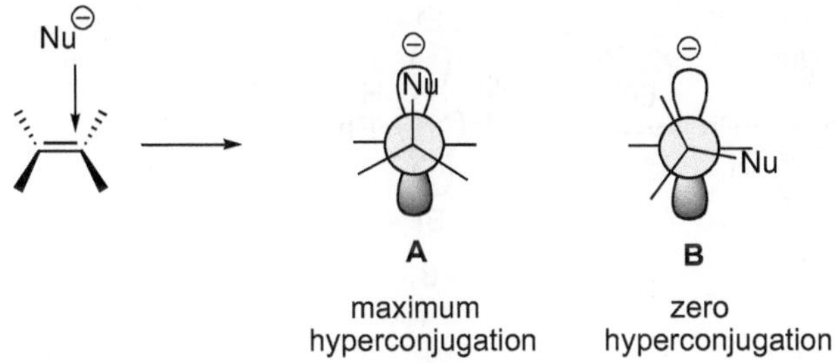

A	B
maximum hyperconjugation	zero hyperconjugation

Scheme 38.6

The elimination step from the highly stable carbanionic intermediate **7** requires the incoming electron pair and the leaving group being antiperiplanar in the transition state. By rotation around the Cα-Cβ bond in **7**, conformer **8** would lead to the product with retained stereochemistry, but inversion would result from the other alternative conformer **9** (Scheme 38.7).

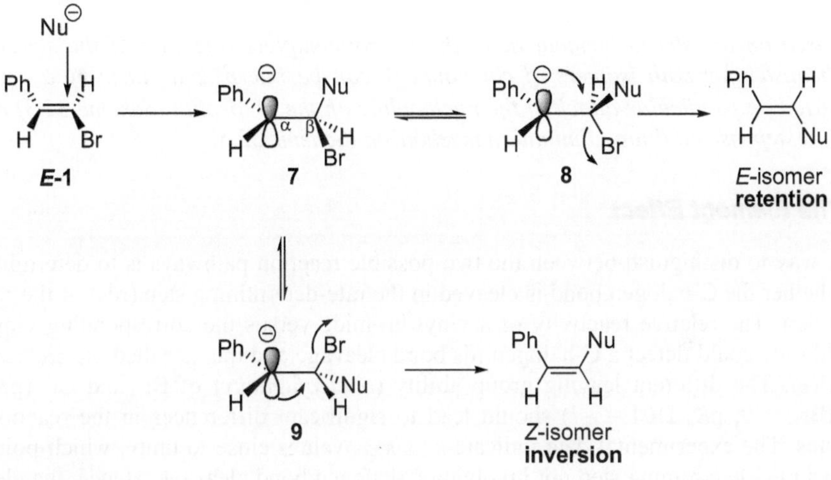

Scheme 38.7

Although both conformers **8** and **9** have similar stability (the degree of hyperconjugation must be comparable, in both cases, see above), calculation of the rotation energy barriers indicate that the 120° rotation required to reach conformer **9** from **7** needs to overcome a higher rotation barrier than in the case of conformer **8** (Scheme 32.8). Calculation of the relative energies of conformers **8**, **9** and **10** indicate that this latter is 10.9 kcal mol^{-1} less stable than the other two. As during the 120° rotation from conformer **7** to **9** conformation of high energy **10** has to be overcome, and the retained substitution products derived from the more favorable conformation **8** are almost exclusively formed (Scheme 38.8).

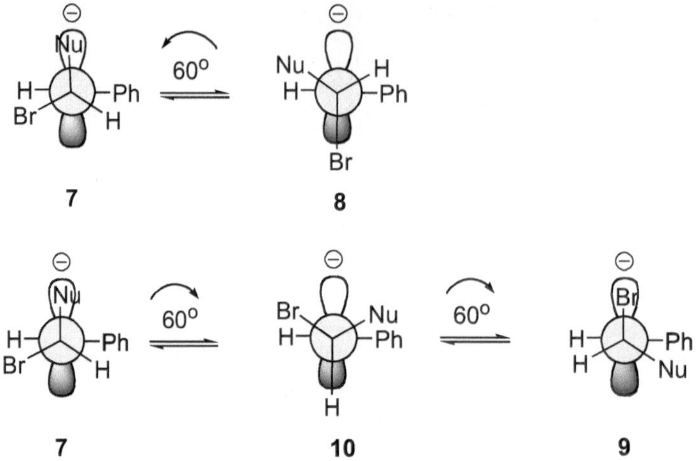

Scheme 38.8

*Based on the stereochemical data, the nearly complete retention of the stereo-chemistry for both isomers of compound **1** can be interpreted either by a con-certed perpendicular attack of the nucleophile on the π^* orbital (Scheme 32.3) or by a stepwise addition-elimination mechanism (Scheme 32.5).*

The Element Effect

A way to distinguish between the two possible reaction pathways is to determine whether the C-halogen bond is cleaved in the rate-determining step (rds) of the re-action. The relative reactivity of a vinyl bromide versus the corresponding vinyl chloride could detect a C-halogen rds bond cleavage and this is called *the element effect*. The different leaving group ability (*nucleofugacity*) of Br$^-$ and Cl$^-$ (pK_a HBr $= -9$; pK_a HCl $= -7$) should lead to significant differences in the reaction rates. The experimental data indicate $k_{Z\text{-}Br}/k_{Z\text{-}Cl}$ values close to unity, which point to a rate-determining step not involving C-halogen bond cleavage. Hence, the ele-ment effect favors the addition-elimination route, in which the nucleophilic attack (formation of the carbanion intermediate) is the slow step of the process. How-ever, based on the absence of element effect, the mechanism of concerted perpen-dicular attack depicted in Scheme 38.3 should be rejected, as the breakage of the C-halogen bond occurs during the slow step of the reaction.

Influence of a p-MeO Substituent on the Aromatic Ring

The experimental results indicate that the reaction rate is reduced when an elec-tron-donating group (MeO) is placed at the *para*-position of the aromatic ring ($k_{Z1}/k_{Z\text{-}p\text{-}MeO} = 1.6 \pm 0.1$). As electron-donating groups should decrease the stability of carbanion intermediates like **6** (Scheme 38.5), this could be an argument in support of the stepwise addition-elimination pathway. However, this data does not discard the concerted mechanism, since the electron-donating group in the aromatic ring makes the styryl C=C double bond less electrophilic, and for that reason, the nucleophilic attack would be also less favored.

Therefore, the retention of the stereochemistry and the lower reactivity of the *p*-methoxy derivative compared with the unsubstituted substrate are not able to distinguish between a concerted and a stepwise mechanism. Only the element effect unequivocally points to an addition-elimination process in which the nucleophilic attack is rate-determining.

Based on the previous discussion, the most likely mechanism for the vinylic substitution of *E*- and *Z*-bromostyrenes **1** can be represented as follows (Scheme 38.9).

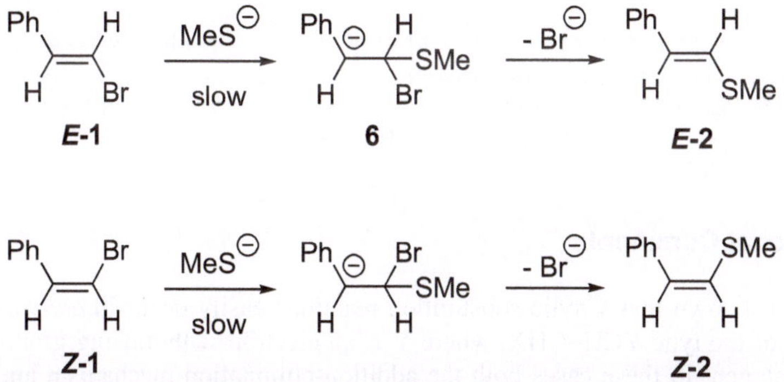

Scheme 38.9

There are two additional points that deserve further discussion. First, the experimental data indicate that the reactions are very fast in *dipolar aprotic solvents* like HMPA, DMSO or DMF, which are excellent cation solvators. The combination of the ability of the solvent in *removing* the cation from the media, with an excellent nucleophile (in fact the MeS⁻ will be *naked* in the presence of such solvents) is ideal to accelerate the process. The addition of solvents that are unable to enhance the reactivity of the nucleophile, either protic (see for example the effect of the addition of CD₃OD in Table 38.1 or aprotic like CHCl₃, will result in an appreciable decrease of the substitution rate.

The second question we should comment is the higher reactivity of the *Z*-isomers compared with the *E*-isomers in all cases studied. Possibly the steric interaction between the vicinal phenyl group and bromine atom is responsible for a less stable (higher in energy) ground state, more prone for a nucleophilic attack that will reduce the steric strain (Scheme 38.10).

These additional points are also in full agreement with the stepwise mechanism proposed above in Scheme 38.9.

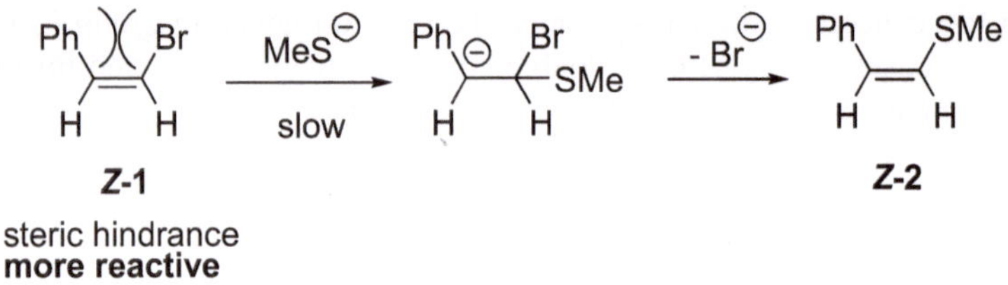

Z-1

steric hindrance
more reactive

Scheme 38.10

In Summary

The vinylic substitution of Z- and E-β-bromostyrenes **1** with MeS⁻ occurs via the stepwise nucleophilic addition-elimination route. The reaction is very fast in dipolar aprotic solvents like HMPA, DMSO and DMF, which due to their ability as cation solvators enhance the nucleophilicity of the reagent.

Additional Comments

It is well known that vinylic substitution reactions easily occur in *activated* substrates, of the type YCH=CHX, where Y is an electron-withdrawing group and X is an halogen. In these cases both the addition-elimination mechanism and direct displacement of the halogen by the nucleophile have been observed. This is why both alternatives have been taken into account for compounds **1**, although strictly they can be considered *unactivated* substrates for nucleophilic vinylic substitutions.

This problem is based on the work by Chen X, Rappoport Z (1998) *J. Org. Chem.* 63:5684-5686.

Subjects of Revision

Nucleophilic substitutions on vinylic systems. Polar and dipolar aprotic solvents: effect on the reaction rates. Parameters of solvent polarity.

Level 3 – Case 39
Periodinane-Mediated Cyclization of Anilides

Key point: *Radical cyclization reaction. SET*

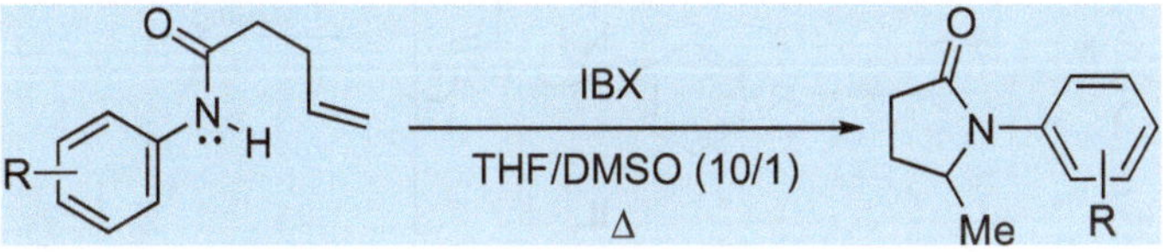

Periodinane-Mediated Cyclization of Anilides

Anilides **1** cyclize to γ-lactams **2** by reaction with an hypervalent iodine reagent (1-hydroxy-1,2-benziodoxol-3(1*H*)-one-1-oxide, IBX) under the conditions shown in Scheme 39.1.

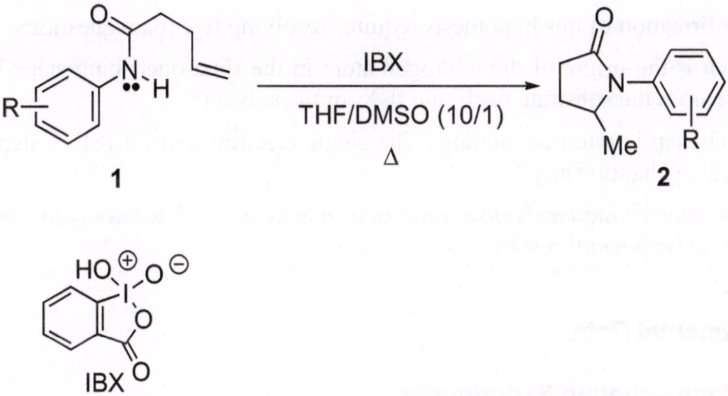

Scheme 39.1

To explain these IBX-mediated ring closures, a stepwise radical mechanism has been proposed. The first step consists of a single electron transfer (SET) process from the substrate to the periodinane, leading to radical cation **I** which is converted into amidyl radical **II** after deprotonation. Cyclization of **II** in a 5-*exo-trig* fashion provides a carbon-centered radical species **III**, which is quenched by the transfer of a hydrogen radical in the termination step (Scheme 39.2).

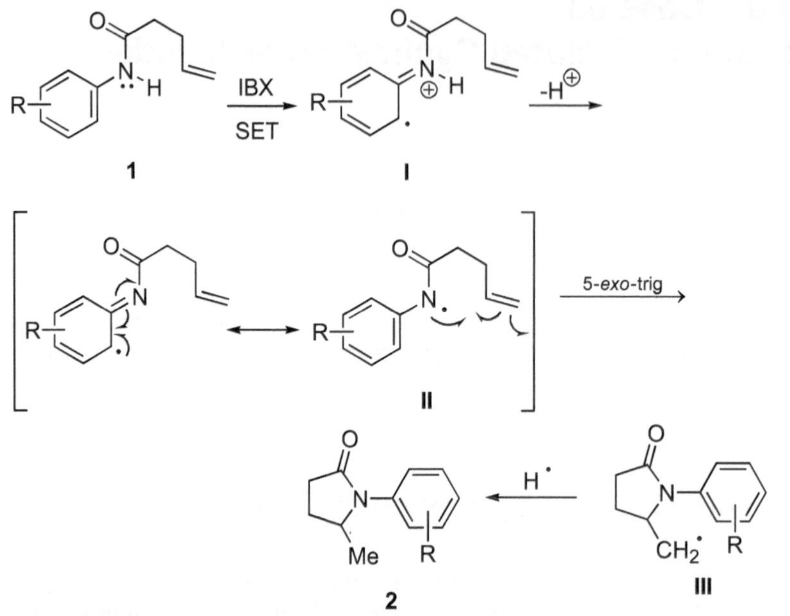

Scheme 39.2

The confirmation of this hypothesis requires resolving two main questions.

1. Which is the origin of the hydrogen atom in the final quenching step? Does it come from the substrate itself, the IBX, or the solvent?

2. Which step is rate-determining? The single electron transfer (SET) step or the radical cyclization step?

Discuss whether the mechanism proposed in Scheme 39.2 is fully consistent with all these experimental results.

Experimental Data

1. Isotope Labeling Experiments

Norbornene-derived anilide **3** was chosen as starting material in the isotope labeling experiments that are collected in Scheme 39.3. When the reaction of **3** with IBX was performed in deuterated media [THF-d_8/DMSO-d_6 (10/1)], deuterated product **4** was obtained.

Carrying out the reaction of **3** in THF/DMSO-d_6 (10/1) led to bicyclic anilide **6** containing no deuterium. The same compound **6** was obtained when labeled anilide **5** was reacted with IBX in THF/DMSO (10/1) (Scheme 39.3). Finally, when the experiment was carried out with **3** and IBX in neat DMSO, no reaction was observed. These results lead to a new question: what is the role of THF in the reaction?

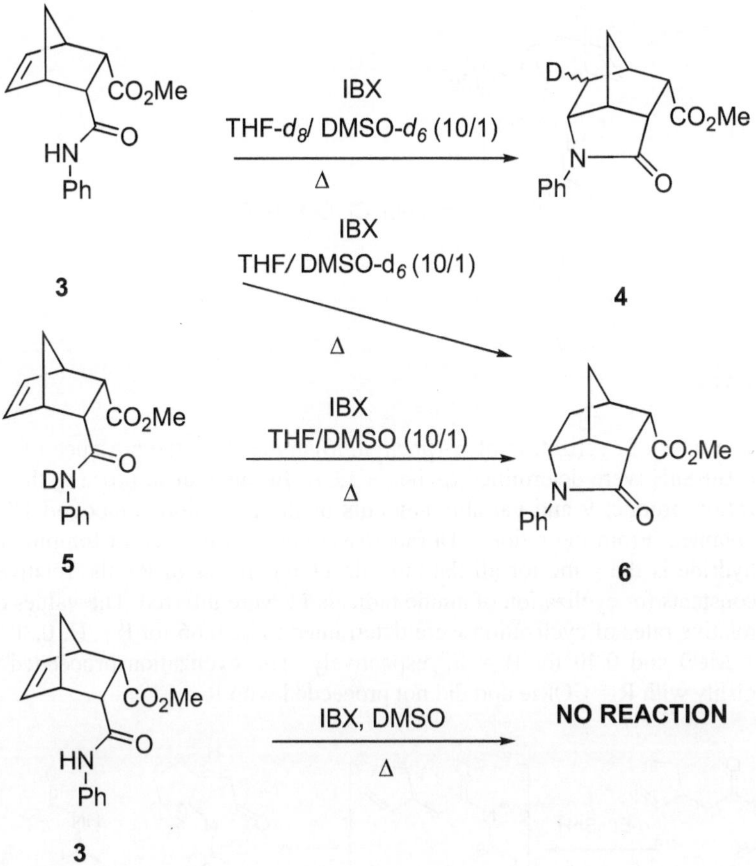

Scheme 39.3

2. Kinetic Studies

2.1 The effect of the substituents on the reaction rates were studied on a series of *p*-aryl-substituted anilides **7**. Electron-donating substituents (R = MeO) resulted in an increase in the reaction rate, whereas electron-withdrawing substituents caused a decrease in the reaction rate (R = Cl, COMe) or had no significant effect on the rate (R = F). For anilides with R = NO_2 or CF_3 no reaction occurred. The Hammett plot gave a linear graph with a negative slope and the better correlations were found with σ^+ parameters (Fig. 39.1).

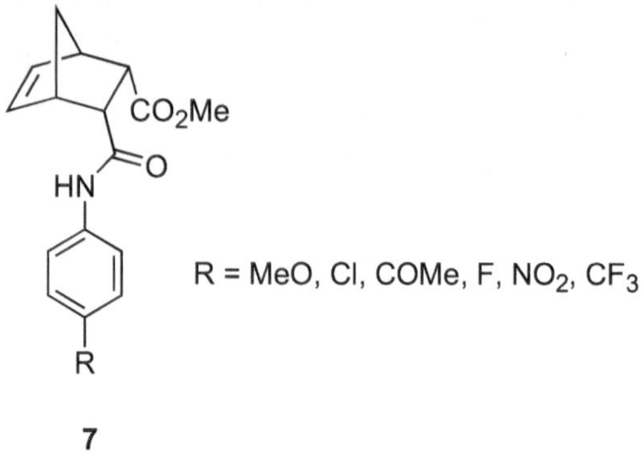
CO2Me
O
HN
R
R = MeO, Cl, COMe, F, NO2, CF3
7

Figure 39.1

2.2 The rates of cyclization of *N*-(phenylthio)amides **8** in the presence of excess *n*-Bu_3SnH were determined (Scheme 39.4). In this radical process, the cyclization product **9** and variable amounts of the reduction compound **10** were obtained. From the ratios **9:10** (and assuming that the rate of trapping of tin hydride is the same for all the radicals **11** regardless of R), the relative rate constants for cyclization of amide radicals **11** were inferred. The values of the relative rates of cyclization were determined to be 0.66 for R = H, 0.30 for R = MeO and 0.40 for R = F, respectively. The cyclization proceeded sluggishly with R = COMe and did not proceeded with R = CF_3.

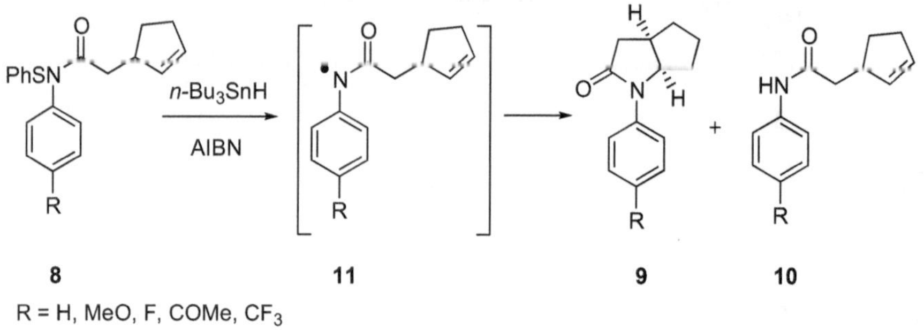
PhSN
O
n-Bu3SnH
AIBN
R
N
O
R
H
O
N
H
R
+
HN
O
R
8
11
9
10
R = H, MeO, F, COMe, CF3

Scheme 39.4

3. Cyclic Voltammetry Measurements

The oxidation potentials of a series of substituted anilides **8** were determined by cyclic voltammetry for CH_2Cl_2 solutions. The observed values were 1.20 V (R = H), 0.82 V (R = MeO), 1.23 V (R = F), 1.48 (R = COMe) and > 1.57 V (R = CF_3).

Discussion

As mentioned above, there are two key questions that have to be discussed in detail to confirm the validity of the stepwise radical mechanism proposed for the periodinane-mediated cyclizations of anilides 1.

Where does the hydrogen atom come from in the termination step?

Termination Step

The isotope labeling experiments give us according information. The experiments performed (Scheme 39.3) have been designed considering all the possible combinations between labeled solvents and substrate. However, if we examine the data in detail it is clear that deuterium is detected in the final product **only when deuterated THF is present in the media**. These results exclude the substrate (anilide) and the DMSO as possible radical quenchers. Finally, the alternative of the IBX being a source of hydrogen is ruled out by the fact that when anilide and IBX are reacted in absence of THF in the media, no reaction was observed.

We can conclude from the deuterium-labeling experiments that the reaction needs to be run in presence of a hydrogen-donating solvent like THF. This solvent is playing an essential role in the reaction, possibly acting as quencher during the termination step.

The peculiar dependence of the reaction with THF could lead us go a step further and think that THF itself is actually playing a role during the process. We should remember that IBX, in DMSO but **in absence of THF,** does not lead to a reaction. That is, **IBX itself is not the reagent of the process**. A very reasonable possibility could involve the coordination of THF to the reagent IBX, leading to an extraordinary oxidant species 12 (Scheme 39.5). This intermediate 12 (and not IBX), could be the electron acceptor in the SET proposed as the first step of the cyclization mechanism.

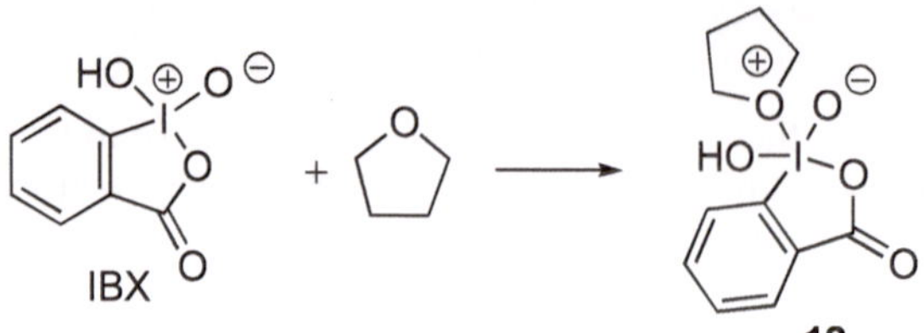

Scheme 39.5

Which is the rate-determining step?

The kinetic experiments were run to decide whether the rate-determining step of the reaction was the SET (formation of radical cation **I**) or the radical cyclization step (transformation of **II** into **III**). The first set of data was obtained from aryl-substituted anilides **7** (Fig. 39.1), in order to get information about the **influence of the electronic effects on the reaction rate** and hence about the possible formation of a radical cation in the slow step.

SET step as rate-determining

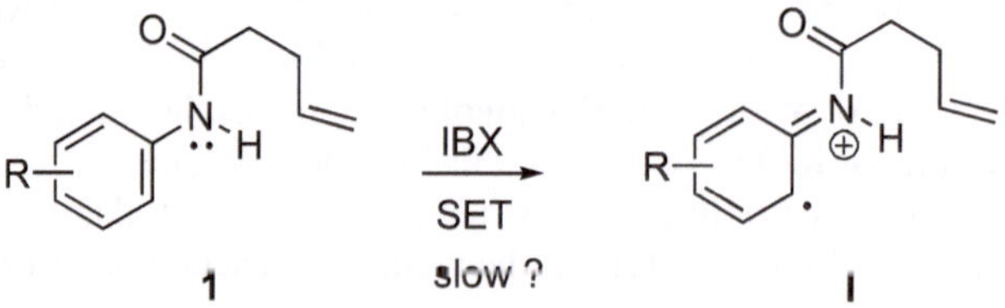

The experimental results indicate that the IBX-mediated cyclization of anilides is sensitive to the electronic character of the substituents in position para of the anilide ring. In fact the reaction rate increases with the presence of electron-donating groups. These data are indicating a lower electron density in the transition state of the slow step of the reaction, relative to the initial ground state of the substrate. In addition, the development of positive charge in the transition state is confirmed by the fact that the Hammett plot shows a negative slope and by the excellent correlation with σ^+ parameters (through conjugation).

We can conclude that the sensitivity of the reaction rate to the electronic effects supports the development of positive charge in the transition state. This is in agreement with the assumption that the SET from the amide aromatic ring to the iodine reagent (leading to radical-cation I) is the rate-determining step of the reaction.

However, is this evidence sufficient to discard the radical cyclization step (**II** to **III**) as rate-determining?

Cyclization step as rate-determining

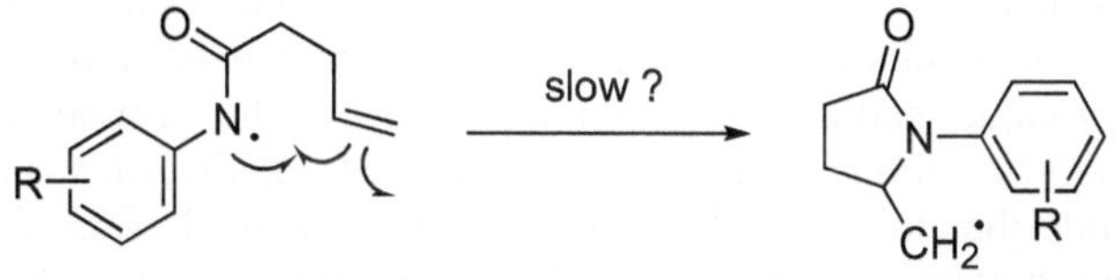

II   **III**

Before making a decision, it would be necessary to confirm whether cyclization step **II** to **III** is sensitive or not to the electronic effects of the substituents in the aromatic ring.

In this case, the information was taken from the study of the n-Bu$_3$SnH-mediated cyclization of substituted *N*-(phenylthio)amides **8**, detailed in Scheme 39.6. The reaction is a radical process in which the nitrogen-centered radical **11** formed in the first instance, cyclizes onto the C=C double bond to give carbon-centered radical species **13** which, after quenching, yields cyclized compound **9**. The cyclization step follows the *exo*-mode in agreement with the Baldwin rules (5-*exo*-trig).

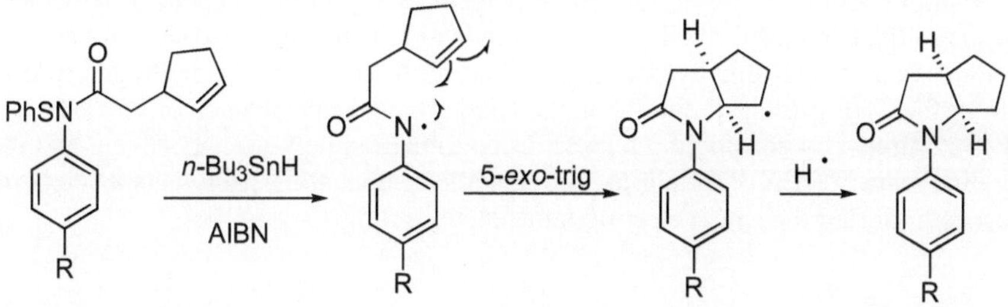

8   **11**   **13**   **9**

R	relative rates
H	0.66
MeO	0.30
F	0.40
COMe	>0.40
CF$_3$	—

Scheme 39.6

From the comparison of the relative cyclization rate values of substituted anilides **8**, it is clear that there is no relationship between the electronic character of the substituents and the reaction rate. Both electron-donating and electron-withdrawing substituents decrease the rate of the cyclization of **8** relative to that of the unsubstituted substrate. Since the opposite trend is experimentally observed in the IBX-mediated cyclization of anilides **1**, we should presume that the transformation of **II** into **III** (a radical cyclization) is not the slow step of the reaction.

Finally, the voltammetry measurements provide data that are in further support of the SET step as rate-determining. If we study the values of the oxidation potentials of substituted anilides **8** we will realize that there exists a direct correlation between their values and the observed relative rates of the reaction. Thus, the anilide with lower oxidation potential (0.82 V) proceeds rapidly, (this is the case of R = MeO), while the anilides with higher oxidation potential (1.23 V, 1.48 V and 1.57 V) proceed only sluggishly (R = F, COMe) or not at all (R = CF$_3$).

Considering the kinetic data, the rate-determining step of the IBX-mediated cyclization of anilides **1** *is the SET process (first step of the reaction). The variations of the measured oxidation potentials with the reaction rates are in good agreement with the initial SET proposed by the mechanism.*

In Summary

The IBX-mediated cyclization of anilides can be interpreted through a stepwise mechanism. The first (slow) step consists of a single electron transfer (SET) from the anilide aromatic ring to the iodine reagent to form radical cation **I**. After loss of a proton, **I** is transformed into amidyl radical **II** that cyclizes to carbon-centered radical species **III**. Finally, **III** is quenched by the transfer of a hydrogen radical from the THF present in the medium. The particular dependence of the THF can be interpreted by considering a previous coordination between the solvent and the iodine reagent (IBX), leading to an oxidant species that probably acts as electron acceptor during the single electron transfer step.

Questions

Based on the SET mechanism previously discussed, explain the transformation of cyclopropylanilide **14** into the novel tetracycle **15** (Scheme 39.7).

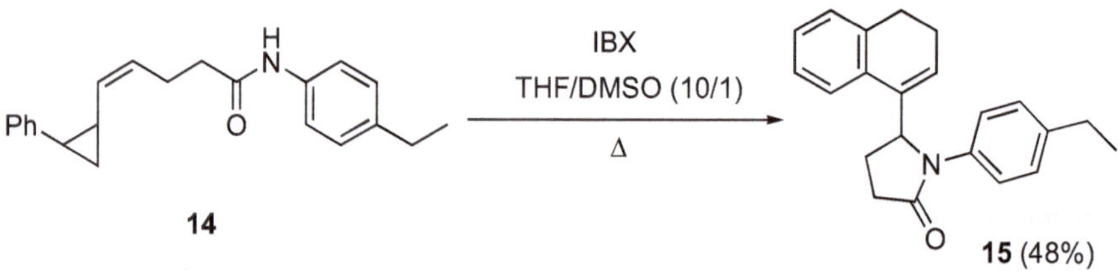

Scheme 39.7

Answer to the Question

To explain the transformation of anilide **14** into tetracycle **15** we should follow the different steps of the mechanism previously discussed. SET from the anilide to the IBX-THF oxidant reagent leads to radical cation **16** that after loss of a proton will be transformed into radical **17**. Cyclization into the C=C double bond will lead to cyclic radical **18**. At this point, the reaction could end by capture of a hydrogen atom. However, in this case, vinylcyclopropane radical **18** is transformed into benzylic radical **19** by rupture of the adjacent cyclopropyl ring. Oxidation of this radical in the reaction medium (IBX-THF) leads to cation **20** which cyclizes to **22** (the cyclization is better understood via canonical form **21**). Finally, rearomatization and loss of a proton in **22** yields the observed product **15**. The overall reaction is a cascade radical cyclization (Scheme 39.8).

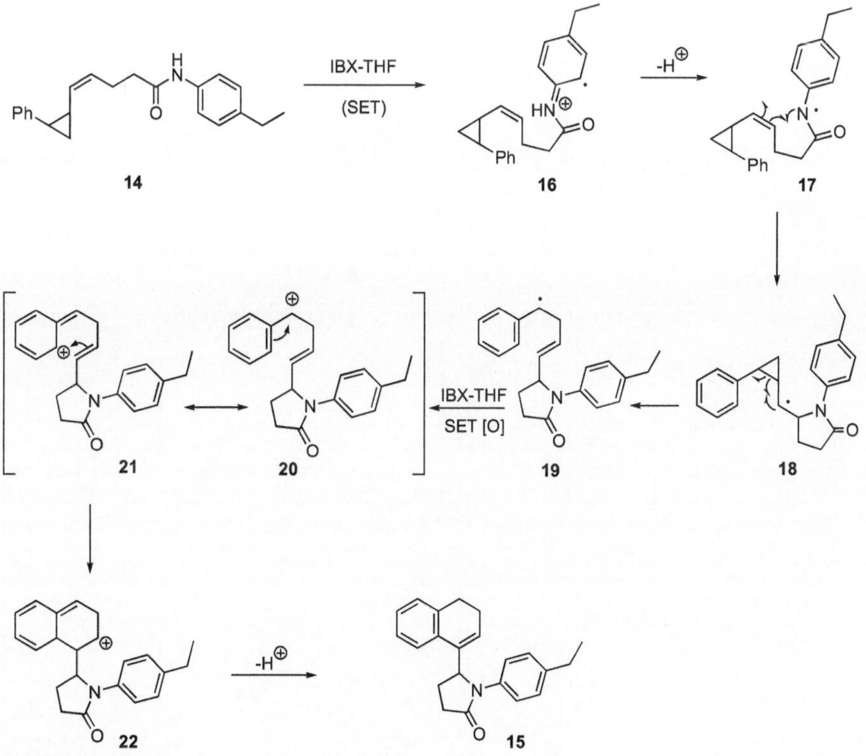

Scheme 39.8

Additional Comments

This problem is based on the work by Nicolaou KC, Baran PS, Kranich R, Zhong Y-L, Sugita K, Zou N (2001) *Angew. Chem. Int. Ed.* 40:202-206.

Subjects of Revision

SET. Radical reactions.

Level 3 – Case 40
Solvolysis of 8-Deltacyclyl Brosylates

Key point: *Nonclassical cations*

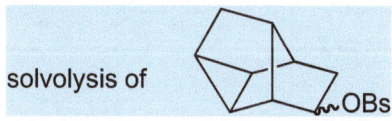

solvolysis of

Solvolysis of 8-Deltacyclyl Brosylates

The deltacyclane system **1** has a fusion of nortricyclane and norbornane skeletons and is an interesting substrate to study skeletal rearrangements or σ-bond participations in solvolysis processes.

1

X = leaving group

nortricyclane   norbornane

In this context, the acetolysis of *exo*- and *endo*-brosylates **2** (brosylate = *p*-toluenesulfonate) has been studied.

exo-2   **endo-2**

Considering the following experimental results discuss the mechanism of acetolysis for both epimers. Which type of cations are involved in the processes?

Experimental Data

1. Acetolysis of *exo*- and *endo*-brosylates **2** gave a single product, *exo*-acetate **3**. No *endo*-substituted products were found (Scheme 40.1).

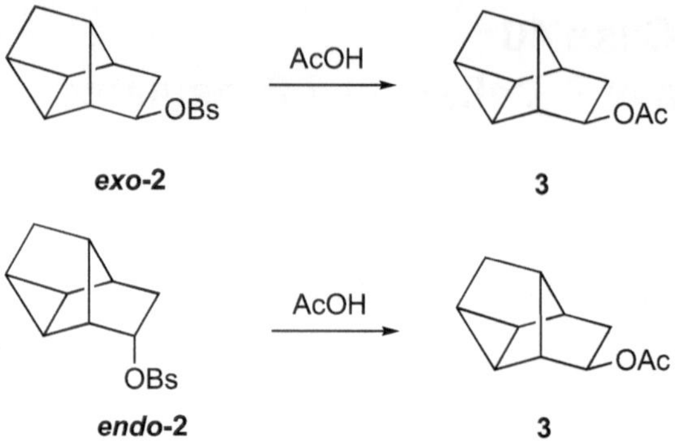

Scheme 40.1

2. The solvolysis rate of *exo*-brosylate **2** was enhanced 57-fold relative to the rate of solvolysis of the *endo*-isomer.

3. A study of the solvolyses of optically active *exo*-and *endo*-brosylates **2** revealed that the *endo*-isomer lost 57% of its optical activity while the *exo*-isomer retained 99% of its optical activity.

4. Independently deuterium-labeled *exo*-brosylates **2** at C8 and C9 positions were employed in separate experiments. When labeled at C8, products from acetolysis of *exo*-**2** exhibited a distribution of deuterium over C4 and C8 positions (compounds **4** and **5**), while when labeled at C9 *exo*-**2** gave scrambling of deuterium over C5 and C9 (compounds **6** and **7**). The distribution of the labels was 50:50 in both cases (Scheme 40.2).

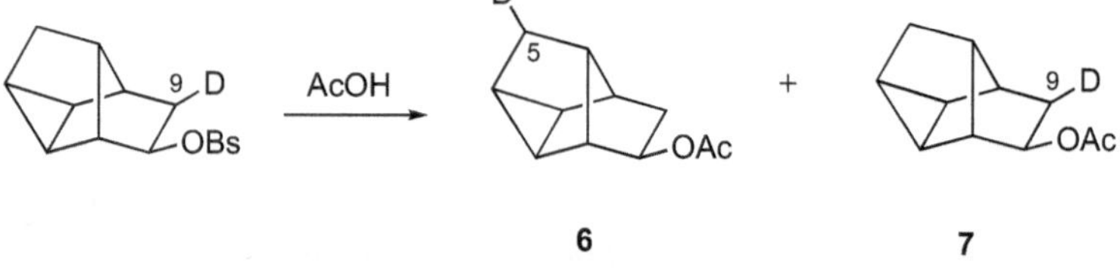

Scheme 40.2

5. Independently deuterium-labeled *endo*-brosilates **2** at C8 and C9 positions gave the same positional scrambling that their *exo*-epimers. In both experiments a 55-60% of the label was retained at the original position (Scheme 40.3).

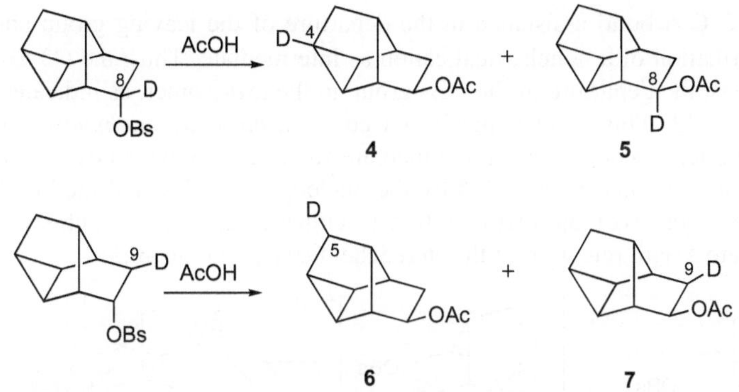

Scheme 40.3

6. A previous study of the acetolysis of *exo-* and *endo*-7-isodeltacyclyl brosylates **8** provided data of great importance to have a complete picture of the solvolysis mechanism of compounds **2**. Thus, the acetolysis of **8** gave *exo*-acetate **3** and *exo*-acetate **9** in 97:3 ratio. It has been proposed that although cation **10** was formed directly upon solvolysis of **8**, it rearranges to nonclassical isodeltacyclyl cation **11** from which the reaction products are formed (Scheme 40.4).

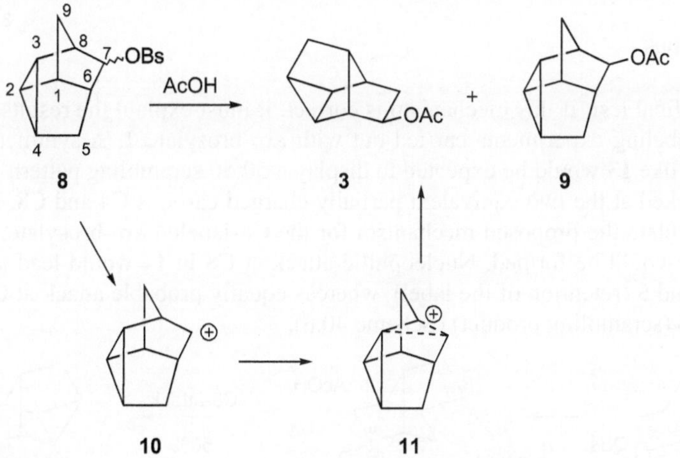

Scheme 40.4

Discussion

Although the same product (*exo*-acetate **3**) is obtained by acetolysis of *exo-* and *endo*-brosylates **2** there are two data that are clearly pointing to a different solvolysis mechanism for each epimer. The first one is *the enhancement of the reaction rate* in the case of the *exo*-isomer. The second is the complete *retention of the optical activity* of the *exo*-brosylate (99%) contrasting to the loss of optical activity observed for the *endo*-isomer. These two data fit well with a mechanism involving C-C σ bond assistance in the departure of the leaving group and subsequent formation of a nonclassical cation as intermediate. Thus, the C3-C4 σ bond could assist the departure of the OBs group in the *exo*-isomer, as indicated in transition state **12**. This bond is nicely placed for a *backside anchimeric assistance* that would lead to nonclassical intermediate **13**. The C4 and C8 positions of **13** are equivalent and could be attacked by the nucleophile with equal facility, but only from the more favorable *exo* direction in either case, leading exclusively to the *exo*-acetate **3** with retention of the stereochemistry (Scheme 40.5).

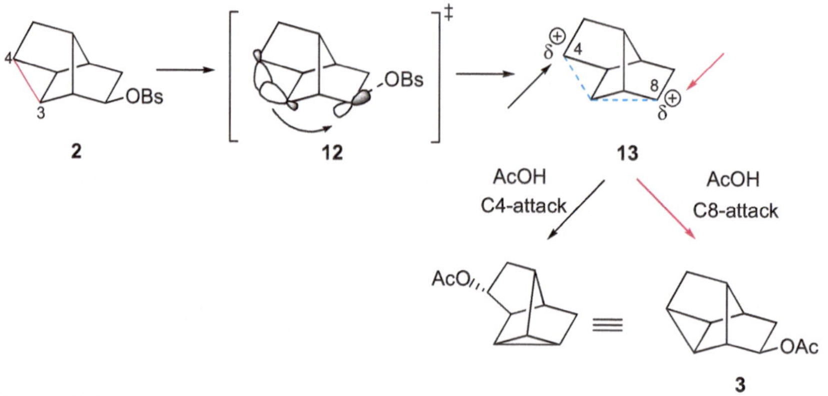

Scheme 40.5

As a final test, if this mechanism is correct, it must explain the results obtained in the labeling experiments carried out with *exo*-brosylate **2**. A symmetric intermediate like **13** would be expected to display a 50:50 scrambling pattern when being attacked at the two equivalent partially charged carbons C4 and C8. Hence, if we formulate the proposed mechanism for the C8-labeled *exo*-brosylate, intermediate **14** would be formed. Nucleophilic attack at C8 in **14** would lead to labeled compound **5** (retention of the label), whereas equally probable attack at C4 would lead to **4** (scrambling product) (Scheme 40.6).

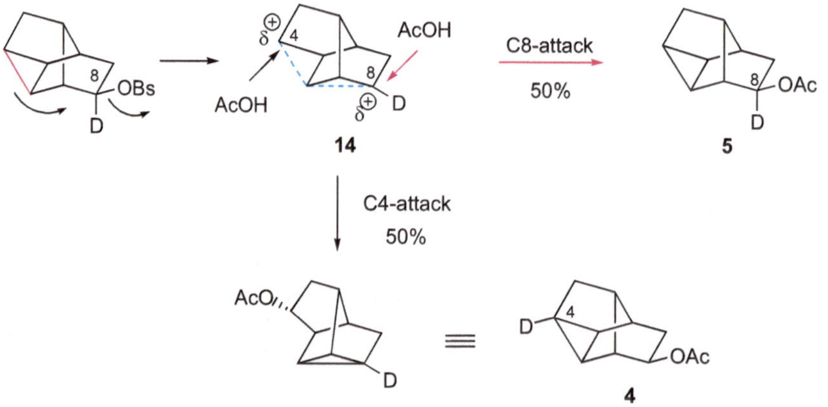

Scheme 40.6

Likewise, we can formulate the mechanistic pathway to explain the acetolysis of the C9-deuterated *exo*-brosylate, this time through intermediate **15** from which the products **6** and **7** are formed with equal probability (Scheme 40.7).

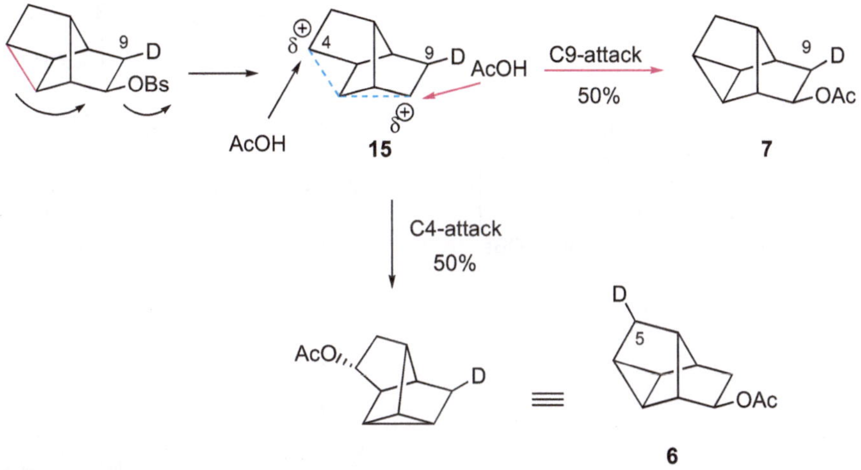

Scheme 40.7

At this point we can conclude that the solvolysis of *exo*-deltacyclyl brosylate **2** proceeds directly through a nonclassical deltacyclyl cation intermediate **13**.[1]

Let us now discuss the acetolysis of the *endo*-isomer. This compound displays a 57% loss of optical activity that cannot be explained by the reaction pathway previously considered for the *exo*-epimer. In fact, the loss of optical activity in solvolysis reactions is more in agreement with the involvement of a classical carbocation intermediate. If we consider the transition state for the departure of the OBs group (**16** in Scheme 40.8) the forming orbital at the C8 position is now poorly lined-up for participation of the C3-C4 bond and a localized carbocation like **17** should be formed (at least at first instance). This intermediate could go directly to product **3** by an *exo* attack of the solvent (path a), or rearrange to form nonclassical cation **13** (path b). The rearrangement of **17** to **13** is very facile, as nonclassical cation **13** is 13.3 kcal/mol more stable than classical cation **17**. Finally, the third alternative for cation **17** is to undergo isomerization to form its enantiomer, cation **18**. The isomerization of **17** to **18** would account for the loss of optical activity (racemization) of *endo*-brosylate **2** during the solvolysis (Scheme 40.8).

---

[1] The nonclassical nature of the deltacyclyl cation **13** was confirmed by NMR and by density functional methods. Deltacyclyl cation **13** is about 5.0 kcal/mol more stable than the known nonclassical 2-norbornyl cation.

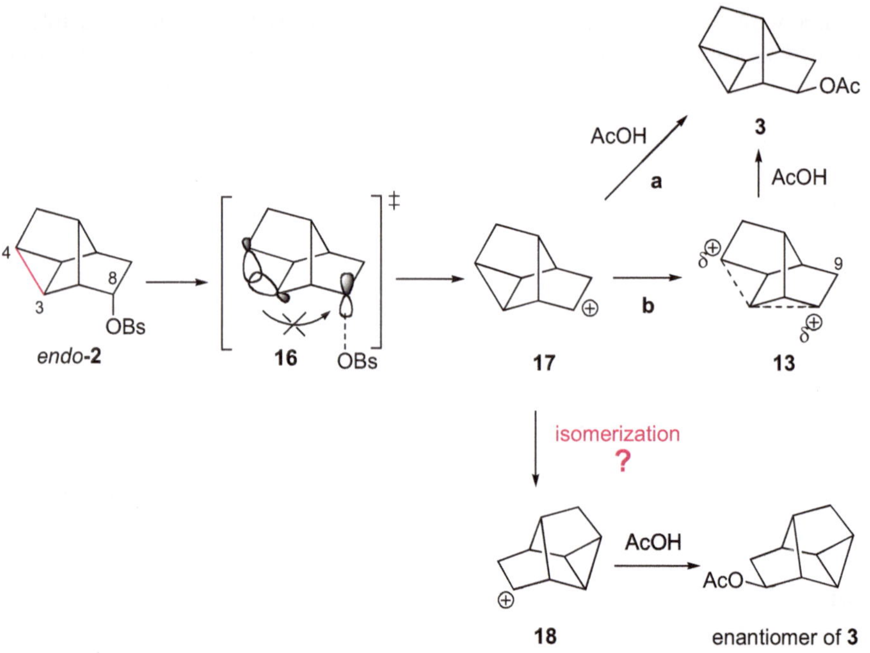

Scheme 40.8

How can we explain the racemization?

Possibly the simplest option is to consider a simple 1,2-hydride shift in cation **17** to form its enantiomer **18** as indicated in Scheme 40.9.

Scheme 40.9

However, this process cannot explain the deuterium scrambling observed during the acetolysis of labeled *endo*-brosylates **2**. Let us discuss this point in detail. We have considered that a classical carbocation like **17** is formed in the first step of the acetolysis of the *endo*-brosylate **2**. Hence, when starting from a C9-labeled brosylate, C9-labeled cation **17** would be formed (Scheme 40.10). If the isomerization of **17** occurs by a 1,2-H(D) shift, a mixture of labeled cations **19** and **20** should be formed. The acetolysis product derived from **19** will be labeled at C9 (observed) but the product derived from **20** will show deuterium scrambling from C9 to C8 positions and this has not been observed in any case (Scheme 40.10).

Remember that a mechanism that only can explain how **some** of the reaction products are formed must be rejected.

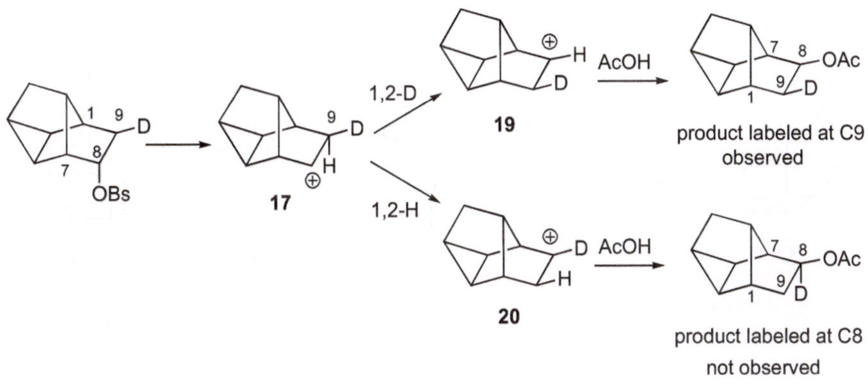

Scheme 40.10

Another possibility for the isomerization step is to consider alkyl (Wagner-Meerwein) shifts, which are frequently proposed to account for the skeletal rearrangements in carbocations. In Scheme 40.11 we have indicated a reasonable series of alkyl shifts that could justify the racemization observed in the solvolysis products. C3 alkyl shift in **17** (from C2 to C8) would lead to cation **20** that by 1,2-carbon shift forms **21**. Cations **20** and **21** could be considered as two extreme canonical forms of nonclassical isodeltacyclyl cation **11**. We know from previous studies (acetolysis of brosylates **8** in Scheme 40.4) that if **11** is formed in the medium, *exo*-acetate **3** is the main solvolysis product. The isomerization of **20** to **22** occurs by 1,2-alkyl migration, as the formation of **23** from **22**. Cations **22** and **23** could also be considered as two extreme canonical forms of nonclassical cation **24**, which would give the enantiomer of the *exo*-acetate **3** by nucleophilc attack of the solvent.[2]

This route justifies the formation of acetate **3** and its enantiomer and hence the observed loss of optical activity during the solvolysis of *endo*-brosylate **2**.

The next step for the reader is to formulate the sequence of Wagner-Meerwein migrations starting from C8 and C9 labeled cations **17** to confirm that the scrambling of deuterium corresponds with the data experimentally observed.

---

[2] The feasibility of the cationic intermediates proposed in Scheme 40.11 was examined by using ab initio and density functional calculations.

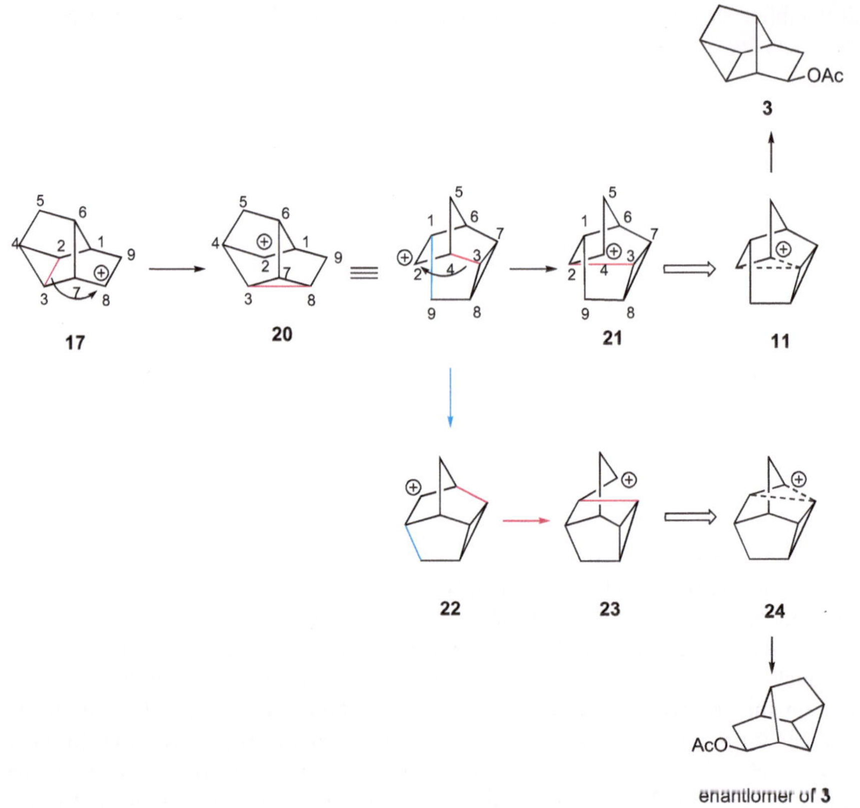

Scheme 40.11

In Summary

The solvolysis of *exo*-8-deltacyclyl brosylate **2** proceeds directly to nonclassical deltacyclyl carbocation intermediate **13**, whereas the *endo*-isomer first forms a classical carbocation **17**, which either generates intermediate **13** or isomerizes by alkyl migrations involving nonclassical isodeltacyclyl cation **11**. The special stability of the nonclassical cations compared with their classical counterparts seems to be the key of the process.

Additional Comments

This problem is based on the work by Freeman PK, Dacres JE (2002) *J. Org. Chem.* 67:3742-3748.

Subjects of Revision

Classical and nonclassical carbocations. Anchimeric assistance. Wagner-Meerwein rearrangements.

Subject Index